TJ

1077

.M457

Metalwor
fluids /
edited by
Jerry P. By

METALWORKING FLUIDS

MANUFACTURING ENGINEERING AND MATERIALS PROCESSING
A Series of Reference Books and Textbooks

FOUNDING EDITORS

Geoffrey Boothroyd
University of Rhode Island
Kingston, Rhode Island

George E. Dieter
University of Maryland
College Park, Maryland

SERIES EDITOR

John P. Tanner
John P. Tanner & Associates
Orlando, Florida

ADVISORY EDITORS

Gary Benedict
Allied-Signal

E. A. Elsayed
Rutgers University

Fred W. Kear
Motorola

Michel Roboam
Aerospatiale

Jack Walker
McDonnell Douglas

1. Computers in Manufacturing, *U. Rembold, M. Seth and J. S. Weinstein*
2. Cold Rolling of Steel, *William L. Roberts*
3. Strengthening of Ceramics: Treatments, Tests, and Design Applications, *Harry P. Kirchner*
4. Metal Forming: The Application of Limit Analysis, *Betzalel Avitzur*
5. Improving Productivity by Classification, Coding, and Data Base Standardization: The Key to Maximizing CAD/CAM and Group Technology, *William F. Hyde*
6. Automatic Assembly, *Geoffrey Boothroyd, Corrado Poli, and Laurence E. Murch*
7. Manufacturing Engineering Processes, *Leo Alting*
8. Modern Ceramic Engineering: Properties, Processing, and Use in Design, *David W. Richerson*
9. Interface Technology for Computer-Controlled Manufacturing Processes, *Ulrich Rembold, Karl Armbruster, and Wolfgang Ülzmann*
10. Hot Rolling of Steel, *William L. Roberts*
11. Adhesives in Manufacturing, *edited by Gerald L. Schneberger*
12. Understanding the Manufacturing Process: Key to Successful CAD/CAM Implementation, *Joseph Harrington, Jr.*

13. Industrial Materials Science and Engineering, *edited by Lawrence E. Murr*
14. Lubricants and Lubrication in Metalworking Operations, *Elliot S. Nachtman and Serope Kalpakjian*
15. Manufacturing Engineering: An Introduction to the Basic Functions, *John P. Tanner*
16. Computer-Integrated Manufacturing Technology and Systems, *Ulrich Rembold, Christian Blume, and Ruediger Dillman*
17. Connections in Electronic Assemblies, *Anthony J. Bilotta*
18. Automation for Press Feed Operations: Applications and Economics, *Edward Walker*
19. Nontraditional Manufacturing Processes, *Gary F. Benedict*
20. Programmable Controllers for Factory Automation, *David G. Johnson*
21. Printed Circuit Assembly Manufacturing, *Fred W. Kear*
22. Manufacturing High Technology Handbook, *edited by Donatas Tijunelis and Keith E. McKee*
23. Factory Information Systems: Design and Implementation for CIM Management and Control, *John Gaylord*
24. Flat Processing of Steel, *William L Roberts*
25. Soldering for Electronic Assemblies, *Leo P. Lambert*
26. Flexible Manufacturing Systems in Practice: Applications, Design, and Simulation, *Joseph Talavage and Roger G. Hannam*
27. Flexible Manufacturing Systems: Benefits for the Low Inventory Factory, *John E. Lenz*
28. Fundamentals of Machining and Machine Tools: Second Edition, *Geoffrey Boothroyd and Winston A. Knight*
29. Computer-Automated Process Planning for World-Class Manufacturing, *James Nolen*
30. Steel-Rolling Technology: Theory and Practice, *Vladimir B. Ginzburg*
31. Computer Integrated Electronics Manufacturing and Testing, *Jack Arabian*
32. In-Process Measurement and Control, *Stephan D. Murphy*
33. Assembly Line Design: Methodology and Applications, *We-Min Chow*
34. Robot Technology and Applications, *edited by Ulrich Rembold*
35. Mechanical Deburring and Surface Finishing Technology, *Alfred F. Scheider*
36. Manufacturing Engineering: An Introduction to the Basic Functions, Second Edition, Revised and Expanded, *John P. Tanner*
37. Assembly Automation and Product Design, *Geoffrey Boothroyd*
38. Hybrid Assemblies and Multichip Modules, *Fred W. Kear*
39. High-Quality Steel Rolling: Theory and Practice, *Vladimir B. Ginzburg*
40. Manufacturing Engineering Processes: Second Edition, Revised and Expanded, *Leo Alting*
41. Metalworking Fluids, *edited by Jerry P. Byers*

Additional Volumes in Preparation

METALWORKING FLUIDS

EDITED BY
JERRY P. BYERS
Cincinnati Milacron
Cincinnati, Ohio

Marcel Dekker, Inc.   New York•Basel•Hong Kong

Library of Congress Cataloging-in-Publication Data

Metalworking fluids / edited by Jerry P. Byers.
    p. cm. — (Manufacturing engineering and materials processing ; 41)
  Includes bibliographical references and index.
  ISBN 0-8247-9201-7
  1. Metal-working lubricants. I. Byers, Jerry P.
II. Series.
TJ1077.M457 1994
671—dc20   94-14919
                                                                                               CIP

The publisher offers discounts on this book when ordered in bulk quantities. For more information, write to Special Sales/Professional Marketing at the address below.

This book is printed on acid-free paper.

Copyright © 1994 by Marcel Dekker, Inc. All Rights Reserved.

Neither this book nor any part may be reproduced or transmitted in any form or by any means, electronic or mechanical, including photocopying, microfilming, and recording, or by any information storage and retrieval system, without permission in writing from the publisher.

Marcel Dekker, Inc.
270 Madison Avenue, New York, New York 10016

Current printing (last digit):
10 9 8 7 6 5 4 3 2 1

PRINTED IN THE UNITED STATES OF AMERICA

Preface

For as long as people have been cutting metal, fluids have been used to aid the process. Water may have been the first fluid, followed by animal fats, vegetable oils, mineral oil, oil-in-water emulsions, and in recent years, by clear synthetic chemical solutions. Today, a broad range of coolants and lubricants for metalworking continue to be key components of the manufacturing process around the world. Popular use of the terms "oils" or "cutting oils" to broadly refer to all these fluids is inaccurate. The term "metalworking fluid" will be used here.

This book was written to serve the current needs of industry by presenting a review of the state of the art in metalworking fluid technology, application, maintenance, testing methods, health and safety, governmental regulations, recycling, and waste minimization. Older texts tended to ignore or give light treatment to important operational aspects of the use of fluids for metalworking. It is hoped that this work will fill these gaps and address new issues that have surfaced in recent years. The contributors are well-known formulators, physicians, college professors, fluid users, industry consultants, and chemical and equipment suppliers.

The 16 chapters that follow summarize the latest thinking on various technologies related to metalworking fluid development, evaluation, and application. Chapter 1 traces the historical development of the use of lubrication in metalworking. Because metalworking fluids are used to shape metal, Chapter 2 covers important aspects of the metallurgy of common ferrous and nonferrous metals. Chapters 3–5 describe fluid application in metal cutting, grinding, and forming. Chapter 6 explains the chemistry of straight oil, soluble oil, semisynthetic, and synthetic (nonoil-containing) fluids. Chapter 7 familiarizes the reader with laboratory methods for evaluating fluid performance. Two aspects of metalworking fluid performance and evaluation are so important and complex that separate chapters have been devoted to them: corrosion control and microbial control (Chapters 8 and 9, respectively). Handling aspects of fluid in a manufacturing facility are covered in Chapters 10–12 on filtration, management and troubleshooting, and recycling. Disposal of the fluid after a long useful life is covered in Chapter 13, which discusses waste treatment processes. Personal concerns of the machine operator are addressed in the chapters on dermatitis (Chapter 14) and health and safety (Chapter 15). Chapter 16 leads the reader through the tangled maze of U.S. government regulations affecting both the manufacture and use of metalworking fluids and explains the impact of these laws on industry. Finally, a comprehensive glossary of over 300 terms is included. These terms were supplied by the contributors.

Metalworking Fluids will appeal to a broad readership including machine operators, plant managers, foremen, engineers, chemists, biologists, and governmental and industrial hygienists, as well as to instructors of manufacturing and industrial disciplines and their students. The authors hope that this text will help modern industry meet the worldwide demand for improved quality and productivity in a cleaner environment.

Several employees of Cincinnati Milacron should be recognized for their contributions to the completion of this work. Mr. Ralph Kelly, Dr. William Lucke, and Dr. Charles Yang reviewed every chapter, offering their comments. Mr. Steven Bridewell spent long hours preparing many of the illustrations. Mrs. Charlene Honroth skillfully and patiently handled communication with the authors and much of the typing. Sincere thanks to these people for their help and dedication.

Jerry P. Byers

Contents

	Preface	iii
	Contributors	vii
1.	Introduction: Tracing the Historical Development of Metalworking Fluids *Jeanie S. McCoy*	1
2.	Metallurgy for the Nonmetallurgist with an Introduction to Surface Finish Measurement *James E. Denton*	25
3.	Metal Cutting Processes *Herman R. Leep*	61
4.	Performance of Metalworking Fluids in a Grinding System *Cornelis A. Smits*	99
5.	Metalforming Applications *Kevin H. Tucker*	135
6.	The Chemistry of Metalworking Fluids *Jean C. Childers*	165

7.	Laboratory Evaluation of Metalworking Fluids *Jerry P. Byers*	191
8.	Corrosion: Causes and Cures *Giles J. P. Becket*	223
9.	Metalworking Fluid Microbiology *L. A. Rossmoore and H. W. Rossmoore*	247
10.	Filtration Systems for Metalworking Fluids *Robert H. Brandt*	273
11.	Metalworking Fluid Management and Troubleshooting *Gregory J. Foltz*	305
12.	Recycling of Metalworking Fluids *Raymond M. Dick*	339
13.	Waste Treatment *Paul M. Sutton and Prakash N. Mishra*	367
14.	Contact Dermatitis and Metalworking Fluids *C. G. Toby Mathias*	395
15.	Health and Safety Aspects in the Use of Metalworking Fluids *P. J. Beattie and B. H. Strohm*	411
16.	Government Regulations Affecting Metalworking Fluids *William E. Lucke*	423
	Glossary	463
	Index	481

Contributors

P. J. Beattie Toxic Materials Control Activity, General Motors Corporation, Detroit, Michigan

Giles J. P. Becket Products Division, Cincinnati Milacron, Cincinnati, Ohio

Robert H. Brandt Brandt & Associates, Inc., Pemberville, Ohio

Jerry P. Byers Product Research and Development, Products Division, Cincinnati Milacron, Cincinnati, Ohio

Jean C. Childers Climax Metals Division, Sales and Marketing Department, Amax, Inc., Summit, Illinois

James E. Denton Metallurgical Engineering Department, Cummins Engine Company, Inc., Columbus, Indiana

Raymond M. Dick Fluid Management Equipment, Products Division, Cincinnati Milacron, Cincinnati, Ohio

Gregory J. Foltz Customer Service, Products Division, Cincinnati Milacron, Cincinnati, Ohio

Herman R. Leep Department of Industrial Engineering, University of Louisville, Louisville, Kentucky

William E. Lucke Laboratory Services, Products Division, Cincinnati Milacron, Cincinnati, Ohio

C. G. Toby Mathias Department of Dermatology, Group Health Associates, Cincinnati, Ohio

Jeanie S. McCoy Cutting Fluid Management Consultant, Jeanie McCoy Technologies, Inc., Lombard, Illinois

Prakash N. Mishra Technical Center, North American Operations, General Motors Corporation, Warren, Michigan

H. W. Rossmoore Department of Biology, Wayne State University, Detroit, Michigan

L. A. Rossmoore Biosan Laboratories, Inc., Ferndale, Michigan

Cornelis A. Smits Advanced Grinding Systems, Products Division, Cincinnati Milacron, Cincinnati, Ohio

B. H. Strohm Toxic Materials Control Activity, General Motors Corporation, Detroit, Michigan

Paul M. Sutton Environmental Engineering Consultant, P. M. Sutton & Associates, Inc., Bethel, Connecticut

Kevin H. Tucker Technical Processes and Engineering, Products Division, Cincinnati Milacron, Cincinnati, Ohio

1

Introduction: Tracing the Historical Development of Metalworking Fluids

JEANIE S. MCCOY
Jeanie McCoy Technologies, Inc.
Lombard, Illinois

I. WHAT ARE THEY?

Metalworking fluids are best defined by what they do. Metalworking fluids are engineering materials that optimize the metalworking process. Metalworking is commonly seen as two basic processes, metal deformation and metal removal or cutting. Comparatively recently, metal cutting has also been considered a plastic deformation process—albeit on a submicro scale and occurring just before chip fracture.

In the manufacturing and engineering communities, metalworking fluids used for metal removal are known as cutting and grinding fluids. Fluids used for the drawing, rolling, and stamping processes of metal deformation are known as metalforming fluids. However, the outcome of the two processes differs. The processes by which the machines make the products, the mechanics of the operations, and the requirements for the fluids used in each process are different.

The mechanics of metalworking govern the requirements demanded of the metalworking fluid. As all tool engineers, metalworking fluid process engineers, and machinists know, the fluid must provide a layer of lubricant to act as a cushion between the workpiece and the tool in order to reduce friction. Fluids must also function as a coolant to reduce the heat produced during machining or forming. Otherwise, distortion of the workpiece and changed dimensions could result. Further, the fluid must prevent metal pick-up on both the tool and the workpiece by flushing away the chips as they are produced. All of these attributes function to prevent wear on the tools and reduce energy requirements. In addition, the metalworking fluid is expected to produce the desired finish on an accurate piece-part. Any discussion of metalworking fluid requirements must include the fact that the manufacturing impetus since the days of the industrial revolution is to machine or form parts at the highest rate of speed with maximum tool life, minimum downtime, and the fewest possible part rejects (scrap), all while maintaining accuracy and finish requirements.

II. CURRENT USAGE IN THE UNITED STATES

The number of gallons of metalworking fluids produced and sold in the United States represents a significant slice of the gross national product as indicated in the 1990 report by the Independent Lubricant Manufacturers Association. Of the 632 million gallons of lubricants produced by independent manufacturers, 92 million gallons were metalworking fluids and 32 million gallons were greases, some of which are used in metal deformation processes [1].

The National Petroleum Refiners Association, in their annual survey on U.S. lubricating oil sales, reported that 2472 million gallons of automotive and industrial lubricants and 56 million gallons of grease sold in 1990, of that, 42% were industrial lubricants. Of the total industrial oil sales, 16% were industrial process oils and 11% were classed as metalworking oils [2].

These statistics indicate the importance and wide usage of metalworking fluids in the manufacturing world. How they are compounded, used, managed, and how they impact health, safety, and environmental considerations will be described in subsequent chapters. This chapter will take the reader through the evolution of metalworking fluids, one of the most important and least understood tools of the manufacturing process.

It is surprising that it is not possible to find listings for metalworking fluids in the available databases. The National Technology Information Services, the Dialog Information Service, the well-known Science Index, the Encyclopedia of Science and Technology Index, and the *Materials Science Encyclopedia* all lack relevant citations. The real story appears to be buried in technical magazines written by engineers and various specialists for other engineers and specialists, and obscured in books on related topics. Clearly, this information needs to be collected and published.

III. HISTORY OF LUBRICANTS: EVIDENCE FOR EARLY USAGE OF METALWORKING FLUIDS

The histories of Herodotus and Pliny, and even the Scriptures, indicate that humankind has used oils and greases for many applications. These include lubrication uses such as hubs on wheels, axles, and bearings, as well as nonlubrication uses such as embalming fluids, unguents, medicines, illumination, waterproofing ships, and setting tiles [3]. However, records documenting the use of lubricants as metalworking fluids are not readily available. Histories commonly report that man first fashioned weapons, ornaments, and jewelry by cold working the metal, then as the ancient art of the blacksmith developed, by hot working the metal. Records show that animal and vegetable oils were used by early civilizations in various lubrication applications. Unfortunately, the use of lubricants as metalworking fluids in the metalworking crafts is not described in these early historical writings [4].

Reviewing the artifacts and weaponry of the early civilizations of Mesopotamia and Egypt, and later the Greek and Roman eras on through the Middle Ages, it is obvious that forging and then wire drawing must have been the oldest of metalworking processes [5]. Lubricants must have been used to ease the wire-drawing process. Since the metalworking fluid is and always has been an important part of the process, it may not be unreasonable to presume that the fluids used then were those that were readily available. These include animal oils and fats (primarily whale, tallow, and lard) as well as vegetable oils from various sources such as olive, palm, castor, and other seed oils [6]. Even today these are used in certain metalworking fluid formulations. Some of the most effective known lubricants have been provided by Nature. Only by inference, since records of their early use have not been found, can we speculate that these lubricants must have been used as metalworking fluids in the earliest of the metalworking processes.

IV. HISTORY OF TECHNOLOGY

A. Greek and Roman Era

The explanation for the lack of early historical documentation might be found by examining the writings of the ancient Greek and Roman philosophers on natural science. It is readily seen that among the "intelligensia" there was little interest in the scientific foundations for the technology of the era.

As Singer points out in his *History of Technology*, the craftsman of this era was relegated to a position of social inferiority because knowledge of the technology involved in the craft process was scorned as unscientific. It was neither studied nor documented, evidently not considered as being worthy of preservation [7]. Consequently, the skills and experience of the craftsman became valuable personal possessions to be protected by secrecy; the only surviving knowledge was handed down through the generations [8].

B. The Renaissance (1450–1600)

During the Renaissance, plain bearings of iron, steel, brass, and bronze were increasingly used, especially da Vinci's roller disk bearings in clock and milling machinery (as early as 1494); Agricola confirmed the wide use of conventional roller bearings in these applications [9]. Although machines were developed to make these parts, there is no record that any type of metalworking lubricant was used in the bearing, gear, screw, and shaft manufacture. It is possible that those parts that were made of soft metals such as copper and brass did not require much if any lubrication in the manufacturing process, but it would seem logical that the finish requirements of iron and steel parts would demand the use of some type of metalworking fluid.

John Schey, in his *Metal Deformation Processes*, points out that metalworking is probably humankind's first technical endeavor and, considering the importance of lubricants used in the process, he was amazed to find no record of their use until fairly recent times [10].

C. Toward the Industrial Revolution (1600–1750)

It was shortly after the turn of the seventeenth century that scientific inquiry into the mechanics of friction and wear became the seed that promoted an appreciation of the value of lubrication for moving parts and metalworking processes. The first scant references to lubrication were in the descriptions of power-driven machinery (animal, wind, and water) by early experimenters on the nature of friction.

In China, Sung Ying-Hsing (1637) wrote of the advantage of oil in cart axles. Hooke (1685) cautioned on the need for adequate lubrication for carriage bearings, and Amontons (1699) elucidated laws of friction in machines through experimentation. The same year, De la Hire described the practice of using lard oil in machinery. Desaugliers (1734) suggested that the role of the lubricant was to fill up the imperfections on surfaces and act as tiny rollers, and Leupold (1735) recommended that tallow or vegetable oil should be used for lubricating rough surfaces [11].

It is interesting to note that although Amontons's endeavors are often considered to be experiments on dry friction, his notes carefully recorded the use of pork fat to coat the sliding surfaces of each experiment. As Dowson points out, Amontons was really studying frictional characteristics of lubricated surfaces under conditions now depicted as boundary lubrication [12], the mechanism operating most frequently in metalworking operations [13]. These concepts were basic to the development of theories of friction and wear which occurred during the eighteenth century, culminating in the profound works of Coulomb, who theorized that both adhesion and surface roughness caused friction.

In the nineteenth century the means to mitigate friction and wear through lubrication were investigated, leading to Reynolds's theory of fluid film lubrication. In the early part of the twentieth century, Hardy with Doubleday introduced the concept of boundary lubrication, which to this day is still a cornerstone of our current foundation of knowledge on the theory of lubrication [14]. It should be noted that William Hardy's works on colloidal chemistry paved the way for the development of water-"soluble" cutting fluids.

However, it was not the development of scientific theory that ultimately led to the explosion of research in this area and especially in the mechanics of metalworking and metalworking fluids in the twentieth century. Rather, it was the wealth of mechanical inventions and evolving technologies that created the need for understanding the nature of friction and wear, and how these effects can be mitigated by proper lubrication.

Interest in craft technologies soared during this period with the founding of the Royal Society of England in 1663 by a group identifying themselves as the "class of new men" interested in the application of science to technology [15]. Their most significant contribution was the sponsorship of *Histories of Nature, Arts or Works*, which for the first time contained scientific descriptions of the craft technologies as practiced in the seventeenth century for popular use. Although the *Histories* published surveys on a wealth of subjects, including long lists of inventions as described by Thomas Sprat, the only reference to a metalworking operation was in the treatise "An Instrument for Making Screws with Great Dispatch." No mention was made of metalworking fluid usage [16].

The lack of early information on machining fluids can only be attributed to a reluctance on the part of the craftsman, seen even today on the part of manufacturers, to disclose certain aspects regarding the compounding of fluids. The revelation of "trade secrets" which might yield a competitive advantage is not done unless the publicity for market value is seen to outweigh the consequence of competitors now knowing "how to do it."

Some information on lubrication in metal deformation processes, however, has been documented. K. B. Lewis relates that in the seventeenth century wire drawing was accomplished with grease or oil, but only if a soft, best quality iron was used. High friction probably caused steel wire to break [17]. About 1650, Johann Gerdes accidentally discovered a method of surface preparation that permitted easy drawing of steel wire. It was a process called "sull-coating" whereby iron was steeped in urine until a soft coating developed. This procedure remained in practice for the next 150 years; later, diluted, sour beer was found to work as effectively. By about 1850 it was discovered that water also worked well [18]. Although the process of rolling was applied to soft metals as early as the fifteenth century—and in the eighteenth century wire rod was regularly rolled—lubricants were not and still are not used for rolling rounds and sections [19].

Since research into the history of lubrication and the history of technology has not yielded documentation on the early use of metalworking fluids, consideration of the elements involved in the metalworking process led to a search through the history of machine tool evolution for answers. A few surprising facts came to light.

V. EVOLUTION OF MACHINE TOOLS AND METALWORKING FLUIDS

L. T. C. Rolt, writing on the history of machine tools, states unequivocally that through all ages, the rate of man's progress has been determined by his tools. Indeed, the pace of the industrial revolution was governed by the development of machine tools [20]. This statement is echoed by R. S. Woodbury who points out that historians traditionally have described the political, social, and economic aspects of human endeavor, including the inventions concerned with power transmission, new materials (steel), transportation, and the textile industry. Most have overlooked the technological development of the machine tool "without which the steam engines and other machinery could not have been built and steels would have little significance" [21].

This same observation could be further extended to include the significance of the technological development of metalworking fluids, without which the machine tool industry could not have progressed to where it is today. The development of metalworking fluids was the catalyst permitting the development of energy-efficient machine tools having the high speed and feed capacities required for today's production needs for extremely fast metalforming and metal cutting operations.

In general, machine tool historians seem to believe that the bow drill was the first mechanized tool as seen in bas relief and carvings in Egypt about 2500 B.C. [22]. The lathe, probably developed from the mechanics of the potter's wheel, can be seen in paintings and woodcuts as early as 1200 B.C. [23]. In the Greek and Roman era (first century B.C. through the first century A.D.) the writings of three authors on technical processes describing various mechanisms have survived:

Hero of Alexandria (50–120 A.D.), whose works include mechanical subjects.
Frontinus (Sextus Julius, 35 B.C.–37 A.D.), who concentrated on water engineering mechanisms.
Vitruvius, whose ten books, *De architectura* (31 B.C.), were the only work of its kind to survive from the Roman world. Book VIII, devoted to

water supplies and water engineering, refers to the use of a metalworking fluid. Vitruvius describes a water pump with bronze piston and cylinders that were machined on a lathe with *oleo subtracti*, indicating the use of olive oil to precision turn the castings [24].

The first record of a mechanized grinding operation that was accomplished by use of a grinding wheel for sharpening and polishing is evidenced in the Utrecht Psalter of 850 A.D., which depicted a grinding wheel operated by manpower turning a crank mechanism [25]. The first grinding fluid probably was water, used as the basic metal removal process in the familiar act of sharpening a knife on a whetstone, as is still done today.

A. Early Use of Metalworking Fluids in Machine Tools

Undoubtedly water was used as the cutting fluid as grinding machines became more prevalent. Evidence for this presumption is seen in a 1575 copper engraving by Johannes Stradanus which is a grinding mill similar to those seen in drawings by Leonardo da Vinci. The engraving depicts a shop set up to grind and polish armor where "the only addition appears to be chutes to supply water to some of the wheels" [26].

It was common practice in Leonardo's day to use tallow on grinding wheels. An indication that oil also was used as a metalworking fluid is illustrated in Leonardo's design for an internal grinding machine (the first hint of a precision machine tool) which had grooves cut into the face of the grinding wheel to permit a mixture of oil and emery to reach the whole grinding surface [27].

The development of machine tools was slow during the following 200 years. In this period the manufacturing of textiles flourished in England with the invention of Hargreaves's spinning jenny and Awkwright's weaving machinery. Carron Ironworks was founded in 1760, no doubt resulting in improvement of iron smelting and steel making. These inventions, plus the introduction of cast iron shafts in machinery, all gave impetus to design machine tools in order to produce these kinds of new machine parts. Still, by 1775 the available machine tools for industry had barely advanced beyond those that were used in the Middle Ages [28].

The troubles between England and the colonies that began in 1718 resulted in a series of events that in time actually promoted machine tool development and the use of metalworking fluids. At that time, American colonial pig iron was exported to England. This alarmed the British iron-masters because they considered the colonies a good market for their iron production. They were successful in getting a ban on the importation of American manufactured iron. In addition, in 1750, the government of England prohibited the erection of steel furnaces, plating forges, and rolling mills in the colonies. In 1785, Britain passed laws that prohibited the export of tools, machines, engines, or persons connected with the iron industry or the trades evolving from it to the newly formed United States [29]. The rationale for this edict was to impact the economy of the colonies by hindering the developing American manufacturing industries and forcing the colonies to purchase English manufactured items. Rather than impeding this American technical development, the British ban stimulated the ingenuity of the American manufacturing pioneers to develop tools, machines, and superior manufacturing skills.

These events encouraged the development of the American textile industry. It was quickened by the inventions of Eli Whitney, first with his cotton gin, permitting the use of very "seedy" domestic cotton, followed by his unique system of rifle manufacturing. The munitions industry began to flourish in America. Whitney developed the system of "interchangeable parts," made possible by more precise machining of castings by which parts of duplicate dimension were effected through measurement with standard gauges. Whitney has been called the father of mass production in that he dedicated each machine to a specific machining operation and then assembled rifles from baskets of parts holding the product of each machine [30]. This system of manufacture was quickly adopted by other American and European manufacturers as well. Whitney continued to be a forerunner of machine tool invention in order to keep pace with the new manufacturing demand. He is credited with the invention of the first milling machine, a multipoint tool of great value [31]. However there is no mention of any metalworking fluid used in any of the machining processes—it was probably known only to the machinist as one of the skills of his trade.

B. Growth of Metalworking Fluid Usage

The practice of using metalworking fluids was concomitant with machine tool development both in the United States and England. R. S. Woodbury relates further evidence for the use of water as a metalworking fluid. In 1838, James Whitelaw developed a cylindrical grinding machine for grinding the surface of pulleys wherein "a cover was provided to keep in the splash of water" [32]. James H. Nasmyth in his 1830, autobiography describes the need for a small tank to supply water or soap and water to the cutter to keep it cool. This consisted of a simple arrangement of a can to hold the coolant supply and an adjustable pipe to permit the coolant to drip directly on the cutter [33].

Woodbury relates that the more common practice of applying cutting fluid during wet grindings (using grinding lathes), was holding a wet sponge against the workpiece. That practice was soon abandoned. A December 1866 drawing shows that a supply of water was provided through a nozzle, and an 1867 drawing shows a guard installed on the slideways of that same lathe to prevent the water and emery from corroding and pitting the slideways [34]. In retrospect, after reviewing the developments in machine tools and machine shop practice, it is obvious that the majority of modern machine tools had been invented by 1850 [35].

C. After the Industrial Revolution (1850–1900)

The next 50 years saw rapid growth in the machine tool industry and concurrently in the use of metalworking fluids. This came about as a result of the new inventions of this period, which in response to the great needs for transportation, saw the development of nationwide railways. The next century saw the development of the automobile and aircraft. In order to build these machines, machine tools capable of producing large heavy steel parts were rapidly designed.

In this period there was growing awareness of the value of metalworking fluids as a solution to many of the machining problems emerging from the new demands upon the machine tools. However, there were four significant happenings that altogether made conditions ripe for rapid progress in the development of compounded metalworking fluids which paralleled the sophistication of machine tools.

1. Discovery of Petroleum in the United States

One of the most important factors in our story was the discovery of huge quantities of petroleum in the United States (1859), which eventually had a profound influence on the compounding of metalworking fluids. Petroleum at that time was largely refined for the production of kerosene used for illumination and fuel. The aftermath of the Civil War with its depressed economic climate led refiners to find a use for oil which was considered a by-product and had been discarded as useless. This caused an environmental problem for the city of Cleveland. The refiners, forced to find a solution to the oil "problem," induced industry to use oil for lubricant applications with the result that mineral oils then began to replace some of the popular animal and vegetable oil-based lubricants [36]. During this period some of today's famous independent lubricant manufacturing companies came into existence offering a variety of compounded lubricants and cutting oils to improve the machining process and permit greater machine output. Some of these original specialty lubricant manufacturers have since been absorbed into the prevailing industrial conglomerates [37].

2. Introduction of Better Alloy Steels

The second factor influencing the development of metalworking fluids was the development of alloy steels for making tools. David Mushet, a Scotch metallurgist, developed methods of alloying iron to make superior irons. One of his sons, Robert Forest Mushet, also a metallurgist, founded a method of making Bessemer's pneumatic furnace produce acceptable steels. Some writers claim that Bessemer's furnace was predated by seven years with the "air-boiling" steels produced by the American inventor, William Kelly [38].

R. F. Mushet made many contributions to the steel industry with his various patents for making special steels. Perhaps his most important legacy is his discovery that certain additions of vanadium and chromium to steel would cause it to self-harden and produce a superior steel for tool making. In the United States, Taylor and White experimented with different alloying elements and also produced famous grades of tool steels. The significance here is that these tough tool steels permitted tools to be run at faster speeds, enabling increased machine output [39].

3. Growth of Industrial Chemistry

The third development that had great impact was the budding petrochemical industry. Chemistry had long been involved in the soap, candle, and textile industries. Chemists' endeavors turned to opportunities that the petroleum industry offered, resulting in the creation of a variety of new compounds; many were used in the "new" lubricants needed for growing industrial and manufacturing applications.

4. Use of Electricity as a Power Source

The fourth factor was the development of electric power stations that permitted the use of the electric motor as a power source. Before the use of electric motors to drive machines, power was transmitted by a series of belts to permit variable gearing, and then replaced by the clutch. The electric motor permitted connection directly to machine drive shafts. This eliminated some of the machining problems caused by restricted and inconsistently delivered power, which had resulted in problems such as "chatter." The introduction of steam turbines to drive Edison's dynamos for the generation of electric power in the 1890s [40] was a boon to machine tool designers. Increased sophistication of design and heavier duty capability in machine tools were required in order to produce the machinery needed for the petroleum and the electrical power industries, and to make the steam engines and railroad cars for the growing railway transportation ventures. Electric power made the design of more powerful machine tools possible, but the stresses between the tool and the workpiece were increasing in heavy duty machining operations. The need to mitigate these conditions brought about the natural evolution of sophisticated metalworking fluids.

This period also heralded the beginnings of the investigation into the scientific phenomena operating in the metal removal process and the effectiveness of metalworking fluids in aiding the process. Physicists, chemists, mechanical engineers, and metallurgists all contributed to unravel the mystery of what happens during metalworking and the effect the metalworking fluid has upon the process.

D. Early Experimentation with Metalworking Fluids

It appears that the first known publication on actual cutting fluid applications was in 1868, in *A Treatise on Lathes and Turning*, by Northcott. He reported that lathe productivity could be materially increased by using cutting fluids [41]. However, the use of metalworking fluids, especially in metal removal operations, was widespread in both England and the United States, as evidenced in a report on the Machine Tool Exhibition of 1873 held in Vienna. Mr. J. Anderson, the superintendent of the Arsenal at Woolrich, England, wrote that in his opinion the machine tools made in continental Europe were not up to the standards of those in England and in America in that there was a conspicuous absence of any device to supply coolant to the edge of the cutting tool [42]. This observation was confirmed by a drawing of the first universal grinding machine which was patented by Joseph R. Brown in 1868 [43] and appeared in a Brown and Sharpe catalogue of 1875. It included a device for carrying off the water or other fluids used in grinding operations [44]. Obviously, the use of metalworking fluids had become standard machine shop practice.

Curiosity regarding the lubrication effect of metalworking fluids in machining had its beginnings in the publication of the *Royal Society of London Proceedings* in 1882. In that publication, Mallock wondered about the mystery of how lubricants appear to mitigate the effects of friction by going between "the face of the tool and the shaving," noting that it was impossible to see how the lubricant got there [45]. In that same time frame, evidence for the use of various types of oil in metal cutting operations appeared in Robert H. Thurston's *Treatise on Friction and Lost Work in Machining and Mill Work* which described various formulas for metalworking. For example, he stated that the lubricants used in bolt cutting must have the same qualities as those required for "other causes of lubrication." He cautioned that the choice of lubricant will be determined by the oil giving the smoothest cut and finest finish with "minimum expenditure of power . . . whatever the market price." His advice was that the best lard oil should be commonly used for this purpose, although he agreed with current practice that mineral oil could be used. Thurston also advised in opposition to "earlier opinions, that in using oil on fast running machinery, the best method is to provide a supply as freely as possible, recovering and reapplying after thorough filtration" [46].

Thurston was an engineer who chaired the Department of Mechanical Engineering at the Stevens Institute of Technology in 1870. His important contributions were in the areas of manufacturing processes, winning him "fame on both sides of the Atlantic" [47]. His well-known lubricant testing machines enabled him to provide advice to machinists. Typically, his studies found that sperm oil was superior to lard oil when cutting steel. In cutting cast iron, he recommended a mixture of plumbago (black lead oxide) and grease, claiming a lower coefficient of friction [48].

It was during this period that chemical mixtures of oils came into usage as metalworking fluids. Most notable was the advent of the sulfurized cutting oils which date back to 1882. The proper addition of sulfur to mineral oil, mineral–lard oil and mineral–whale oil mixtures was found to ease the machining of difficult metals by providing better cooling and lubricating qualities and prevented chips from welding onto the cutting tools. Sulfur has the ability to creep into tiny crevices to aid lubrication [49].

About this same time, another famous engineer was engaged in an endeavor that forever changed the way machining was done and how machine shops were managed. Thurston's contemporary, Fredrick W. Taylor, was a tool engineer in the employ of the Midvale Steel Company, Philadelphia. As foreman of the machine shop, he aspired to discover a method to manage the cutting of metals so that by optimizing machine speeds and work feed rates, production rates could be significantly increased. In 1883, his various experiments in cutting metal proved that directing a constant heavy stream of water at the point of chip removal so increased the cutting speed that the output of the experimental machine rose 30 to 40% [50]. This was a discovery of prime importance when it is considered that it contradicted Mushet, who insisted that as standard practice his "self-hardening" tools must be run "dry" [51]. Taylor's experiments revealed that the two most important elements of the machining processes were left untouched by earlier experimenters, even those in academia [52]. Those two elements were the effect of cooling the tool with a rapid cooling fluid and the contour of the tool.

Taylor published his findings in an epochal treatise, *On the Art of Cutting Metal*, in 1907, based on the results of 50,000 tests in cutting 800,000 pounds of metal. He reported that the heavy stream of water, which cooled the cutting tool by flooding at the cutting edge, was saturated with carbonate of soda to prevent rusting. The cutting fluid was termed "suds." This practice was incorporated onto every machine tool in the new machine shop built by the Midvale Steel Works in 1884. At Taylor's direction, each machine was set in a cast iron pan to collect the suds which were drained by piping into a central well below the floor. The suds were then pumped up to an overhead tank from which the coolant was returned to each machine by a network of piping [53]. This was the first central coolant circulation system, the forerunner of those huge 100,000 plus gallons central coolant systems in use today for supplying cutting fluids to automated machine transfer lines in machining centers.

No secret was made of Taylor's coolant system, and by 1900 the idea of a circulating coolant system was copied in a machine designed by Charles H. Norton. It had a built-in suds tank and a pump capable of circulating 50 gallons of coolant per minute, evidence that Norton appreciated the need to avoid heat deformation at high cutting rates [54].

E. Status of Metalworking Fluids (1900–1950)

As a result of engineers seeking more productive machining methods in upgrading the design of machine tools, and metallurgists producing stronger and tougher alloy steels, the compounding of metalworking fluids likewise improved. At the turn of the century, the metalworking fluids industry provided machinists with a choice of several metalworking fluids: straight mineral oils, combinations of mineral oils and vegetable oils, animal fats (lard and tallow), marine oils (sperm, whale, and fish), mixes of free sulfur and mineral oil used as cutting oils, and of course "suds."

The lubricant manufacturers of this era were well versed in the art of grease making, having learned the value of additives as early as 1869 with E. E. Hendrick's patented "Plumboleum," a mixture of lead oxide and mineral oil. Grease, in many cases, was the media of choice used for metal deformation. They were simple compounds, mixtures of metallic soaps, mineral or other oils and fats, and sometimes fibers [55].

World War I had a significant effect on the course of metalworking fluid development. In the early stages of the European involvement, white oil could no longer be imported from Russia. An American entrepreneur, Henry Sonneborn, who had made petroleum jelly and white oil for the pharmaceutical industry since 1903, found his white mineral oil and related products in great demand by lubricant manufacturers [56]. Chemists entered the endeavor by using a chemical process, the acidification of neutral oil with sulfuric acid, which resulted in a reaction product of a mixture of white oil and petroleum sulfonate. The white oil was extracted with alcohol. The sulfonate was discarded until it was discovered to be most useful as a lubricating oil additive and also in compounding metalworking fluids. Sulfonates eventually were found to combine with fatty oils and free fatty acids to make emulsions [57].

1. Development of Compounded Cutting Oils

As tougher alloy steels became more common and as machine tool and cutting tool speeds increased, the stresses incurred in the machining process tended to overwork the cutting oils. These were mostly combinations of mineral oils and lard oil, or mixtures of free sulfur and mineral oil. Overworking caused a chemical breakdown resulting in objectionable odors, rancidity, and very often dermatitis.

The disadvantages of those cutting oils had to be addressed. In 1918, no doubt spurred by the demands of the munitions industries and the need for greater precision in machining, serious research into better compounding of sulfurized cutting oils began and continued into the late 1920s. The problem to overcome was to extend the limits of sulfur combined with mineral oil by effecting a means of chemically reacting sulfur with the hydrocarbon molecules. This inhibited the natural corrosiveness of sulfur, yet gained the maximum benefit for sulfur for the machining process. In 1924 a special sulfochlorinated oil was patented by one of the oldest lubricant compounding companies in the United States and marketed as Thread-Kut 99. It is still used today for such heavy duty machining operations as thread cutting and broaching on steels [58].

But these chemically compounded oils did not solve all cutting difficulties. The new, highly sulfochlorinated cutting oils could not be used for machining brass or copper since sulfur additives stained those metals black and contributed to eventual corrosion [59].

2. Development of "Soluble Oils"

The worth of Taylor's experience was not lost on the engineering and manufacturing community. His demonstration of the profound effect that an aqueous chemical fluid had on machine productivity began the search for water/oil/chemical-based formulas for metalworking fluids. W. H. Oldacre has written that although "water-mixed oil" emulsions were used extensively in the first quarter of the twentieth century, and the wide range of formulations made a very important contribution to machine shop practice, it is not clear when the first crude emulsions were made by mixing "suds" (soda water) with fatty lubricants. History has neglected the commercial development of soluble oils [60].

Around 1905, when chemists began to look at colloidal systems, the scientific basics of metalworking fluid formulation began to unfold. Industrial chemists focused their attention upon emulsions—colloidal systems in which both the dispersed and dispersing phases are liquids. Two types of emulsions were recognized: a dispersion of oil or hydrocarbon in aqueous material such as milk and mayonnaise; and dispersions of water in oil such as butter, margarine, and oil field emulsions. Theories of emulsification began with the surface tension theory, the adsorption film theory, the hydration theory, and the orientation theory put forth by Harkins and Langmuir. These theories explained the behavior of emulsifying agents, which eventually found a direct application in the formulation of cutting fluids.

It has been reported that an English chemist, H. W. Hutton, discovered a way to emulsify oil in water in 1915. What it comprised and how it was made is not described [61]. However, in the United States in 1915, an early brochure (Technical Bulletin 16, still available from the Sun Refining Co., Tulsa, Oklahoma) by one of the oldest oil companies claimed the innovation of the first "all petroleum based (naphthenic) soluble oil." This was first marketed under the name of Sun Seco during World War I.

The growing body of knowledge on colloid and surfactant chemistry led to the compounding of various "soluble oils" using natural fatty oils. H. W. Hutton was granted a patent for the process of producing water-soluble oils by compounding sulfonated and washed castor oil with any sulfonated unsaponified

Historical Development

fatty oil (other than castor oil), and then saponifying the sulfonated oils with caustic alkali [62].

After World War I, new developments in lubrication science through the work of Hardy and Doubleday (1919–1933) ellucidated the mechanism of boundary lubrication [63]. The petrochemical industry began to flourish while applications for new synthetic chemicals such as detergents and surfactants found many commericial and industrial uses. The automobile industry recovered. The effort to speed up mass production of cars required stronger machine tools capable of faster cutting speeds. Oil-in-water emulsions were the preferred fluids except in heavy duty machining operations such as broaching, gear hobbing, and the thread cutting of tough alloy steels where chemically compounded oils were used.

The need for stable emulsions in the food, cosmetics, and soapmaking industries, as well as by the metalworking fluid manufacturers, maintained high interest in oil/water emulsions. The research of B. R. Harris, expanding upon the orientation theory of emulsions, focused on the synthesis of many new compounds relating their chemical structure to various types of surface modifying activity. Reporting in *Oil and Soap*, Harris established that all fatty interface modifiers have two essential components, a hydrophilic part which makes the compound water soluble and a lipophilic part which makes the compound fat soluble. These must be in balance to effect a good emulsion [64]. As research in this area continued, many emulsifying agents were developed for the previously mentioned industries. Some, the amine soaps, wetting agents, and other special function molecules, were compounded with mineral and/or vegetable oils by metalworking fluid compounders to effect stable "soluble oils" [65].

3. Influence of World War II

With the growth of the aircraft industry, exotic alloys of steel and nonferrous metals were introduced, creating the need for even more powerful machine tools having greater precision capability. Better metalworking fluids to effectively machine these new tough metals were also needed. The circumstances of World War II, which demanded aircraft, tanks, vehicles and other war equipment, began a production race of unknown precedent. Factories ran 24 hours daily, never closing in the race to produce war goods. The effort centered on new machine tool design to shape the new materials and to make production parts as fast as possible. The cover of the February 24, 1941 edition of *Newsweek* featured a huge milling machine carrying the title "The Heart of America's Defense: Machine Tools." In fact, metalworking fluids along with machine tools are at the heart of the cutting process. The demand for more effective war production translated into faster machining speeds. Higher feed rates using the available fluids led to problems such as poor finishes, excessive tool wear, and part distortion. The need to satisfy the war production demand mandated inquiry into the mechanics of the machining process in both Europe and the United States.

4. Mechanisms of Cutting Fluid Action

In 1938 Schallbroch, Schaumann, and Wallichs in Germany tested machinability by measuring cutting temperature and tool wear, and in so doing derived an empirical relationship between tool life and cutting tool temperature [66]. In the United States in about the same period, H. Ernst, M. E. Merchant, and M. C. Shaw studied the mechanics of the cutting process. Ernst studied the physics of metal cutting and determined that a rough and torn surface is caused by chip particles adhering to the tool causing a built-up edge (BUE) on the nose of the cutting tool due to high chip friction. Application of a cutting fluid lowered the chip friction and reduced or eliminated the BUE [67]. This confirmed Rowe's opinion that the BUE was the most important consideration to be addressed in the machining process [68]. Many studies were made but the researchers who made the most important discoveries affecting the course of metalworking fluid development were employed by one of the largest machine tool builders in the United States.

Ernst and Merchant, seeking to quantify the frictional forces operating in metal cutting, developed an equation for calculating static shear strength values [69]. Merchant, in another study, was able to measure temperatures at the tool–chip interface. He found that in this area heat evolves from two sources, the energy used up in defoaming the metal and the energy used up in overcoming friction between the chip and the tool. Roughly two-thirds of the power required to drive the cutting tool is consumed by deforming the metal, and the remaining third is consumed in overcoming chip friction. Merchant found that the right type of cutting fluid could greatly reduce the frictional resistance in both metal deformation and in chip formation, as well as reduce the heat produced in overcoming friction [70].

Ernst and Merchant began a three-year study to scientifically quantify the friction between the cutting tool and the chip it produced. They found temperatures at the tool–chip interface ranging between 1000 and 2000°F (530 and 1093°C) and the pressure at the point was frequently higher than 200,000 psi (1,380,000 kPa) [71]. Bisshopp, Lype, and Raynor also investigated the role of the cutting fluid in machining experiments to determine whether or not a continuous film existed in the tool–chip interface. They admitted that in some experiments the cutting fluid did appear to penetrate as indicated by examination of the tool and the workpiece under ultraviolet light. They concluded that a continuous film, as required for hydrodynamic lubrication, could not exist in the case where a continuous chip was formed. Neither was it possible for fluid to reach the areas where there was a tool–chip contact in the irregularity of the surfaces [72]. Other researchers, A. O. Schmidt, W. W. Gilbert, and O. W.

Historical Development

Boston, investigated radial rake angles in face milling and the coefficient of friction with drilling torque and thrust for different cutting fluids [73]. Schmidt and Sirotkin investigated the effects of cutting fluids when milling at high cutting speeds. Depending on which of the various cutting fluids were used, tool life increased approximately 35 to 150% [74].

Ernst and Merchant further studied the relationship of friction, chip formation, and high-quality machined surfaces. Their research belied the conclusions of Bisshopp et al. They found that cutting fluid present in the capillary spaces between the tool and the workpiece was able to lower friction by chemical action [75]. Shaw continued this study of the chemical and physical reactions occurring in the cutting fluid and found that even the fluid's vapors have constituents that are highly reactive with the newly formed chip surfaces. The high temperatures and pressures at the contact point of the tool and chip effect a chemical reaction between the fluid and tool–chip interface resulting in the deposition of a solid film on the two surfaces which becomes the friction reducing agent [76,77].

Using machine tool cutting tests on iron, copper, and aluminum with pure cutting fluids, Merchant demonstrated that this reaction product, which "plated out" as a chemical film of low shear strength, was indeed the friction reducer at the tool–chip interface. He stated that materials such as free fatty acids react with metals to form metallic soaps and that the sulfurized and sulfochlorinated additives in turn form the corresponding sulfides and chlorides acting as the agents that reduce friction at the tool–chip interface. However, he quickly cautioned that as cutting speeds increase, temperature increases rapidly and good cooling ability from the fluid is essential. At speeds of over 50 ft/min (254 mm/s) the superior cutting fluid must have the dual ability to provide cooling as well as friction reduction capacity [78].

Having learned which chemical additives are effective as friction reducers, Ernst, Merchant, and Shaw theorized that if they could combine these chemicals with water in the form of a stable chemical emulsion, a new cutting fluid having both friction reducing and cooling attributes could be created. In 1945, as a result of this research, their company compounded a new type of "synthetic" cutting fluid [79]. The new product, described as a water-soluble cutting emulsion with the name of CIMCOOL, appeared as a news item in a technical journal in October 1945 [80]. Two years later, the first semisynthetic metalworking fluid was introduced by this same company at the 1947 National Machine Tool Builders Show. It was a preformed emulsion very similar to a soluble oil but with better rust control and chip washing action [81]. This research was one of the important developments in metalworking fluid formulation in that it provided the impetus for a whole new class of metalworking fluids, facilitating the new high-speed machining and metal deformation processes developed in the next quarter century.

5. Metalworking Fluids and the Deformation Process

During the same period of investigation into cutting fluid effects upon the metal removal process, many papers appeared in technical journals on the ameliorating effects of lubricants and "coolants," as the aqueous-based fluid came to be termed. In the next decade, much research appeared in the technical literature on the theories of metalforming and how the lubricants used affected the metal deformation processes of extrusion, rolling, stamping, forging, drawing, and spinning. Notable among them is the often-quoted work by Bowden and Tabor on the friction and lubrication of solids [82], Nadai's theory of flow and fracture of solids [83], Bastian's works on metalworking lubricants discussing their theoretical as well as the practical aspects [84], theories of plasticity by Hill in 1956 [85] and by Hoffman and Sacks [86], followed by Leug and Treptow's discussion of lubricant carriers used in the drawing of steel wire [87]. Also notable are the investigations of Billingmann and Fichtl on the properties and performance of the new cold-rolling emulsions [88] and Schey's investigations on the lubrication process in the cold rolling of aluminum and aluminum alloys [89].

Metalworking deformation processes involve tremendous pressures on the work metal. Consequently, very high temperatures are produced demanding a medium to effect friction reduction and cooling. If these stresses are not mitigated, there is the imminent danger of wear and metal pick-up on the dies, producing scarred work surface finishes [90]. To prevent these maleffects of metalforming, a suitable material must be used to lubricate, cool, and cushion both the die and the workpiece. In general, metal deformation processes rely upon the load carrying capacity and the frictional behavior of metalworking lubricants as their most important property. In some cases, however, friction reduction is critical, as in rolling operations. Insufficient friction would permit the metal to slide edgewise in the mill and cause the rolls to slip on the entering edge of the sheet or strip. Lack of friction also causes a problem in forging, a condition known as "flash," which prevents sufficient metal from filling the die cavities [91].

VI. METALWORKING FLUIDS TODAY

At midcentury, metalworking fluids had acquired sufficient sophistication and proved to be the necessary adjunct in high-speed machining and in the machining of difficult material: the exotic steels and specially alloyed nonferrous metals. They began to be regarded as the "corrector" of many machining problems and sometimes, by the uninitiated or inexperienced, were expected to be a cure-all for most machining problems. In the next decade many cutting fluid companies sprang into existence, offering a multitude of metalworking formulations to ameliorate machining problems and increase rates of production. Listings of metalworking fluids are to be found in a great number of publications of technical papers and handbook publications of various societies that cater to the lubrication engineering, tool making, and metallurgical communities.

Considering the many processes and the myriad of products available, there was and is confusion and controversy as to the best choice of fluid in any given situation. In the 1960s the literature published by various technical organizations on the subject of how and what to use in metalworking processes was profuse. It was recognized by the metalworking community that direction was desirable, but there seems to be an isolation of those involved in the metalworking process from those involved in metalworking lubrication. As Schey has pointed out, the province of the metalworking process has traditionally been within the sphere of mechanical engineers and metallurgists; while the area of metalworking lubrication was within the expertise of chemists, physicists, and manufacturing process engineers. The National Academy of Science, observing this division, realized the need for communication among these specialists to integrate current knowledge and further the expansion of metalworking fluid technology and metalworking processes. They directed their Materials Advisory Board to institute the "Metalworking Processes and Equipment Program," a joint effort of the Army, Navy, Air Force, and NASA. One of the outcomes of this program was a comprehensive monograph containing the interdisciplinary knowledge of metalworking processes and metalworking lubricants to serve both as a text and a reference book [92].

This brief history of the evolution of metalworking fluids shows that the dynamics of metalworking fluid technology are dependent upon the dynamics of metalworking processes as created by the parameters of machine tool design. These dynamics are mutually dependent parts of the total process and can only be investigated jointly. The body of knowledge evolving from metalworking fluid technology developed by these "cross culture" [93] engineers and scientists contributed significantly to the growing body of science and technology in the area of friction, lubrication, and wear. In the late 1960s this technology blossomed into the new science of *tribology*. A "veritable explosion of information" has followed since 1970 [94].

REFERENCES

1. E. Cleves, "Report on the volume of lubricants manufactured in the United States by independent lubricant manufacturers in 1990." 1991 Annual Meeting Independent Lubricant Manufacturers Association, Alexandria, VA, pp. 1–5 (1991).
2. National Petroleum Refiners Association, "1990 Report on U.S. lubricating oil sales," Washington, D.C., p. 5 (1991).
3. J. G. Wills, *Lubrication Fundamentals*, Marcel Dekker, New York, pp. 1–2 (1980).

4. J. A. Schey, *Metal Deformation Processes: Friction and Lubrication*, Marcel Dekker, New York, p. 1 (1970).
5. E. L. H. Bastian, *Lubr. Eng. 25*(7): 278 (1968).
6. D. Dowson, *History of Tribology*, Longmans Green, New York, p. 253 (1979).
7. C. Singer et al., *A History of Technology*, Vol. III, Oxford Univ. Press, London, p. 668 (1957).
8. C. Singer et al., *A History of Technology*, Vol. III, Oxford Univ. Press, London, p. 663 (1957).
9. C. Dowson, *History of Tribology*, Longmans, Green, New York, p. 126 (1979).
10. J. A. Schey. *Metal Deformation Processes: Friction and Lubrication*, Marcel Dekker, New York, pp. 1–2 (1970).
11. D. Dowson, *History of Tribology*, Longmans, Green, New York, pp. 177–178 (1979).
12. D. Dowson, *History of Tribology*, Longmans, Green, New York, p. 154 (1979).
13. E. L. H. Bastian, *Metalworking Lubricants*, McGraw–Hill, New York, p. 3 (1951).
14. J. A. Schey, *Metal Deformation Processes: Friction and Lubrication*, Marcel Dekker, New York, p. 1 (1970).
15. C. Singer et al., *A History of Technology*, Vol. III, Oxford Univ. Press, London, p. 663 (1957).
16. C. Singer et al., *A History of Technology*, Vol. III, Oxford Univ. Press, London, p. 669 (1957).
17. K. B. Lewis, *Wire Ind. 1*: 4–8 (1936).
18. K. B. Lewis, *Wire Ind. 2*: 49–55 (1936).
19. J. A. Schey, *Metal Deformation Processes: Friction and Lubrication*, Marcel Dekker, New York, p. 5 (1970).
20. L. T. C. Rolt, *A Short History of Machine Tools*, MIT Press, Cambridge, MA, p. 11 (1965).
21. R. S. Woodbury, *Studies in the History of Machine Tools*, MIT Press, Cambridge, MA, p. 1 (1972).
22. D. Dowson, *History of Tribology*, Longmans, Green, New York, pp. 21–23 (1979).
23. R. S. Woodbury, *History of the Lathe to 1850*, MIT Press, Cambridge, MA, p. 23 (1961).
24. J. G. Landels, *Engineering in the Ancient Worlds*, Univ. of California Press, Berkeley, CA, pp. 77, 199–215 (1978).
25. R. S. Woodbury, *History of the Grinding Machine*, MIT Press, Cambridge, MA, p. 13 (1959).
26. R. S. Woodbury, *History of the Grinding Machine*, MIT Press, Cambridge, MA, p. 23 (1959).
27. R. S. Woodbury, *History of the Grinding Machine*, MIT Press, Cambridge, MA, p. 21 (1959).
28. K. B. Gilbert, C. Singer et al., *History of Technology*, Vol. IV, Oxford Univ. Press, London, p. 417 (1958).
29. L. T. C. Rolt, *A Short History of Machine Tools*, MIT Press, Cambridge, MA, p. 138 (1965).
30. O. Handlin, *Eli Whitney and the Birth of Modern Technology*, Little, Brown, Boston, MA, pp. 119–143 (1956).

31. O. Handlin, *Eli Whitney and the Birth of Modern Technology*, Little, Brown, Boston, MA, p. 170 (1956).
32. R. S. Woodbury, *History of the Grinding Machine*, MIT Press, Cambridge, MA, p. 41 (1959).
33. J. Naysmyth, *James Naysmyth, An Autobiography*, Samuel Stiles, ed., Harper and Brothers, Franklin Square, London, p. 437 (1883).
34. R. S. Woodbury, *History of the Grinding Machine*, MIT Press, Cambridge, MA, p. 41 (1959).
35. K. B. Gilbert, C. Singer et al., *History of Technology*, Vol. IV, Oxford Univ. Press, London, p. 417 (1958).
36. D. Dowson, *History of Tribology*, Longmans, Green, New York, pp. 286–287 (1979).
37. *Million Dollar Directory, America's Leading Public and Private Companies*, Dunn and Bradstreet, Parsippany, NJ (1991).
38. F. M. Osborn, *The Story of the Mushets*, Thomas Nelson, London, pp. 38–52 (1952).
39. F. M. Osborn, *The Story of the Mushets*, Thomas Nelson, London, pp. 89–95 (1952).
40. R. B. Morris, ed., *Encylopedia of American History*, Harper & Row, New York, pp. 562–565 (1965).
41. W. H. Northcott, *A Treatise on Lathes and Turning*, Longmans, Green, London (1868).
42. W. H. Steed, *History of Machine Tools 1700–1910*, Oxford Univ. Press, London, p. 91 (1969).
43. R. S. Woodbury, *History of the Grinding Machine*, MIT Press, Cambridge, MA, p. 167 (1959).
44. R. S. Woodbury, *History of the Grinding Machine*, MIT Press, Cambridge, MA, p. 65 (1959).
45. A. Mallock, Action of cutting tools, *Royal Society of London Proceedings*, London, Vol. 33, p. 127 (1881).
46. R. H. Thurston, *A Treatise on Friction and Lost Work in Machining and Millwork*, Wiley, New York, p. 141 (1885).
47. D. Dowson, *History of Tribology*, Longmans, Green, New York, p. 551 (1979).
48. R. H. Thurston, *A Treatise on Friction and Lost Work in Machining and Millwork*, Wiley, New York, p. 284 (1885).
49. *Lubrication*, The Texas Oil Co., New York, Vol. 39, p. 10 (1944).
50. F. W. Taylor, *On the Art of Cutting Metals*, Society of Mechanical Engineers, New York, pp. 138–143 (1907).
51. L. T. C. Rolt, *A Short History of Machine Tools*, MIT Press, Cambridge, MA, p. 198 (1965).
52. F. W. Taylor, *On the Art of Cutting Metals*, Society of Mechanical Engineers, New York, pp. 14, 137–138 (1907).
53. F. W. Taylor, *On the Art of Cutting Metals*, Society of Mechanical Engineers, New York, p. 9 (1907).
54. L. T. C. Rolt, *A Short History of Machine Tools*, MIT Press, Cambridge, MA, p. 213 (1965).

55. *Lubrication*, The Texas Oil Co., New York, Vol. 39, p. 73 (1944).
56. *The Oil Daily*, August 6, pp. B7–B8, (1960).
57. H. W. Hutton, Improvement in or relating to the acid refining of mineral oil, British Patent No. 13,888/20, May 20 (1920).
58. W. H. Oldacre, Cutting fluid and process of making the same, U.S. Patent No. 1,604,068, October 19, 1926, assigned to D. A. Stuart Oil Co. (1923).
59. *Lubrication*, The Texas Oil Co., New York, Vol. 39, pp. 9–16 (1944).
60. W. H. Oldacre, *Lubr. Eng. 1*(8): 162 (1944).
61. R. Kelly, *Carbide Tool J. 17*(3): 28 (1985).
62. H. W. Hutton, *Process of producing water-soluble oil*, British Patent No. 13,999, accepted Aug. 20 (1923).
63. D. Dowson, *History of Tribology*, Longmans, Green, New York, p. 351 (1979).
64. B. R. Harris, S. Epstein, and R. Cahn, *J. Am. Oil Chem. Soc. 18*(9): 179–182 (1941).
65. *Emulsions*, 7th ed., Carbide and Carbon Chemicals, pp. 28–39, (1946).
66. H. Schallbock, H. Schaumann, and R. Wallichs, Testing for machinability by measuring cutting temperature and tool wear, *Vortrage der Hauptversammlung, Der Deutsche Gesellschaft fur Metalkunde*, VDI Verlag, pp. 34–38 (1938).
67. H. Ernst, *Machining of Metals*, American Society of Metals, Cleveland, OH, pp. 1–34 (1938).
68. G. W. Rowe, *Introduction of Principles of Metalworking*, St. Martin's Press, New York, pp. 265–269 (1965).
69. H. Ernst, M. E. Merchant, "Surface friction of clean metals," *Proceedings of Massachusetts Institute of Technology Summer Conference on Friction and Surface Finish*, pp. 76–101 (1940).
70. M. E. Merchant, *J. Appl. Phys. 16*: 267–275 (1945).
71. R. Kelly, *Carbide Tool J. 17*(3): 26 (1985).
72. K. E. Bisshopp, E. F. Lype, and S. Raynor, *Lubr. Eng. 6*(2): 70–74 (1950).
73. A. O. Schmidt, W. W. Gilbert, and O. W. Boston, *Trans. ASME 164*(7): 703–709 (1942).
74. A. O. Schmidt and G. V. B. Sirotkin, *Lubr. Eng. 4*(12): 251–256 (1948).
75. H. Ernst and M. E. Merchant, *Surface Treatment of Metals*, American Society of Metals, Cleveland, OH, pp. 299–337 (1948).
76. M. C. Shaw, *Met. Prog. 42*(7): 85–89 (1942).
77. M. C. Shaw, *J. Appl. Mech. 113*(3): 37–44 (1948).
78. M. E. Merchant, *Lubr. Eng. 6*(8): 167–181 (1950).
79. *1884 Cincinnati Milacron 1984*, Cincinnati Milacron, Cincinnati, OH, p. 88 (1984).
80. *Lubr. Eng. 6*(6): 79 (1950).
81. R. Kelly, *Carbide Tool J. 17*(3): 29 (1985).
82. F. P. Bowden and F. Tabor, *Friction and Lubrication of Solids*, Clarendon Press, Oxford, England (1950).
83. A. Nadai, *Theory of Flow and Fracture of Solids*, McGraw–Hill, New York (1950).
84. E. L. H. Bastian, *Metalworking Lubricants*, McGraw–Hill, New York (1951).
85. R. Hill, *The Mathematical Theory of Plasticity*, Oxford Univ. Press, London (1956).
86. O. Hoffman and G. Sachs, *Theory of Plasticity*, McGraw–Hill, New York, (1957).
87. W. Leug and K. H. Treptow, *Stahl u. Eisen 76*: 1107–1116 (1956).

88. J. Billigman and W. Fichtl, *Stahl u. Eisen 78*: 344–357 (1958).
89. J. Schey, *J. Inst. Met. 89*: 1–6, (1960).
90. G. W. Rowe, *Wear 7*: 204–216 (1964).
91. L. B. Sargent, *Lubr. Eng. 21*(7): 286 (1965).
92. J. A. Schey, *Metal Deformation Processes: Friction and Lubrication*, Marcel Dekker, New York (1970).
93. K. C. Ludema, *Lubr. Eng. 44*(6): 447–452 (1988).
94. J. Schey, *Tribology in Metalworking*, American Society for Metals, Metals Park, OH, p. v (1983).

2
Metallurgy for the Nonmetallurgist with an Introduction to Surface Finish Measurement

JAMES E. DENTON
Cummins Engine Company, Inc.
Columbus, Indiana

I. INTRODUCTION

In one sense, it is unfortunate that metallurgy is such a mature science. Metals have been in use for so long and have been so effective in performing their role that they tend to be taken for granted. In the modern "high-tech" world of materials, it is ceramics, engineered polymers, and fiber-reinforced composites that monopolize the scientific literature. Meanwhile, metals continue to quietly go about their business of supporting the infrastructure of our society.

Metalworking fluids, as the name implies, exist solely to facilitate the shaping of a useful metal object by a cutting, grinding, stamping, or drawing process. The ultimate objective is the useful metal part, and the lubricating/cooling fluid is only a process aid. Before developing or selecting the proper metalworking fluid for an operation, it is necessary to understand something about the metal that will be encountered. Is the metal easy or difficult to machine?

How does mild steel differ from stainless steel? What do the numerical metal alloy designations indicate? Does the metal corrode readily, or is it resistant to corrosion? Can the same fluid be used to machine both cast iron and aluminum? Will this fluid affect tool steel or carbide tooling? Surface finish may definitely be affected by the choice of metalworking fluid; how is surface finish expressed?

In this chapter, a short course in metallurgy and surface finish measurements is presented for the nonmetallurgist. It is the intent of the chapter to give the reader an appreciation for the technology of metals and an understanding of the fundamentals of metallurgy with special focus on the ways in which these fundamentals influence the behavior of metals during the various forming and fabrication processes they undergo on their way to becoming useful products. We will begin the chapter with a discussion on the general topics of structure and properties that are common to all metals. Later in the chapter, we will concentrate on the more specific topics of composition and thermal treatments. A short section on the subject of surface texture measurement concludes the chapter.

II. THE STRUCTURE OF METALS

As is all matter in our universe, metals are composed of atoms. Further, metals are crystalline in structure. By crystalline we understand that the individual atoms in metals are arranged in a regular and predictable three-dimensional array. This array is called the crystal lattice. The metallic bonding between the neighboring atoms in the crystal lattice is somewhat different than chemical bonding and gives rise to the inherent strength and malleability of the metals, properties not exhibited by other chemical compounds. It is also a special property of this metallic bond that the outer electrons of the atoms are generally shared by all the atoms within the structure and the electrons are free to circulate throughout the whole of the metal. This feature gives rise to the property of electrical conductivity, one of the properties that distinguish metals from nonmetals. Further, the atoms vibrate about their nominal position within the crystal lattice giving rise to the thermal properties of conductivity, thermal expansion, specific heat, and, ultimately, the melting point of the metal.

While there are a large number of possible crystalline arrangements, there are four that are particularly important in understanding most of the metals we come into contact with in our daily work. These are termed body-centered cubic, face-centered cubic, body-centered tetragonal, and close-packed hexagonal. When these structures are depicted, the arrangement of only the smallest number of atoms that completely describe their spatial relationship, the so-called unit cell, need be shown. Repetition of these unit cells in all three dimensions builds up the total structure. The unit cell should be thought of as the basic building block. Figure 1 shows the basic unit cells of the four important crystal structures.

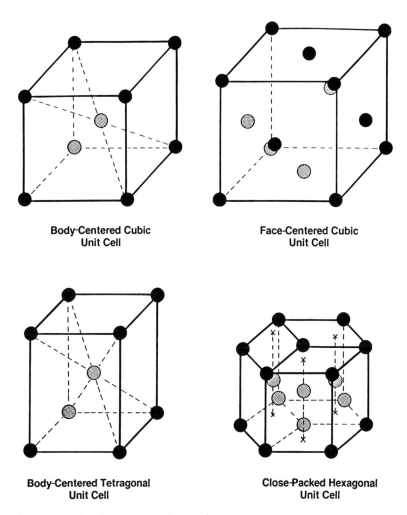

Figure 1 Crystal structures in metals.

In Fig. 1 the atom sites are shown as small points which might be considered as representing the nucleus of the atom. In reality, the atoms are more nearly spheres and the outer electron shells of neighboring atoms actually touch each other. As might be deduced, the body-centered cubic structure consists of a cube with atoms located at each corner and one atom at the very center of the cube. The face-centered cubic structure consists of a cube with atoms located at each corner and at the center of each face, like dice with a five showing on every face. The body-centered tetragonal cell is like the body-centered cubic cell except that it is stretched in one direction. The top and bottom faces are squares while the four side faces are rectangles. The hexagonal cell is most easily recognized as a hexagonal prism with atoms located at every corner and the center of the top and bottom ends.

Since the hexagon is symmetrical about its center, the simplest unit cell is one whose end faces are parallelograms arrived at by slicing the hexagon into three pieces. This is the simple hexagonal structure. There is an additional potential atom site in each of the three unit cells making up the hexagon. If this site is occupied, the unit cell is termed close-packed hexagonal. The crystal structures are frequently referred to by their acronyms: body-centered cubic is abbreviated BCC, face-centered cubic is abbreviated FCC, body-centered tetragonal is abbreviated BCT, and close-packed hexagonal is abbreviated HCP.

Visualizing these four bodies is, at first, difficult but there are very good reasons for understanding the structures. Malleability, or the ability of a metal to undergo deformation without breaking into fragments, is explained by the slipping of the various planes of atoms past one another. Slip tends to occur most readily on planes that have the highest atomic density. For example, in the face-centered cubic structure, a diagonal plane that intersects three corners, as shown in Fig. 2, has the densest atom population of any of the possible planes in the FCC system.

It is called the 111 plane, accounting for most of the slip occurring when a FCC-structured metal is deformed. The fact that slip is confined to certain discrete crystallographic planes explains the development of strain lines and texture on the surface of formed sheet metal components. It also accounts for "earing" which is the concentration of excess material at specific points on the edges of drawn or ironed metal components [1]. Table 1 shows some common metals and their normal crystal structures.

It is necessary to specify conditions when identifying the crystal structure of a metal, because many metals can exist in more than one crystal structure at different temperatures. This phenomena is called allotropism. Iron, for example, has three equilibrium crystal structures. Up to 1670°F (910°C) iron is body-centered cubic. From 1670 to 2552°F (910–1400°C) iron is face-centered cubic. From 2552°F (1400°C) to the melting point, it is body-centered cubic again but with different distances between the neighboring atoms than in the room-temperature form. The transition between crystal structures is accompanied by a change in volume. This allotropic change in crystal structure is the basis for heat treatment strengthening of metals which will be discussed in some detail later in this chapter.

All is not perfect in the world of crystal structure. Several types of defects in the normal atomic arrangement are frequently encountered. There may be atoms missing from a site and this type of defect is called a vacancy. Occasionally an extra plane of atoms may cause a disruption in the otherwise normal lattice

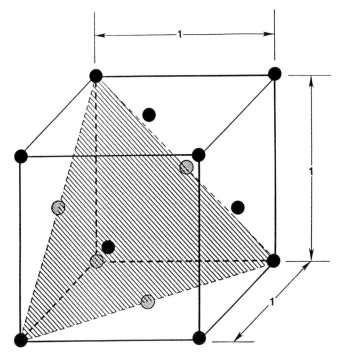

Figure 2 [111] Primary slip plane in face-centered cubic unit cell.

TABLE 1 Normal Crystal Structure of Some Common Metallic Elements

Metal	Crystal system
Aluminum	FCC
Chromium	BCC
Copper	HCP
Iron	BCC
Nickel	FCC
Tin	BCT
Zinc	HCP

Source: Ref. 1, p. 84.

and this is called a dislocation. There may also be foreign or alloy atoms substituting in a normal atom site in the parent structure. Foreign atoms are likely to have a different diameter than the matrix atoms giving rise to a local distortion in the lattice. A special form of foreign atom defect is the interstitial atom. Elements such as carbon and nitrogen have very small atomic diameters and typically have the correct size to fit into interstitial lattice sites. While these defects may be thought of as imperfections in the crystal structure they have many beneficial effects and therefore are not necessarily detrimental. Figure 3 illustrates some of the common crystal defects.

Up until this point we have discussed crystal structure within the context of a single crystal. To be sure, there are examples in nature, as well as intentionally prepared mechanical components that exist as single crystals. Very large, naturally occurring quartz crystals are quite common. Some jet engine turbine blades, for example, are directionally solidified to result in the entire component being composed of a single crystal. The silicon crystals used to construct transistors are single crystals in which extreme care has been taken in their preparation to create as perfect a crystal structure as possible. However,

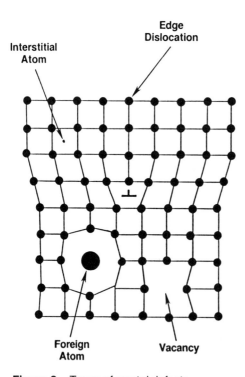

Figure 3 Types of crystal defects.

most metallic bodies are polycrystalline or composed of many crystals. Within each individual crystal the arrangement of the atoms is near perfect, as we have previously described them. Adjacent crystals, however, may have completely different orientations as shown in Fig. 4, so that where two adjacent crystals meet, their atom planes do not line up exactly.

The individual crystals in a polycrystalline metal are called grains, and the contact regions between adjacent crystals are called grain boundaries. In a properly manufactured and processed metal, the grain boundaries are stronger than the grains themselves; and when they are broken, failure occurs transgranular or through the grain. Under some adverse conditions the grain boundaries may develop problems which diminish their strength. When such a metal is fractured, the failure is intergranular or within the grain boundaries. The strength and ductility of the metal are significantly reduced in this type of failure. The fracture in this case may appear grainy and faceted giving rise to the old folklore conclusion that "the metal crystallized" and caused the failure. As we now know, the metal was always crystalline but the intergranular fracture made the crystallinity more visually apparent.

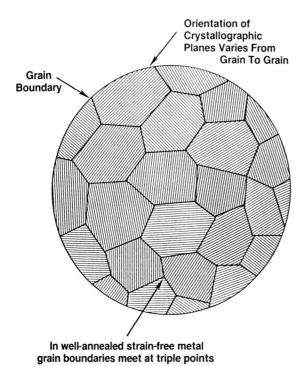

Figure 4 Polycrystalline grain structure.

III. THE PROPERTIES OF METALS

The properties of metals are divided into two categories: physical and mechanical. Physical properties include those inherent characteristics that remain essentially unchanged as the metal is processed. Such characteristics as density, electrical conductivity, thermal expansion, and modulus of elasticity are some of the more typical examples of physical properties.

Mechanical properties are those that can be drastically changed by processing and thermal treatment. Such characteristics as hardness, strength, ductility, and toughness are the properties that govern the performance of the metal in use. A common and easily determined mechanical property of a metal is hardness. One of the earliest hardness tests developed was a scratch test based on the Mohs scale. This test involves a scale of one to ten with talc as one and diamond as ten. The test is based on a determination of which material will scratch the other and is somewhat subjective and not particularly quantitative. Most hardness testing is based on indentation where an indentor of a specified geometry is pressed into the sample under a specified load. The projected area of the indentation or depth of penetration is determined and the hardness is expressed in the dimensions of pressure, for example, kilograms per square millimeter. In the United States, Brinell and Rockwell tests are the most common. In Europe, the Vickers test is more common.

The Brinell test uses a 10 mm diameter ball with a load of 3000 kg for ferrous metals or 500 kg for softer nonferrous metals. The Rockwell test uses either a conical diamond indentor, a 1/16 in. diameter ball or a 1/8 in. diameter ball with loads of 60, 100, or 150 kg. The combinations of three different loads with three different indentors provides nine possible Rockwell scales that are useful for metals ranging from very soft to very hard. There is also a special Rockwell test for thin or fragile samples called the Superficial test that uses lighter loads of 15, 30, or 45 kg. The Vickers test is similar to the Rockwell test but uses a pyramidal-shaped diamond indentor with equal diagonals. There are numerous other hardness tests that have been devised for special purposes but the ones described earlier are the most frequently encountered. In specifying a hardness value it is necessary not only to give the numerical value but also to indicate the scale or type of test used. Recognized abreviations appended to the hardness number are of a format that begins with H for hardness followed by additional letters and numbers indicating the particular test. HV indicates the Vickers test. HV and DPH, for diamond pyramid hardness, are used interchangeably. HBN stands for the Brinell hardness test. If a particular test is conducted with a variety of loads, the load may be subscripted after the abbreviation, e.g., HBN_{500}. HRC is Rockwell hardness on the C scale, HRB is Rockwell hardness on the B scale, etc. Table 2 shows a summary of these various tests and the hardness ranges over which they are used [2].

TABLE 2 Approximate Hardness Conversions[a]

Brinell		Vickers	Rockwell				Rockwell Superficial					Applications
500 kg	3000 kg		A 60	B 100	C 150	15N	30N brale	45N	15T	30T 1/16 in. ball	45T	
—	—	1076	87	—	70	—	86	—	—	—	—	Carbide
—	757	860	84	—	66	92	83	70	—	—	—	
—	682	737	82	—	62	91	79	68	—	—	—	
—	560	605	79	—	56	88	74	62	—	—	—	Hard Steel
—	496	528	77	—	51	86	69	56	—	—	—	
—	429	455	73	—	45	83	64	49	—	—	—	
—	372	393	70	—	40	80	60	43	—	—	—	
—	332	353	68	—	36	78	56	38	—	—	—	
—	290	309	66	—	31	75	51	32	—	—	—	
—	265	278	64	—	27	73	48	28	—	—	—	
195	234	247	61	—	21	70	42	21	93	83	72	Soft Steel
179	216	222	59	99	16	—	—	—	92	80	69	
163	195	202	56	96		—	—	—	90	78	65	
151	176	182	54	92		—	—	—	89	75	61	Hard Brass
135	156	163	51	88		—	—	—	87	71	55	
118	135	140	46	82		—	—	—	85	66	47	
110	125	131	44	74		—	—	—	83	63	43	Bronze
104	117	113	42	70		—	—	—	82	60	39	
95	107	107	40	68		—	—	—	80	56	33	Aluminum
85	—	100	—	60		—	—	—	76	51	25	Copper
75	—	93	—	52		—	—	—	74	48	14	
				41								

[a] Approximate conversions for a variety of metals based on ASTM E-140.

There is a proportional relationship that exists between hardness and tensile properties of steel that is useful in estimating strength. If the Brinell hardness number is multiplied by 500, the result is approximately equal to the tensile strength in pounds per square inch. If the Brinell hardness number is multiplied by 400, the result is approximately equal to the yield strength in pounds per square inch. This relationship is reasonably valid over a wide range of hardness including annealed and hardened steels. The proportionality, however, is not so good for very soft or very hard steels.

The tensile test is another common test used to describe the properties of a metal and illustrates several important characteristics. There are a large variety of tensile test specimens but all share a design that has a controlled cross section and enlarged ends for gripping. During testing the specimen is subjected to an increasing pull load at a carefully controlled rate of increase called the strain rate. Most metals are sensitive to strain rate and exhibit different properties when tested at different strain rates. Figure 5 shows the various points depicted by a tensile test.

First, it may be noted that this is a plot of load versus extension or in engineering terms, stress versus strain. It is an expected characteristic of all metals that they behave elastically. That is, as stress is applied, the metal elongates in a linearly proportional response. The first part of the tensile curve,

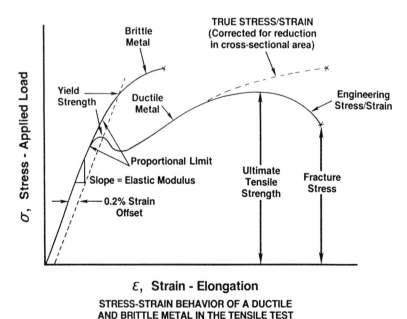

Figure 5 Tensile test stress–strain diagram.

then, is a straight line. The slope of this straight line portion, stress divided by strain, is called the elastic modulus or Young's modulus of elasticity. In the elastic portion of the curve, the elongation is completely recoverable. The metal behaves as a rubber band. If the load is removed, the metal returns exactly to its original size. If, however, the load is continually increased, a point is reached where the metal is no longer capable of stretching elastically and permanent deformation is produced. This point is called the proportional limit, above which stress is no longer linearly proportional to strain. If the proportional limit is exceeded, the metal undergoes plastic deformation. When the load is removed, the sample will be found to be longer than its original size. The proportional limit is difficult to accurately discern on the tensile curve, so an arbitrary point is defined and called the yield strength. To determine the yield strength, a line offset 0.2% from the origin is drawn parallel to the straight line portion of the tensile curve. Where it intersects the curve is called the yield point and the stress corresponding to this point is called the yield strength. There are other methods of defining the yield point but the 0.2% offset method is, by far, the most common. The appearance of the stress/strain curve above the yield point varies dramatically as a function of the thermal and mechanical history of the metal, as well as the inherent characteristics of the metal itself.

For soft ductile metals the curve may take a small dip after yield before continuing upward to the maximum load, which is called the ultimate tensile strength. While the load is being applied, the sample is responding by elongating and becoming smaller in diameter. Local necking down eventually occurs so the sample assumes an hourglass shape. Because of the reduction in diameter, the stress/strain curve shows a dropoff in load. When the stress/strain curve is corrected for the reduction in diameter, it may be seen that the work-hardening effect persists up to final failure. The corrected stress/strain curve is called the true stress/true strain curve. This is the normal behavior of metals that have a capacity to work harden.

In very hard or brittle metals the curve bends only slightly at yield and ultimate failure occurs soon after. The point of ultimate failure is called the fracture strength. A metal that exhibits a lot of plastic deformation after the yield point is called a ductile metal, while a metal that fails with very little deformation is called a brittle metal. The area under the curve is a product of force times distance and may be thought of as a work function whose physical equivalent is toughness. High toughness, then, is exhibited by a material that combines high strength with the ability to deform in order to maximize the area under the stress/strain curve.

A more common procedure for determining toughness is the impact test. This procedure rigidly supports the test specimen which is then struck by a falling pendulum, thus producing a relatively high strain rate representative of the many collision events a component may experience during its service life. In the Charpy impact test the specimen is supported at both ends as a simple beam and struck in the center by the pendulum. In the Izod impact test the specimen is clamped in a vise as a cantilever beam and struck at the unsupported end by the pendulum. In both tests the specimens are sometimes notched to concentrate the impact energy from the falling pendulum in a smaller volume of material.

Metals are susceptible to a peculiar and insidious failure mode called fatigue, where fracture occurs at a stress level substantially below the yield strength of the material. This type of failure requires cyclical stress or repeated loading and unloading of the component. Cracks are initiated and grow until the remaining unfractured area is insufficient to support the applied load, at which point failure occurs catastrophically. There is no observable deformation or bending so there is little or no warning that failure is imminent. Inspection of the fractured surface of a fatigue failure shows a series of so-called "beach marks" or approximately concentric lines whose focus point coincides with the fracture initiation site. The beach marks or lines are actually crack arrest fronts. The crack grows in stages during load application and stops between cycles, leaving a track of the characteristic beach marks. Since the effects of fatigue can be so serious, a measure of fatigue strength is an important property of a metal. A graph plotting number of cycles to failure vs. applied load yields a curve of the shape shown in Fig. 6.

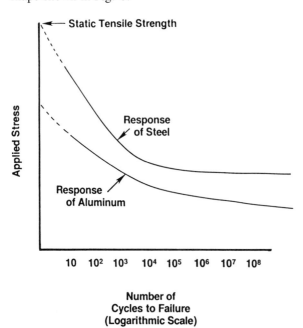

Figure 6 Fatigue test endurance limit curve.

At lower stress levels, the slope of the curve levels out and approaches a flat line for some metals, like steel. Other metals, such as aluminum never really approach horizontal and do not have a well-defined fatigue limit.

This threshold level, below which failure does not occur, is called the endurance limit. Fatigue tests are normally terminated at 10 million load cycles, and samples that survive this long are assumed to have infinite life. Fatigue life is drastically affected by numerous factors including surface finish, geometry, residual stresses, and environmental conditions [3].

IV. FERROUS METALS

The use of iron and steel as structural materials is so dominant that all metals are classified into two broad categories: ferrous and nonferrous. Ferrous metals include all the alloys whose major alloying element is iron. This broad category includes cast iron, carbon steel, alloy steel, stainless steel, and tool steel. Nonferrous materials include, basically, everything else. The more common materials in this group are aluminum, magnesium, titanium, zinc, copper- and nickel-based alloys.

A. Steel

In the simplest form, steel is an alloy of iron and carbon. The melting and reduction of iron ore was traditionally done using charcoal or coal as a fuel. As the molten iron was in contact with the carbon-rich fuel, it absorbed excess carbon into its structure resulting in an inherently brittle product called pig iron. To develop the desired ductile and tough properties of steel it was necessary to reduce the carbon content. A technological breakthrough in the production of low carbon steel from high carbon pig iron occurred in 1856 when Henry Bessemer developed the Bessemer converter which blew compressed air through the molten iron to convert the excess carbon to carbon dioxide. The basic differentiation between cast iron and steel is the carbon content, ranging up to approximately 1.5% in steel and up to 4% in cast iron.

Steel comes in a very wide range of alloys and carbon contents. The most common specifying bodies for steel composition are the American Iron and Steel Institute, AISI, The Society of Automotive Engineers, SAE, and the American Society for Testing and Materials, ASTM. AISI doesn't actually write standards but, acting as the voice for the steel industry, determines which grades are manufactured and sold in such quantities as to be considered standard grades. In the format of AISI/SAE standard alloy steels, the composition is represented by a four-digit number. The first two numbers designate the major metallic alloys present and the last two numbers designate the carbon content in hundredths of a percent. A frequently used alloy steel grade is 4140. According to this designation system it is possible to tell that the material contains chromium and molybdenum as the principal alloys along with 0.40% carbon. Table 3 shows the alloying elements associated with each of the standard grades.

The alloying elements used in steel have specific functions and their concentration ranges have been developed for the various grades to produce specific properties.

Carbon is the essential element that determines the ultimate hardness a steel is capable of achieving. Steels with low ranges up to 0.20% carbon do not respond well to heat treatment and can achieve maximum hardness up to about 35 HRC (Rockwell C hardness). When used for purposes requiring high hardness, these steels must be carburized, a case-hardening heat treatment. Steels with medium carbon, up to about 0.50%, may be fully hardened to as high as 60+ HRC. This is about the limit of martensitic hardness in medium-alloyed steels. Carbon in excess of 0.50% has little additional effect on hardness. The excess carbon above 0.50% forms carbide particles which, while they do not increase the hardness, do have a beneficial effect on wear resistance and compressive strength. The high carbon steels, notably 52100 and 1095, find application in bearings and cutting edge use.

TABLE 3 AISI/SAE Carbon and Alloy Steel Designation System

Alloy series	Nominal alloy content	Alloy series	Nominal alloy content
10XX	Plain carbon	43XX,47XX 81XX,86XX 87XX,88XX 93XX,94XX 97XX,98XX	Nickel plus chromium plus molybdenum
11XX	Resulfurized		
12XX	Resulfurized and rephosphorized		
13XX,15XX	Manganese	46XX,48XX 97XX,98XX	Nickel plus molybdenum
23XX,25XX	Nickel	50XX,51XX 51XXX[a]	Chromium
31XX,32XX 33XX,34XX	Nickel plus chromium	52XXX[a]	
40XX,44XX 41XX	Molybdenum Chromium plus molybdenum	61XX	Chromium plus vanadium

[a]These alloys have carbon in excess of 1.0%.
Source: Ref. 4.

Manganese is the most effective alloy at increasing hardenability. Simply defined, hardenability governs the rate at which a steel must be cooled from the hardening temperature to achieve full hardness. Steels with low hardenability must be cooled rapidly, usually requiring a water quench. Steels with high hardenability can be cooled more slowly and may be quenched in oil or air. Since cooling rate is also a function of mass, steels with high hardenability may also through-harden in very thick section sizes. Steels with low hardenability will harden only at the surface or not at all in thick section sizes. In resulfurized steels, some of the manganese combines with sulfur to form manganese sulfide inclusions.

Vanadium, chromium, and molybdenum are also very effective at increasing hardenability, but unlike manganese, are also strong carbide formers. These elements promote increased wear resistance and also resist softening upon exposure to elevated temperatures.

Nickel generally improves hardenability but its primary effect is as an austenite stabilizer. Austenite is the allotropic form of iron that exists at the elevated temperatures used for heat treating. Steels that contain appreciable amounts of nickel can typically be hardened from lower heat-treating temperatures. Nickel also promotes toughness and is frequently used for applications requiring high impact resistance.

Sulfur and phosphorus are usually thought of as contaminants in steel but when intentionally added, as in the 11XX and 12XX series, they promote machinability. They are essentially insoluble in iron and form nonmetallic stringer-type inclusions having relatively low melting points. During machining operations these stringers lubricate the cutting tool and also act as chip breakers. Both sulfur and phosphorus have a negative effect on strength and toughness. Lead is another alloy that falls into the category of a free machining additive but due to its toxicity is being used less and less. In an attempt to mitigate the negative effects of sulfur on strength, the so-called shape-controlled steels were developed. Calcium and tellurium are being added to some proprietary, free machining steels to reshape the stringer-type sulfides into more globular inclusions that have less of a detrimental effect on properties [4]. The alloy steels contain a maximum of about 5% alloying elements. Steels for more demanding applications such as the tool steels require much higher alloy additions.

B. Tool Steels

The obvious definition of this class of steels hardly fits. In addition to tools, these steels are used for molds, bearings, wear parts, and a wide variety of structural components. They are likewise difficult to categorize. Tool steels evolved in a highly proprietary market where a diverse population of specialty steel mills developed materials for unique applications. Each manufacturer puts its own special twist in the chemical composition to provide some real or perceived commercial market advantage. Even today, more than a few brand names still persist and are widely recognized and specified by name. Attempts to organize the extensive array of tool steels eventually settled on a system of classification by function. In the AISI designation system, tool steels are identified by a letter followed by one or two numbers. The letters classify the grades by function or application while the numbers were assigned chronologically within the grade.

1. High-Speed Steels

In the late 1800s and early 1900s, all metal cutting was done with straight high carbon steels. This grade is capable of developing high hardness but tends to soften rapidly when it heats up. For this reason cutting speeds were restrictively low to prevent the tool from overheating. In the late 1800s, it was discovered that the addition of tungsten and chromium to cutting steel made it much more resistant to softening when heated and, therefore, made it possible to increase the cutting speed to a remarkable degree. These steels came to be known as high-speed steel. A classical composition of 18% tungsten, 4% chromium, and 1% vanadium evolved and became the basis of the tungsten-type high speeds now designated as T1. The tungsten-type high-speed steels dominated the metal cutting industry until World War II when, because of a shortage of tungsten, molybdenum was substituted for most of the tungsten. The lower cost of molybdenum led to the growing popularity of the M-type high-speed steels.

2. Hot Work Die Steels

This group of tool steels, designated the H series, was developed basically for die casting of zinc, magnesium, and aluminum or for such high-temperature forming operations as extrusion and forging dies. It has three subgroups: the H1–H19 chromium type, the H20–H39 tungsten type, and the H40–H59 molybdenum type.

3. Cold Work Steels

This group of steels is restricted to forming operations that do not exceed 500°F owing to the lack of refractory alloys that resist softening. There are three subgroups: the O-type oil hardening grades, the A-type medium alloy air hardening grades, and the D-type high carbon, high chromium series.

4. Shock-Resisting Steels

This group of steels, designated the S type, is used for chisels, punches, and other applications requiring extreme toughness at high hardness levels.

5. Mold Steels

This group of steels, designated the P series, is used primarily for plastic molds. Types P2 to P6 are carburizing steels that have very low hardness in the annealed condition permitting the mold cavity to be generated by hubbing. Hubbing is a cold work process where a tool having the geometry of the desired cavity is pressed into the mold blank rather than creating the cavity by conventional machining operations. The mold is subsequently carburized and hardened for long-term durability. Types P20 and P21 are normally supplied in the preheat treated condition in the 30 HRC hardness range. Following final machining, the molds are ready for service without further heat treatment.

6. Water-Hardening Steels

This group of steels, designated the W series, are low alloy high carbon grades that have very little resistance to softening at elevated temperature. When used for cutting tools, they are restricted to woodworking tools and slow cutting speeds. When heat treated in moderate section thicknesses, they must be water quenched and develop full hardness only at the surface, the core remaining somewhat softer and tougher.

7. Special Purpose Steels

This group of steels, designated the L series, are low in alloy and carbon content compared with the other tool steels. The characteristics of good hardness, wear resistance, and high toughness make this grade useful for machinery parts such as collets, cams, and arbors.

Since tool steels normally contain large amounts of expensive alloying elements and are required to withstand very severe service conditions, particular care is devoted to their production. Double-melting practice is common wherein the steel is produced by electric furnace or vacuum induction melting and then subjected to a secondary refining process such as vacuum arc or electroslag remelting. An alternate manufacturing procedure uses high-purity powders to produce moderate size billets by the powder metallurgy (PM) process. The PM billets are then consolidated by hot isostatic pressing and rolling to final size. An added advantage of this process is that it prevents segregation of the alloying elements that naturally occurs during solidification from the molten stage and also produces fine carbide particle size.

These factors make tool steels a premium-priced commodity. A highly alloyed grade in a double melted or PM form can cost well over $10/lb [4,5].

C. Stainless Steels

In the AISI numbering system the primary stainless steels are designated by a three-digit number: 2XX, 3XX, 4XX, and 5XX. In addition, there is one other class called the precipitation-hardening grades which are designated by their chromium and nickel contents. For example, 17-7 PH is a common precipitation-hardening grade containing 17% chromium and 7% nickel along with minor amounts of other elements that promote the precipitation-hardening behavior. As one can imagine, there are also a never-ending variety of proprietary and special purpose grades beyond the scope of this discussion. Table 4 shows a generalized format of these compositional classifications.

Aside from the compositionally based classification, stainless steels are also categorized by their crystallographic structures: ferritic, martensitic, and austenitic. Both the ferritic and martensitic stainless steels are magnetic while the austenitic grade is nonmagnetic. The most corrosion-resistant grades are to be found in the austenitic series followed by the ferritic series and the precipitation-hardening grades. The martensitic grades, which are hardenable by a quench and temper heat treatment, generally have the poorest corrosion resistance.

The primary mechanism by which stainless steels gain their corrosion resistance is through the development of a stable, protective surface oxide film. It is generally accepted that a minimum chromium content of 12% is necessary to form the protective oxide film. To a great extent, stainless steels form this film naturally by reaction with oxygen in the atmosphere. However, a denser, more stable oxide may be forced through a process called passivation. This process subjects the material to a strong oxidizing acid, usually concentrated nitric acid, at an elevated temperature. During passivation, minute particles of nonstainless metals which may have become embedded in the surface during machining and contact with nonstainless forming equipment are dissolved while the oxide film is being generated. This treatment leaves the metal in its most corrosion-resistant condition.

Stainless steel is subject to a serious reduction in corrosion resistance through a mechanism called sensitization. Chromium has a very strong affinity

TABLE 4 AISI/SAE Stainless Steel Designation System

Alloy series	Nominal alloy content
2XX	Chromium plus manganese plus nickel
3XX	Chromium plus nickel
4XX	Chromium (12–20%)
5XX	Chromium (5%)
PH	Chromium plus nickel plus molybdenum or aluminum or copper

Source: Ref. 6.

for carbon and tends to form a very stable carbide. Moreover, a small amount of carbon can combine with a large amount of chromium, thereby effectively negating the effect of the chromium in promoting corrosion resistance. When heated in the range of 1000 to 1200°F (540 to 650°C), a reaction between carbon and chromium occurs at the grain boundaries resulting in the sensitization. Since a significant portion of the chromium has been tied up, corrosion can readily proceed at the grain boundaries in a form of corrosion called intergranular attack. One common fabrication process that can induce sensitization is welding. At some point along the edges of the weld a temperature in the sensitization range will occur rendering the stainless subject to intergranular attack in the heat-affected zone of the weld. Sensitization can be reversed by a heat-treating process which consists of a high-temperature cycle that redissolves the precipitated chromium carbides followed by rapid cooling through the sensitization range. While this corrective treatment is effective, it can be expensive and result in unacceptable distortion of welded fabrications. Specific grades have been developed that minimize sensitization by controlling the carbon content to very low limits as in types 304 and 316L or by incorporating elements that have a stronger affinity for carbon than chromium such as in type 347 [6,7].

D. Cast Iron

Cast iron is a very important class of engineering materials. The relatively low melting point and fluidity of cast iron makes it readily castable into complex shapes. It has good mechanical properties and is easily machinable due to its unique microstructure. In its broadest description, cast irons are alloys of iron, carbon, and silicon. The carbon is usually present in the range of 2.0 to 4.0% and the silicon in the range of 1.0 to 3.0%. This composition results in a microstructure that has excess carbon present as a second phase. The form taken by the excess carbon is the basis of the three major subdivisions of cast irons: gray iron, white iron, and ductile iron.

The most common variety of cast iron has the excess carbon present in the form of graphite flakes and is called gray iron. This terminology derives from the appearance of a freshly fractured surface which has a dull gray texture owing to the presence of the graphite flakes. During fracture the crack propagates along the graphite flakes since graphite has practically no strength.

In irons with the carbon and silicon content minimized and where a very rapid solidification rate was attained, the excess carbon is present as a carbide and there is no free graphite. This type is called white iron because a freshly fractured surface has a smooth, white appearance. White iron is very hard and brittle in the as-cast condition. It is frequently used in applications where extreme wear resistance is required.

White iron can be converted to so-called malleable iron by a heat treatment which causes the carbide to decompose into compact clumps of graphite called temper carbon. The compact clump form of graphite interrupts the crack path more effectively than the flake form found in gray iron thereby providing some measure of ductility.

Another method of producing a ductile form of cast iron is by a nodulizing inoculation. If magnesium or rare earth metals are added to the molten iron, the excess carbon forms spheroidal nodules of graphite rather than the flake form found in gray iron. The nodular graphite structure results in a substantial increase in strength and ductility. Ductile iron castings can compete with steel castings or forgings in some near net shape applications.

Commercial iron castings are seldom melted to strict chemical composition. It is more common that the required mechanical properties are specified and the foundry selects the composition that will meet the specification. The mechanical properties of cast iron are drastically affected by cooling rate during solidification, which is a function of metal section thickness. In order to meet the specified mechanical property requirements, the foundry must have the latitude to adjust the chemical composition to suit the weight and section thickness of the particular casting. The parameter that is most often used for controlling mechanical properties is the carbon equivalent. This parameter is the sum of the total carbon plus one-third of the silicon content plus one-third the phosphorus content. These three elements in the given ratios affect the rejection of carbon from the melt during solidification and determine the resulting graphite size and distribution.

Gray cast iron, therefore, is designated by its tensile strength. Class 25 iron has a tensile strength of 25,000 lb/in.^{2} (172,000 kPa), class 30 has 30,000 lb/in.^{2} (206,000 kPa), etc. Since gray cast iron has essentially no ductility, there is no measurable yield strength.

Ductile iron is designated by three numbers: the tensile strength, yield strength, and percent elongation from the tensile test. Common varieties of ductile iron are 60-40-18, 80-55-06, 100-70-03, and 120-80-02. The minimum mechanical properties of 60-40-18 would be 60,000 lb/in.^{2} (415,000 kPa) ultimate tensile strength, 40,000 lb/in.^{2} (275,000 kPa) yield strength, and 18% elongation [4].

V. HEAT TREATMENT AND MICROSTRUCTURES OF IRON AND STEEL

There are several strengthening mechanisms in the metallurgy of iron and steel. Steel exhibits a substantial increase in hardness and strength as the result of cold work or plastic deformation. Distortion of the crystal lattice and creation of extensive dislocations that occur during plastic deformation results in strengthening the metal. Addition of alloying elements can also cause strengthening through two separate mechanisms. Alloys that form substitutional solid solutions strengthen through inhibiting crystallographic slip. Atoms of solid solution alloying elements replace the base metal at some atom sites. Since they have a different atomic diameter, they cause protuberances or depressions in the atomic planes they occupy and interfere with slip, thereby strengthening the metal.

Another mode of alloy strengthening is through the formation of a second phase. The iron/carbon system is a striking example of this mechanism. Carbon is essentially insoluble in iron at room temperature. In steel, carbon forms a second phase called cementite having the composition Fe_3C. From the composition it may be noted that one carbon atom unites with three iron atoms and so a small addition of carbon forms a lot of the second phase, cementite. The normal microstructure of iron that has no carbon is the single phase body-centered cubic form of iron, called ferrite. As carbon is added, grains of a second phase involving cementite appear in the ferrite. The typical form that cementite takes in steel is a microconstituent called pearlite. Pearlite is a mixture of ferrite and layers of cementite arranged in a lamellar or plywoodlike sandwich. Within the grains of the pearlite phase, the equilibrium carbon content is 0.8 weight percent. As additional carbon is added to the steel, the amount of the pearlite phase increases. The ferrite phase is essentially free of carbon. Therefore, a medium carbon steel such as 1040 would have a microstructure of approximately 50% ferrite phase and 50% pearlite. At an overall carbon content in the steel of 0.8%, the entire microstructure is composed of pearlite. As the carbon content is increased above 0.8%, discrete particles of cementite appear in the microstructure. Figure 7 illustrates the effect of increasing carbon content on the pearlite content.

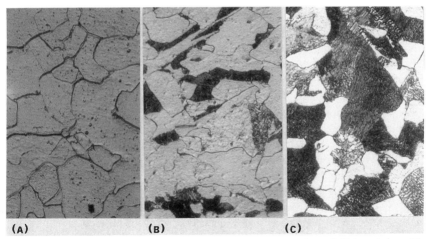

Figure 7 Effect of carbon content on the microstructure of steel. Original magnification of all micrographs was 500×, 2% nital etchant. (A) 1005 Steel with 0.05% carbon. Microstructure is completely ferritic. (B) 1045 Steel with 0.45% carbon. Microstructure is a mixture of ferrite (white grains) and pearlite (dark etching phase). (C) 1075 Steel with 0.75% carbon. Microstructure is predominantly pearlite (dark etching phase) with lesser amounts of ferrite (white grains).

If steel is heated above its critical temperature, the body-centered cubic crystal structure of ferrite changes to the face-centered cubic form called austenite. The cementite that was insoluble in the body-centered cubic ferrite is readily soluble in the face-centered cubic austenite and quickly forms a single phase solution of high carbon austenite when heated above the critical temperature. If the steel was to be cooled slowly back to room temperature, the structure would revert to its original microstructure of the ferrite and cementite mixture. However, if the steel is rapidly cooled or quenched, the dissolved carbon is trapped in solution in the austenite and transforms to the metastable body-centered tetragonal form called martensite. The martensite crystal structure is the hard, strong form in steel.

In summary, there are three criteria for hardening steel. First, the material must contain adequate carbon. Second, it must be heated above its critical temperature where it transforms to the face-centered cubic, austenite form and absorbs the carbon into solution. Third, it must be cooled at a rate that is sufficiently fast to prevent reversion to ferrite, trapping the crystal structure in the hard martensite form. Freshly transformed martensite has a highly stressed crystal lattice. Although it is very hard, it also is very brittle. A degree of toughness can be restored with only a slight sacrifice in hardness by performing a tempering operation. Tempering is carried out by heating in the range of 300 to 1000°F (150 to 540°C), which is below the critical temperature where the steel would again transform to austenite. The higher the tempering temperature, the lower the resulting hardness. At 300°F (150°C) the reduction in hardness is only one or two points on the Rockwell C scale. At 1000°F (540°C) the reduction in hardness may be as much as 30 Rockwell C points. Hardened parts are never used in the as-quenched condition without tempering.

Manipulation of these three criteria, carbon plus heating above the critical temperature followed by rapid cooling, forms the basis for all of the various heat treatments commonly used for martensitic hardening of steel. The same principles apply to cast iron, the matrix of which, exclusive of the graphite particles, may be thought of as high carbon steel. The most straightforward hardening technique is called neutral or through hardening and subjects the entire part being hardened to all three of the criteria as previously described. This process produces uniform hardness throughout the part.

Frequently it is desired to selectively harden only a small area of the part, the end of a shaft for example. In this case it is possible to heat only that portion of the shaft to be hardened. When quenched, only the heated area will be transformed to martensite. Alternatively, it would be possible to heat the entire shaft but only quench the end to be hardened. The remainder of the shaft would be allowed to cool slowly thus reverting to the soft ferrite plus cementite form. Another method, called carburizing, begins with a low carbon steel that has insufficient carbon to harden. When heated above its critical temperature in the presence of an atmosphere containing a source of carbon, such as methane, the surface of the part absorbs carbon. When quenched, only the high carbon case transforms to hard martensite while the center or core of the part, where carbon did not reach, will remain softer and tougher. A similar type of case hardening may be accomplished using a nitrogen atmosphere instead of methane in a process called nitriding. A hybrid process using both methane and nitrogen is called carbonitriding.

There are other reasons for heat treatment besides hardening. Normalizing is a conditioning heat treatment applied to steel bars, plate, castings, and forgings usually during or after the rolling operation at the steel mill, forge shop, or foundry. Its purpose is to soften or prepare the steel for machining or further processing. It consists of heating the metal above the critical temperature and allowing it to cool normally in the open air under ambient conditions. Since the cooling rate is not controlled, the results are not uniformly predictable. Another softening process is called annealing. This process is carried out like normalizing except the cooling rate is controlled, usually inside the annealing furnace. The results are more predictable and the material may be put in its softest condition by this process. Another process called stress relieving is applied to precision manufactured components to remove residual stresses that may have been induced by the manufacturing processes. The purpose is usually to render the part dimensionally stable. Although normalizing and annealing are also effective at relieving stresses, the resulting change in hardness and microstructure may not always be desirable. For this reason stress relieving is usually carried out below the critical temperature so there is no transformation in crystal structure.

The heat-treating processes described above are generally applicable to steel, cast iron, and the martensitic grades of stainless. Exceptions are that cast iron is not a candidate for carburizing since it already contains excess levels of carbon, and stainless is not usually carburized because of its adverse effects on corrosion resistance. The precipitation hardening grades of stainless are hardened by a completely different mechanism, one that is analogous to the heat treatment applied to aluminum.

VI. NONFERROUS METALS

It is implicit in the term nonferrous that this category of materials includes a great number of different metals. Omission of a detailed discussion of the metallurgy of the pure elemental metals as well as the alloys of nickel, titanium, etc., does not imply that these metals are unimportant, but is requisite simply because it is beyond the scope of this chapter in introductory metallurgy. For the sake of brevity we will concentrate on the two most common nonferrous alloy systems, aluminum and copper.

A. Aluminum

The most striking property of the aluminum alloys is density, which is only about one-third that of steel. Other important properties of aluminum are high thermal and electrical conductivity and good corrosion resistance and ease of fabrication. This unique collection of properties makes aluminum well suited to a variety of commercial applications. Aluminum is available in practically all wrought product forms such as plate, sheet, foil, bar, rod, wire, tubing, forgings, and complex cross-section extrusions. In addition, its low melting point and fluidity make aluminum ideal for casting. Sand, plaster, permanent mold, and pressure die castings are readily available.

There are two main classification systems for aluminum alloys; one system for wrought products and one for castings. The major specifying body is the Aluminum Association. The designation system for wrought products is based on a four-digit system while the cast form is designated by a three-digit number. Both product forms typically carry a suffix of one letter and one to four numbers which describe the temper or strengthening process applied to the product. Table 5 shows the Aluminum Association designation system for wrought aluminum products while Table 6 shows a similar designation system for cast aluminum products.

Pure aluminum is a chemically reactive metal and will form many chemical compounds. The good corrosion resistance of aluminum is due to the natural tendency of aluminum to form a tenacious oxide, Al_2O_3, on the surface when exposed to air. Even if scratched, the oxide will quickly renew itself. In addition to this natural oxide forming tendency, a thicker and more protective oxide layer can be induced by an electrochemical process called anodizing. In this process, the aluminum part is made the anode in an electrolyte of a strong oxidizing acid such as sulfuric or chromic acid. The cathodes are inert lead bars. Passing a

TABLE 5 Aluminum Association Designation System for Wrought Aluminum Products

Alloy series	Nominal alloy content
1XXX	99.00% purity, <1% alloy
2XXX	Copper plus minor additions of manganese and magnesium
3XXX	Manganese plus magnesium
4XXX	Silicon
5XXX	Magnesium plus minor additions of manganese and chromium
6XXX	Silicon and magnesium plus minor additions of copper, manganese, or chromium
7XXX	Zinc and magnesium plus minor additions of copper and chromium

Source: Ref. 8.

TABLE 6 Aluminum Association Designation System for Cast Aluminum Products

Alloy series	Nominal alloy content
1XX	99.0% purity, <1.0% alloy
2XX	Copper
3XX	Silicon with minor additions of copper and/or magnesium
4XX	Silicon
5XX	Magnesium
6XX	Unused series
7XX	Zinc
8XX	Tin

Source: Ref. 8.

current through this system develops a heavy, porous, aluminum oxide coating. After rinsing, the still porous coating may be dyed a variety of colors by immersion in a dye bath. When the desired color has been achieved, the anodized coating is sealed by immersing the part in hot water which hydrates the oxide, thus sealing the porosity. If the acid anodizing bath is refrigerated during the anodizing process, exceptionally thick and ceramiclike oxide coatings can be developed. This process is called hard anodizing and produces very wear-resistant coatings. Aluminum alloys containing appreciable quantities of silicon are difficult to anodize, especially for decorative purposes [8].

B. Heat Treatment of Aluminum Alloys

Certain wrought alloys in the 2XXX, 6XXX, and 7XXX series and the 2XX, 3XX, and 7XX series casting alloys are heat treatable by a process called precipitation hardening. The alloys 2024, 6061, and 7075 are common high-strength wrought alloys and 319, 355, and 356 are common casting alloys which are subject to hardening through a precipitation or age-hardening heat treatment. The mechanism as described here also applies to the precipitation hardening grades of stainless, albeit the temperatures required are substantially higher. In addition to strengthening, precipitation hardening provides the most machinable condition in most aluminum alloys.

Copper, magnesium, and zinc form intermetallic compounds or secondary phases with the aluminum microstructure as shown in Fig. 8. In the equilibrium or slowly cooled condition, these intermetallic compounds form and grow to relatively large microconstituents within the metal. As such, they have a relatively

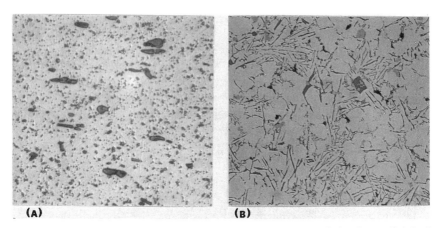

Figure 8 Comparison of wrought and cast microstructure of aluminum. Original magnification of all micrographs was 500×, 2% HF Etchant. (A) Wrought alloy 3003. Microstructure is particles of $MnAl_6$ in a solid solution matrix. Particles are elongated and oriented in the rolling direction (left to right). (B) Die cast alloy 308. Microstructure is particles of $CuAl_2$ and silicon eutectic in a solid solution matrix. There is no orientation as in the wrought alloy.

minor effect on hardening or strengthening the alloy. If the alloy is heated to a point where the intermetallic compounds are redissolved into the matrix and then rapidly cooled at a rate faster than would allow them to reform, the alloy is said to be solution annealed. A low-temperature heating then encourages the precipitation of the intermetallics in very fine particles that are practically undetectable by conventional optical microscopy. An alloy processed in this manner is said to be precipitation or age hardened. The extensive distribution of the very fine precipitates inhibits slip on the critical crystallographic slip planes and results in the strengthening effect. Typical solution annealing is done by heating in the range of 1000°F (540°C) followed by water quenching. This leaves the alloy in its softest and most formable condition. A few alloys will allow the precipitate to form at room temperature and these are said to be naturally aging grades. Other alloys must be heated to the range of 350°F (180°C) to induce the precipitation to occur and these are said to be the artificial aging grades. The heat-treated condition is described by a system of suffixes, called the temper designation, which is appended to the alloy number. Table 7 shows the common temper designations used for aluminum products [7].

C. Copper and Copper Alloys

Copper is one of the few metals found in its metallic form in nature. The metal, along with its principal alloys, is an important group of engineering materials.

TABLE 7 Commonly Specified Temper Designations for Aluminum Products

Temper designation	Process description
F	As fabricated, no additional thermal processing.
O	Annealed to lowest strength for maximum ductility and dimensional stability.
H	Strain hardened (applies to wrought products only). May be followed by one or more numbers that further describe the strain hardening process used.
W	Solution heat treated, applies only to alloys that naturally age.
T	Heat treated to produce a stable temper. Usually requires artificial aging. Always followed by one or more numbers that define the actual process used.

Source: Ref. 8.

Among the outstanding characteristics of copper and copper alloys are excellent thermal and electrical conductivity, formability, castability, and corrosion resistance. Certain of the bronze and brass alloys have excellent tribological compatibility with steel and are frequently used in bearing applications. Copper alloys protect themselves from corrosion by forming a tenacious oxide which is the familiar green patina seen on exposed architectural elements and marine fittings. The oxide coating on copper is electrically conductive and as a result, copper wiring does not have the problem with contact resistance that aluminum wiring does. Copper alloys are susceptible to stress corrosion cracking, especially in the presence of ammonia. Much of the classification of copper alloys was done by the Copper Development Association. The CDA numbers were adapted into the Unified Numbering System which is now the most widely recognized designation system. Table 8 shows the UNS designation system for copper alloys.

The significance of the solid solubility limits shown in Table 8 is that copper alloys having less than the indicated limit of alloying element will have a single phase microstructure. In these systems the alloying element is present in the form of a substitutional solid solution. Such alloys exhibit substantially increased strength in combination with good ductility and formability. When the alloy content exceeds the limit of solid solubility, a second phase appears in the microstructures, and even higher strengthening results.

Two-phase alloys lose some of their ductility, and the capability for rolling thin sheets may be diminished or completely eliminated. Lead, tellurium, and selenium are elements added to copper alloys to promote machinability. These elements are insoluble in the matrix, produce a separate phase in the microstruc-

TABLE 8 Unified Numbering System for Copper Alloy Designation

Alloy Group	UNS designation	Principal alloy element	Solid solubility (at. %)
Copper and high copper alloys	C10000	None	
Brasses	C20000, C30000, C40000, C66400–C69800	Zn	37
Phosphor bronze	C50000	Sn	9
Aluminum bronze	C60600–C64200	Al	19
Silicon bronze	C64700–C66100	Si	8
Copper nickel	C70000	Ni	100

Source: Ref. 8.

ture, and behave much as sulfur does as a free-machining additive to steel. Lead also imparts a self-lubricating characteristic to bearing alloys.

Additional hardening and strengthening above that produced by alloying can be obtained by cold working. The hardness of cold-worked, wrought copper and brass is usually expressed as a fraction related to the degree of cross-sectional area reduction accomplished during the rolling process. Table 9 shows the temper designations commonly produced in wrought copper alloys [8].

VII. MEASUREMENT OF SURFACE FINISH

Specifying surface finish is basically a process of describing the topography and texture of the boundary surface of a solid body in quantifiable terms. Surface finish is an important parameter of a component part. It determines how the part will respond to sliding friction, how well it will retain a lubricant, the wear rate that will be experienced, and how well it will retain a coating, such as electroplating or painting to name just a few. Surface finish measurement is also closely linked to dimensional tolerancing. It would be irrational to reference a very precise dimension from a very rough surface.

A. Terminology

Although there are various noncontact methods of measuring surface texture, such as electrical capacitance and laser interferometry, the most common and widespread method currently in use is by the contact stylus method. In this technique a diamond-tipped stylus having a radius typically 0.0004 in. (0.01 mm) is dragged across the surface being measured and the up and down motion of the stylus is tracked electronically. Much of the terminology in this discussion is appropriate to either contact or noncontact measuring methods. The language

TABLE 9 Temper Designations for Rolled Copper Alloys

Temper designation	Reduction in thickness/area (%)
1/4 Hard	10.9
1/2 Hard	20.7
3/4 Hard	29.4
Hard	37.1
Extra hard	50.1
Spring	60.5

Source: Ref. 8.

of surface finish measurement contains a number of unique terms and before going much further it would be well to provide definitions for these terms. Figure 9a serves to illustrate the physical significance of these terms.

Nominal surface is a hypothetical surface that defines the shape of the body such as would be depicted by an engineering drawing. The nominal surface is smooth and serves only as a reference for dimensioning and assigning the allowable tolerance for deviations from the surface.

Surface topography, also called *texture*. This parameter is the composite of all the deviations from the nominal surface. The irregularities comprising the texture are several, including the following:

Form: Deviations from the specified surface geometry such as taper, concavity, convexity, twist, etc.

Waviness: Relatively long-range periodic deviations which may have resulted from such sources as cutter chatter, machine vibrations, or an out-of-balance grinding wheel, which alter the path of the cutter from that actually intended.

Roughness: Relatively closely spaced deviations which result from the interaction of the tool and workpiece such as tears, gouges, cutter marks, and built-up edge sloughing. The actual cutter path need not vary from the path intended.

Flaws: Defects not necessarily related to the cutting process. Scratches or dents occurring after the surface was cut and cracks or porosity in the material are examples of flaws. These defects are random in orientation and spacing with respect to other surface texture features.

Lay Surfaces that have been generated by a cutting process such as lathe turning, milling, planing, or grinding have an obvious directional quality. Processes like flame cutting and welding produce directionality on a very gross

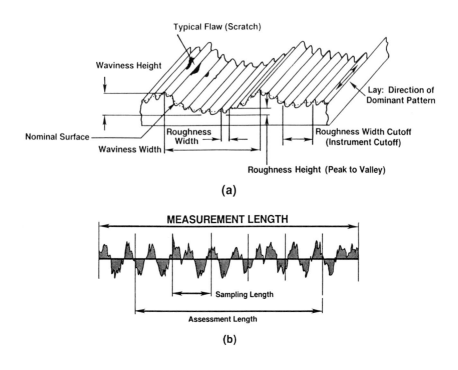

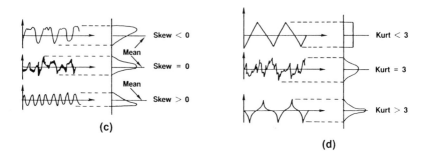

Figure 9 Surface finish terminology. (Reprinted with permission from SAE AS291D ©1991 Society of Automotive Engineers, Inc.)

scale. This directional orientation is called lay and the lay direction is parallel to the major lines defining the lay. Some processes such as casting, shotblasting, or electrical discharge machining produce surfaces with no discernible directional characteristics.

Nominal profile is the hypothetical line created by the intersection of a plane at a right angle with the nominal surface. The nominal profile serves as the reference baseline for superposition of the measured profile. The nominal profile through a flat surface will be a straight line, through a cylindrical surface will be a circle, etc.

Measured profile is the wavy, zigzag line that is defined by the intersection of a plane at a right angle to the measured surface and at a right angle to the lay of the finish. It is generally this profile that the stylus of the surface measurement device tries to trace.

Peaks and valleys are points or ridges that protrude above the plane of the surface or holes or troughs that lie beneath the plane of the surface.

Mean line is a straight constructed line that divides the measured profile line such that the area enclosed by the peaks is equal to the area enclosed by the valleys.

Sampling length, also called *cutoff length*. This is a preselected distance over which the measurements are taken and parameters computed. By judicious selection of the cutoff length, longer-scale effects of waviness and form errors can be filtered out. A commonly used cutoff length is 0.030 in. (0.76 mm).

Assessment length is the stylus travel distance over which the average profile is determined. The stylus stroke should encompass at least five cutoffs plus some amount of overtravel on each end of the assessment length.

B. Concepts and Parameters

The reader has likely seen traces from surface finish measuring equipment and noted that the profile is very sharp with the slope of the peaks and valleys being very steep. The deviations of actual surfaces on real bodies are much more gently undulating hills, bumps, and valleys. Indeed, if actual bodies had such sharp-sloped peaks and valleys, friction would be infinite and sliding of two mating surfaces past each other would likely be impossible. The reason the surface finish trace looks different from the actual surface is that the vertical scale of the trace must be greatly magnified to resolve the minute fluctuations of the surface. If the profile were equally magnified in both the vertical and horizontal directions the profile trace would be a proportional magnification of the surface but the trace would require yards and yards of chart paper. Typical magnifications are 200× in the horizontal direction and 5000× in the vertical direction and this makes the profile look much sharper and steeper than it actually is.

Once the profile trace had been acquired, a number of mathematical manipulations of the data are made to generate the following roughness parameters.

R_a is the average surface roughness computed as the arithmetic mean of the absolute value of the distance between the baseline to the maximum peak or valley height. The average roughness is easier to visualize if the bottom half of the trace (the valley part) is flipped up onto the top half (the peak part) of the trace as shown in Fig. 9b. The line defining the mean height of this flipped trace is the R_a roughness. R_a is the most universally used surface roughness parameter. The units of surface roughness are microinches in the English system and micrometers in the metric system.

R_q is the equivalent of R_a except the root mean square method is used as the averaging technique instead of arithmetic averaging.

R_{sk} Skewness refers to the distribution of peaks and valleys about the mean line as shown in Fig. 9c. Surfaces that have peaks and valleys of equal height and depth have zero skew. If the valleys are deeper than the peaks are high, the surface has negative skew. If the peaks are higher than the valleys are deep, the surface has positive skew.

R_{ku} Kurtosis is a parameter that describes the sharpness of the profile as shown in Fig. 9d. A typical surface has a kurtosis of approximately three. If the points of the peaks and valleys are more obtuse or flatter than average the kurtosis is less than three. If the points on the peaks and valleys are very acute or sharper than average, the kurtosis is greater than three.

t_p is the percent bearing ratio. It is determined by drawing a construction line parallel to the mean line at a specified height above the mean line thus cutting off the peaks. It simulates the effect of wearing away the peaks and projects the resulting bearing area that would be in contact with a perfectly flat mating surface.

Various methods of creating surfaces produce their own characteristic surface textures. Figure 10 shows the range of surface roughness that can be expected for a number of common commercial production processes. Under ideal conditions actual surface roughness may be controlled to higher or lower values, but this chart gives a good approximation of the surface finish that can be achieved by the various manufacturing methods [9–11].

C. Interpretation of Engineering Symbols

Engineering drawing symbols are used to convey the designer's intentions to the machinist or manufacturer who must create the actual component. As shown in Fig. 11, surface finish requirements are indicated by a checklike mark which may sometimes have an extension bar extending from the top of the long leg of the check.

The point of the symbol touches the surface to which it applies or the dimension extension line from that surface. The required surface roughness appears as a number just above the short leg of the check. If only one number appears here it indicates maximum allowable roughness. If the roughness must be controlled between a maximum and minimum value, two numbers appear. The units of the numbers are consistent with the other dimensions of the drawing. The maximum allowable waviness height is shown on the extension bar to the

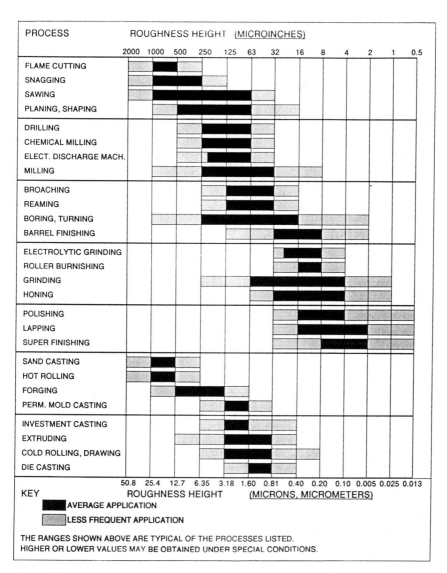

Figure 10 Surface finish roughness produced by common production methods. (Reprinted with permission from SAE AS291D ©1991 Society of Automotive Engineers, Inc.)

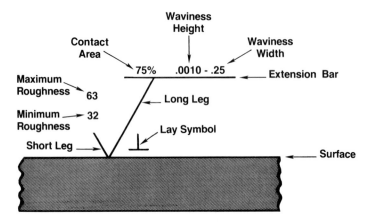

Figure 11 Engineering symbols for surface texture.

right of the intersection of the long leg of the check. Waviness width, when required, is placed to the right of the waviness height separated by a dash. Desired contact area with the mating surface is expressed as a percentage and appears directly above the intersection of the extension bar and long leg of the check. The lay direction is shown to the right of the point of the check and may specify perpendicular, parallel, angular, multidirectional, circular, or radial lay direction.

The precision and functional performance of any manufactured component is strongly dependent on its surface texture. Understanding the fundamentals of specifying surface texture parameters will help ensure that this finished component will perform as intended.

REFERENCES

1. A. G. Guy, *Elements of Physical Metallurgy*, Addison–Wesley, Reading, MA, pp. 72–78 (1959).
2. V. E. Lysaght and A. DeBellis, *Hardness Testing Handbook*, Wilson Instrument Div., American Chain and Cable, Reading, PA, pp. 25–55 (1969).
3. C. A. Keyser, *Basic Engineering Metallurgy*, Prentice–Hall, Englewood Cliffs, NJ, pp. 59–78 (1959).
4. W. H. Cubberly et al., *Metals Handbook, Vol. 1, Properties and Selection: Irons and Steels*, 9th ed., American Society of Metals, Metals Park, OH, pp. 3–56, 114–143 (1978).
5. G. A. Roberts and R. A. Cary, *Tool Steels*, 4th ed., American Society for Metals, Metals Park, OH, pp. 227–229 (1985).
6. W. H. Cubberly et al., *Metals Handbook, Vol. 3, Properties and Selection: Stainless Steels, Tool Materials and Special-Purpose Metals*, 9th ed., American Society for Metals, Metals Park, OH, pp. 3–6, 421–433 (1980).

7. D. Peckner and I. M. Bernstein, *Handbook of Stainless Steels*, McGraw–Hill, New York, pp. 1.1–1.10 (1977).
8. W. H. Cubberly et al., *Metals Handbook, Vol. 2, Properties and Selection: Nonferrous Alloys and Pure Metals*, 9th ed., American Society for Metals, Metals Park, OH, pp. 3–44, 140–143, 239–251 (1979).
9. H. Amstutz, *Surface Texture: The Parameters*, The Sheffield Measurement Div., Warner Swasey (1978).
10. Aerospace Standard AS 291D, Society of Automotive Engineers, 485 Lexington Ave., New York, NY.
11. Surface Texture Standard ANSI B46.1, The American Society of Mechanical Engineers, United Engineering Center, 345 East 47th Street, New York, NY 10017 (1978).

3
Metal Cutting Processes

HERMAN R. LEEP
University of Louisville
Louisville, Kentucky

Most manufactured goods require some type of metal cutting during production. The basic metal cutting or machining processes include drilling, turning, milling, and broaching. Before describing the metal cutting processes, the mechanism of chip formation will be discussed.

I. MECHANISM OF CHIP FORMATION

The basic metal cutting processes remove material by generating chips. The most widely used model for explaining the formation of chips was proposed by Merchant in 1945 [1,2]. Most machining operations involve three-dimensional (oblique) cutting, but to simplify the mechanism, a two-dimensional (orthogonal) model is assumed. The diagram in Fig. 1 shows the cutting tool moving across the surface of a stationary workpiece. This model shows the cutting tool moving at velocity V (cutting speed) and depth of cut t. Photomicrographs have been used to show that chips form along a shear plane making an angle ϕ (shear angle) with the workpiece surface.

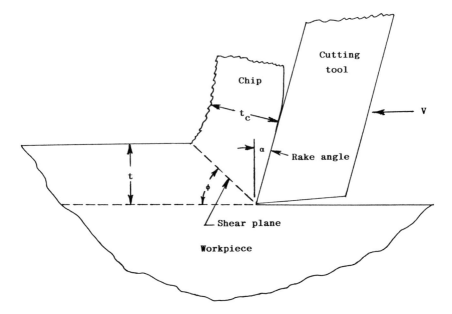

Figure 1 Schematic diagram of two-dimensional orthogonal cutting.

Usually, the chip thickness t_c is greater than the depth of cut. The chip thickness depends on the rake angle α of the tool, shear angle, and cut depth.

An equation can be developed for calculating the shear angle in terms of the rake angle and the chip thickness ratio, r_c, defined as t/t_c. The shear angle is important in predicting cutting forces, power requirements, heat generated by the work of deformation, and vibrations. By making the proper assumptions, the equation for the chip thickness ratio is as follows:

$$r_c = \frac{t}{t_c} = \frac{\sin \phi}{\cos(\phi - \alpha)} \tag{1}$$

In order to solve Eq. (1) for the shear angle, the cosine term must be expanded. After simplifying, the following expression results:

$$\tan \phi = \frac{r \cos \alpha}{1 - r \sin \alpha} \tag{2}$$

Numerous methods have been used to determine the shear angle. The "quick stop" method makes it possible to interrupt the cutting process with the chip attached to the workpiece and the cutting tool retracted [3,4].

II. TOOL WEAR

Factors that cause cutting tools to wear include the high, localized stresses between the workpiece and cutting edges, the high temperatures which develop in the cutting zone, and the sliding of chips along and over the cutting edges [5]. Tool wear can decrease the quality and accuracy of machined surfaces and increase the costs of machining operations. Tooling costs increase when tools must be replaced, indexed, or resharpened and reset too frequently [6].

Tool life can be defined as a measure of the length of time a tool will cut satisfactorily. The following methods may be used to determine the life of a cutting tool: (a) the tool fails to cut, (b) the surface finish of the workpiece deteriorates due to a change in the tool geometry, (c) the force or power required for the same metal removal rate increases, (d) the dimensions of the workpiece increase, or (e) the amount of tool wear exceeds a predetermined limit [7].

There are three broad categories of tool failure as follows:

1. *Catastrophic failure*: Fracture or chipping of the cutting tool due to forces that are too large.
2. *Gradual wear*: Rate depends on factors such as materials of the cutting tool and workpiece, tool geometries, cutting fluids, cutting conditions (such as cutting speed, feed rate, and depth of cut), and characteristics of the machine tool.
3. *Plastic deformation*: Due to the elevated temperatures and high compressive stresses on the cutting tool [8].

The most common mechanisms used to explain wear on cutting tools for machining metals are abrasion, adhesion, and diffusion. Chemical reaction, electrolysis, and oxidation can also affect wear on cutting tools. These tool-wear mechanisms can be described as follows:

1. *Abrasion.* Small, hard particles in the workpiece materials can abrade the cutting tool. Examples of these particles from steels include carbides, nitrides, and oxides.
2. *Adhesion.* The high pressures and temperatures in localized areas on a cutting tool surface can cause chips to deform and weld to the cutting tool. As these tiny welds are fractured, the surface of the cutting tool will deteriorate. Adhesion is common when machining soft metals such as annealed aluminum and annealed copper.
3. *Diffusion.* The high temperatures can cause atoms in the metallic crystal of the cutting tool to move from one lattice point to another. For example, the diffusion of carbon in a high-speed steel (HSS) cutting tool can make it softer.

4. *Chemical.* Additives in a cutting fluid can cause a chemical reaction between the cutting tool and workpiece which could lead to wear on the cutting tool.
5. *Electrolytic.* Possible galvanic corrosion between the cutting tool and workpiece could cause the cutting tool to wear.
6. *Oxidation.* The high temperatures can oxidize the carbide in the cutting tool, decreasing its strength and causing the edges to wear [6].

There are two types of wear on single-point tools used in turning operations: flank wear and crater wear. These types of wear are illustrated in Fig. 2. On the flank or relief face of the cutting tool, abrasion causes a small land to develop. This flank wear will extend from the cutting edge to approximately 0.4 to 1.5 mm downward, depending on the cutting tool material [5]. On the rake face of the cutting tool, a small crater will develop behind the cutting edge. This crater is caused by abrasion when the chips slide over the tool face. Crater wear is also subject to the diffusion mechanism since crater wear will increase with temperature.

The developer of HSS cutting tools, F. W. Taylor, published in 1907 the results of the first extensive study of tool life. Taylor established the following relationship between tool life and cutting speed for turning steels:

$$VT^n = C \tag{3}$$

where

V = cutting speed, cm/s
T = tool life, s
n = exponent depending on cutting conditions and materials of tool and work
C = constant equal to cutting speed for a tool life of 60 s

Since Taylor first published the results of his study, an extensive amount of research related to tool wear has been performed. Results from some of the more recent research projects are discussed in the last section of this chapter.

III. FUNCTIONS OF CUTTING FLUIDS

Cutting fluids are also called coolants and lubricants. The term coolant was coined by researchers soon after F. W. Taylor reported that tool life could be improved by applying water. The term lubricant originated with the introduction of oils.

The primary function of a cutting fluid is to serve as a coolant [3]. In this role, the coolant dissipates the heat in the cutting zone by absorbing the heat in the workpiece, chip, and tool. In order to increase metal removal rates, feeds and speeds must be increased. These increases cause elevated temperature which

Metal Cutting Processes

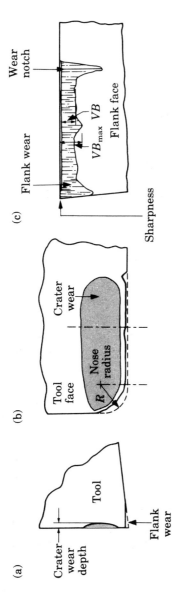

Figure 2 (a) Flank and crater wear in a cutting tool. Tool moves to the left. (b) View of the rake face of a turning tool, showing nose radius R and crater wear pattern on the rake face of the tool. (c) View of the flank face of a turning tool, showing the average flank wear VB and the wear notch or depth-of-cut line. (From Ref. 5 with permission.)

is the most critical limitation to tool life. Heat is carried away by the cutting fluid which circulates through the system. Consequently, the temperatures of the tool and workpiece are reduced. The reduced temperature of the cutting tool will increase its life since the hardness is retained, or it will allow the machining to be performed at a higher cutting speed for the same tool life. The reduced temperature of the workpiece reduces thermal distortion and provides better dimensional control and surface finish [9].

Almost all the work that goes into machining is dissipated as heat. Approximately 75% of the total heat generated in machining is due to the heat of deformation and the other 25% is attributed to the heat of friction. On the average, about 80% of the total heat produced goes into the chip, 10% into the tool, and 10% into the workpiece [8]. Hence, it is advantageous to produce small chips which carry the heat away via a coolant. For a typical shear angle between the base of the chip and the workpiece, approximately 60% of the heat is generated in the shear plane. As the shear angle is increased, the heat generated in the shear plane will decrease since the plastic flow of the metal occurs over a shorter distance. The rest of the heat is produced in the friction plane between the chip and the cutting tool (approximately 30%) and in the surface plane between the cutting tool and the newly machined surface (approximately 10%). The shear angle can be increased by reducing the friction at the tool–chip interface with a cutting fluid and by selecting the proper tool geometry [6].

The secondary function of a cutting fluid is to serve as a lubricant. In this role, the lubricant reduces the coefficients of friction between the cutting tool and the chip, and between the cutting tool and the workpiece. In processes such as drilling and tapping, the body of the tool rubs against the workpiece and increases the power required unless a good lubricant is used. By reducing friction and wear, tool life and surface finish are improved. Improved surface finish is accomplished by restraining the formation of a built-up edge (BUE) on the cutting tool. A lubricant reduces the BUE by producing a thinner, less deformed, and cooler chip [3].

The tertiary function of a cutting fluid may be to wash the chips away from the cutting region. In certain drilling applications, coolant is forced under pressure through holes in the drill to provide cooling at the tip, but as the coolant flows out through the flutes of the drill, the coolant helps in removing chips from the hole.

Another function of a cutting fluid may be to provide protection against environmental corrosion for the newly machined surfaces of the workpiece and the machine tool. Some coolants have additives that will provide a protective coating.

A cutting fluid's effectiveness depends on factors such as the method used to apply the cutting fluid, temperatures encountered, cutting speed, and type of machining process [9]. The role of a cutting fluid as a coolant or lubricant is very sensitive to the cutting speed. For example, in high-speed cutting operations such as turning and milling where the tool–work interface is small, the cooling characteristic of a coolant is extremely important. Conversely, in low-speed cutting operations such as broaching, threading, and tapping, lubricity is more important since it tends to reduce the formation of a BUE and improves surface finish.

As the severity of a machining process increases, the need for an effective cutting fluid becomes more critical. The degree of severity depends on the temperatures developed in the cutting region, the forces experienced by the cutting tool, the tendency to form a BUE, and the ease with which chips are removed from the cutting region [9]. The relative severity of the machining processes discussed in this chapter, the relative cutting speeds used, and the relative need for an effective cutting fluid are shown in Table 1.

Even though some solids and gases are used as coolants, most coolants are liquids. These liquid cutting fluids are easily recirculated and may be directed to the point in the machining process that makes them most effective.

The most common theory for the method used by a cutting fluid to penetrate the tool–chip interface is capillary action. This model was used by Merchant to explain the results from his studies using a transparent sapphire cutting tool. His experiments showed that the cutting fluids entered the interface by seeping in from the sides of the chip. In order to penetrate this tiny interlocking network of surface asperities, the molecules of the cutting fluid should be small and have good wetting properties. As cutting speeds increase, penetration may be improved since the high temperatures can convert the cutting fluid into smaller gaseous molecules, but since these processes take time, the capillary action is probably less effective. At high speeds, the cooling function of the cutting fluid is more important.

TABLE 1 Severity of Machining Process

Machining process	Process severity	Cutting speed	Cutting fluid activity
Broaching (internal)	High	Low	High
Tapping	↑	│	↑
Broaching (surface)	│	│	│
Threading (general)	│	│	│
Deep drilling	│	│	│
Drilling	│	↓	│
Milling	│	│	│
Turning	Low	High	Low

Source: Ref. 9.

A different theoretical model for the cutting fluid action was proposed by Smith, Naerheim, and Lan [10]. Their model was based on the capillary flow theory, but the model assumed that fissures in the chip and along the tool–chip interface allowed the cutting fluid to enter into the chip and along the interface. Carbon tetrachloride, with small molecules, a high vapor pressure, and the affinity of chlorine for machined surfaces, was used to validate this model.

Other researchers have suggested that a cutting fluid can reduce the shear strength of the workpiece material by interacting with the dislocations near the surface to promote plastic deformation. This phenomenon is called the Rebinder effect and numerous experiments have been performed to verify it. Researchers have also suggested diffusion of the cutting fluid through the chip to the interface to form a film between the chip and tool.

The flood method is the most common method for applying cutting fluids in turning, drilling, and milling processes. The proper methods of applying cutting fluids in various machining processes are illustrated in Fig. 3. Figures 4 and 5 demonstrate some additional methods used in milling processes. Flow rates can be as low as 10 l/min for turning and as high as 200 l/min for face milling.

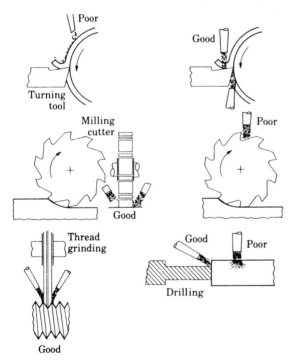

Figure 3 Schematic illustration of proper methods of applying cutting fluids in various machining processes. (From Ref. 5 with permission.)

Metal Cutting Processes

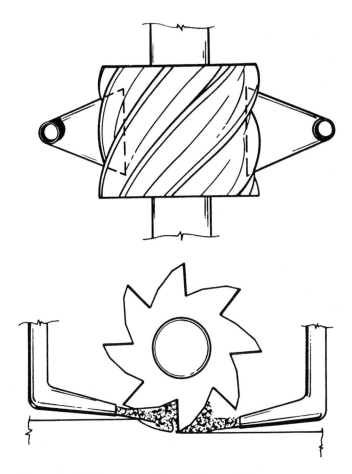

Figure 4 Fluid is pumped from right nozzle by cutter teeth while fluid from left nozzle washes away chips as they emerge. (From J. Silliman, ed., *New Cutting and Grinding Fluids: Selection and Application*, 1992, by permission of the Society of Manufacturing Engineers.)

Chips can be washed away from the cutting region in deep-hole drilling and end milling by using fluid pressures ranging from 700 to 14,000 kPa [9].

When selecting a cutting fluid, consideration should be given to its effects on workpiece materials and machine tools. Cutting fluids with sulfur and chlorine additives can cause stress-corrosion cracking on some workpiece materials. Some cutting fluids will stain aluminum and copper workpieces. Cutting fluid residues should be removed from these parts after machining. Machine tool components

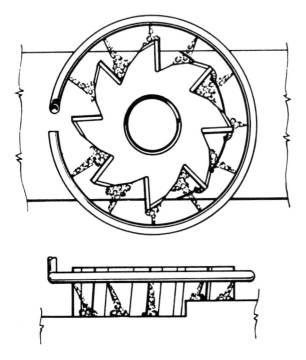

Figure 5 Ring distributor used with face mill directs fluid against the entire circumference of the cutter for even cooling. (From J. Silliman, ed., *New Cutting and Grinding Fluids: Selection and Application*, 1992, by permission of the Society of Manufacturing Engineers.)

such as bearings and slideways are susceptible to certain cutting fluids, and care should be taken in choosing fluids which are compatible with these materials.

Neat oils are used when lubricity is more crucial than cooling. Good examples of these applications include broaching, tapping, and some milling processes. Broaching and tapping are normally performed at low cutting speeds but the chip loads and cutting forces are high, especially when high-strength or heat-resistant alloys are machined. Some milling operations require intermittent cuts which create heavy forces between the milling cutter and the workpiece. An oil-based cutting fluid can work better than a water-based cutting fluid in this application since rapid cooling could cause thermal shock which would result in chipped cutting edges.

IV. DRILLING PROCESSES

Probably the most common shape found in a manufactured part is a circular hole, and many of these holes are produced by drilling. Since the chips are formed within the part, the flutes or grooves that spiral around the drill body normally serve two purposes. In addition to providing a conduit for the removal of chips, they also allow the cutting fluid to reach the tool–workpiece interface. Because of the geometric constraints inherent in the design of drills, most drills are made from HSS which demands the application of a cutting fluid [11].

Most drills have two cutting edges. The cutting speeds recorded in standard reference books apply to the outer ends of the cutting edges. The following equation is used to calculate the spindle speed of the machine tool for a given drill size and cutting speed:

$$N = \frac{10V}{\pi D} \quad (4)$$

where

N = spindle speed, rev/s
V = cutting speed, cm/s
D = drill diameter, mm

The tabulated feed rate is given in mm/rev and the programmed feed rate is given by the following equation:

$$f_p = f_r N \quad (5)$$

where f_p = programmed feed rate, mm/s, and f_r = feed rate, mm/rev.

The most common drill design is the twist drill shown in Fig. 6. Rotary motion is transferred from the machine tool spindle to the drill shank, which may be straight or tapered. Straight shanks are common for drill diameters up to 12 mm and tapered shanks are used for larger sizes. Drills usually have two flutes. Typical helix angles are in the range from 24 to 32° [5]. The recommended helix angle depends on the workpiece material. A low helix angle is used for copper alloys but a high helix angle is suggested for aluminum alloys. The drill point generally has an included angle of 118°. An included angle of 135° is better for drilling high-strength stainless steels and titanium alloys.

The chisel edge functions as a wedge causing the material to deform and slide up the cutting edges. Since the relatively long twist drill can flex, it has a tendency to "walk" on the workpiece surface [3]. If tight tolerances are specified, center drilling or the application of a bushing is recommended.

Most conventional drills have a chisel edge at the end of the web which connects the cutting edges. This design is not effective when drilling difficult-to-machine materials such as stainless steels and titanium alloys. The high thrust on conventional drills can be reduced by using a design with split points and a thinned web. Figure 7 shows the end of a drill with split points on the cutting edges and the thickness of the web. Web thinning is done with a narrow grinding wheel which removes part of the web, making the chisel edge shorter.

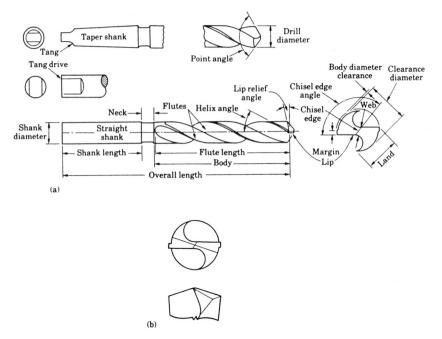

Figure 6 (a) Standard chisel-point drill indicating various features. (b) Crankshaft-point drill. (From Ref. 9 with permission.)

Many deep holes are drilled with gun drills. Various features of a gun drill are shown in Fig. 8. The name of this cutting tool came from its original application, that is, drilling the bores of rifle barrels. Holes that have a depth of at least three diameters are classified as deep holes. A hole in the drill allows cutting fluid under high pressure to reach the tip of the drill. High cutting speeds and low feed rates are used.

A typical range of diameters for gun drills varies from 1.4 to 25 mm. The depth of the hole is restricted by the torsional rigidity of the shank. Latinovic and Osman studied the frictional losses in coolant flow through the passages in a gun drill [12]. They concluded that a kidney-shaped passage in the tip was superior to both single-circular and double-circular passages. The researchers presented a method of calculating the pressure drop for a given shank size and hole depth.

When drilling deep holes that have a large diameter (e.g., 12 to 200 mm), a method that can be used is BTA (from the Boring-Trepanning Association) drilling. A high-pressure fluid system is used to remove the chips and cool the cutting tool. The coolant also eliminates the formation of a BUE [13]. BTA drilling uses carbide inserts and has internal chip removal. The coolant flows

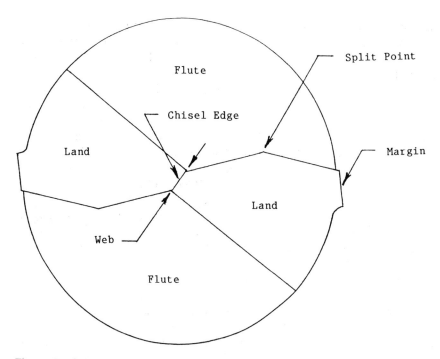

Figure 7 Schematic diagram of a split-point drill.

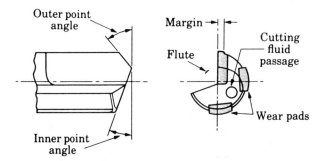

Figure 8 A gun drill showing various features. (From Ref. 5 with permission.)

between the boring bar and the wall of the hole to the machining zone. The chips are forced out through the boring head and boring bar. Since the chips do not rub against the workpiece, the accuracy or surface finish of the hole is not affected.

Many machine tool builders are providing spindles which allow the coolant to be delivered directly to the cutting tool. If this feature is not available, it can be added to the machine tool as a retrofit. A tool supplier has designed and manufactured a special coolant gland for the toolholder which has a connector that lines up with a coolant port mounted on the face of the spindle [14]. Feeding coolant through the tool is recommended as the most effective method in an application such as drilling deep holes. A higher feed rate or a longer tool life is possible with coolant-fed drills than without coolant-fed drills.

The coolant pressure recommended in pressurized-coolant drilling depends on several factors. The most important factors include the following: (a) workpiece hardness, (b) feed rate, (c) hole diameter, (d) hole depth, (e) hole tolerance, and (f) hole finish. As the coolant pressure is increased, containing the splash and recirculating the coolant could become a problem [15].

Drilling times can be reduced by using internally cooled drills with inserts. Tungsten carbide inserts and inserts with titanium nitride coatings are commonly used in these applications. An efficient coolant system supplies that coolant at the proper pressure and flow rate. An inefficient coolant system can lead to poor surface finish inside the drilled holes. The drill diameter is the most important factor in determining the coolant pressure and flow rate [16]. Chips can block the channels that deliver the coolant to the inserts if the coolant system is inefficient. This condition can cause tool breakage.

V. TURNING PROCESSES

A lathe or turning center uses turning processes to produce parts which are round in shape. Cylindrical and conical surfaces are generated from a rough cylindrical blank which rotates about its longitudinal axis and against a cutting tool. The ends of a long workpiece are held between centers in the machine tool, the left end in the chuck and the right end in the tailshock. Short workpieces are clamped into the chuck on the spindle only.

Most turning operations use single-point tools. Normally, a tool holder contains an indexable insert with multiple cutting edges (e.g., a square insert can have eight cutting edges).

The part programmer specifies the rotational speed of the workpiece, the feed rate of the tool, and the depth of cut. Roughing cuts, which remove metal at the maximum rate, are followed by finishing cuts at a higher cutting speed, a lower feed rate, and a smaller depth of cut [3]. These machining conditions depend on the workpiece and tool materials, the surface finish and dimensional accuracy desired, and the machine tool capacity. Each combination of workpiece and tool materials has an optimal set of tool angles. Many of the recommended values of the controllable variables have been determined by experimentation and are tabulated in machining handbooks.

Tool geometry (including tool angles) will affect the direction of chip flow. The objective is to avoid long continuous chips which can interfere with the machine tool operation or damage the part surface. Chips will break when they hit the workpiece or tool holder. The following methods have been used to break chips: (a) determine the optimal tool angles, (b) include grooves in tool inserts, and (c) add obstructions to the tops of the inserts [17].

The major turning processes performed on a lathe are straight or cylindrical turning, taper turning, facing, and boring. These processes are shown in Fig. 9. In turning and boring processes, the cutting tool normally is fed from the operator's right to the left or toward the chuck. The tool is fed perpendicular to the axis of rotation, and usually toward the center of the part, in a facing cut.

Turning operations occur on the external surface of a part. The tool

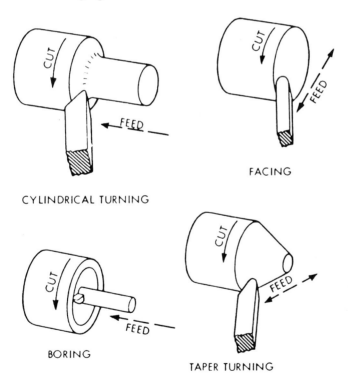

Figure 9 Major turning operations performed on a lathe. (From Ref. 7 with permission.)

penetrates beneath the part's surface, and is fed parallel to the axis of rotation in straight turning. Taper turning involves feeding the tool at an angle to the axis of rotation in order to produce a truncated cone.

Facing is used to produce a flat surface which is perpendicular to the axis of rotation. The part normally is clamped into the chuck.

Boring is an internal turning process in which metal is removed by a tool held in the proper position. Boring may be done for one of the following reasons: (a) to enlarge a hole made by a previous process, (b) to enlarge the inside diameter of a hollow tube, or (c) to machine internal grooves. As in facing, the workpiece usually is clamped into the chuck. If the machining is done deep into the part, a long boring bar which is sufficiently stiff must be used in order to avoid vibrations and chatter. These undesirable conditions can cause poor surface finish and premature tool failure.

VI. MILLING PROCESSES

A variety of milling processes are available. Common examples include end milling, slab milling, and face milling. These examples of milling processes are shown in Fig. 10. The workpiece is clamped to the table or held in a fixture which is clamped to the table. High metal removal rates are possible in milling since milling cutters have multiple teeth and each tooth produces a chip. In most applications, the workpiece is fed into a rotating cutter. The feed motion usually is perpendicular to the cutter axis, and cutting occurs on the circumference of the cutter.

One way of classifying milling cutters is according to the way the cutting tool is mounted into the machine tool [3]. A shank cutter is inserted directly into the machine tool spindle or into a tool holder which goes into the spindle. The cutting teeth are ground into the shank. An end mill is a good example of a

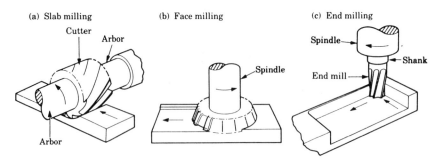

Figure 10 Basic types of milling cutters and processes. (From Ref. 5 with permission.)

shank cutter. An arbor cutter has a center hole which allows it to be mounted onto an arbor which is inserted into the spindle of the milling machine. A facing cutter which is bolted to the end of a stub arbor, or short shaft, is a good example of an arbor cutter.

In end milling, the axis of cutter rotation is perpendicular to the workpiece surface to be milled. Several types of end-milling cutters are available. Most end mills are relatively long compared to their diameter and the shank usually is an integral part of the cutter. Smaller cutters normally have straight shanks, but larger cutters have tapered shanks. The cutter is designed so that the teeth on the end allow it to be plunged into the workpiece like a drill. Then, the teeth on the periphery of the cutter do the majority of the machining when the milling machine table moves. End mills commonly have two, three, or four flutes which are helical. End mills with hemispherical ends, called ball mills, can be used to generate curved surfaces.

Typical applications for end mills include flat surfaces and recess cuts for making dies. Grooves that extend partially through the material and profiles around thin parts are other common operations.

Shell end mills and hollow end mills are examples of cutters designed for special applications [3]. The shell end mill is a larger cutter for heavy duty machining. Since the shell end mill does not have a shank, one stub arbor can be interchanged with several cutters to reduce tooling costs. Hollow end mills are used on automatic screw machines. The internal cutting teeth will machine a cylindrical surface to an accurate diameter from round bar stock.

In slab milling, or peripheral milling, the cutter axis is parallel to the machined surface. The cutter has several teeth along its periphery. Cutting speed at the surface of the cutter (V) is given by the following equation:

$$V = \pi D N \tag{6}$$

where D is the cutter diameter and N is cutter rotational speed.

Face milling is similar to end milling. However, a face-milling cutter is relatively large in diameter compared to its length. Face-milling cutters are designed to machine flat surfaces. The geometry of the cutter and the major application are appropriate for carbide inserts which can be indexed. The cutter usually is mounted to the spindle of the machine tool. Due to the relative motion between the cutting tool and the workpiece, tool marks, such as those produced in turning, are found on face-milled surfaces also.

The direction of cutter rotation in milling can produce different effects. In the traditional method, called up milling or conventional milling, the cutter rotates against the direction in which the workpiece is fed. In down milling or climb milling, the cutter rotates with the feed direction. Up milling produces a chip that gets thicker, but the chip gets thinner in down milling. These milling methods are shown in Fig. 11.

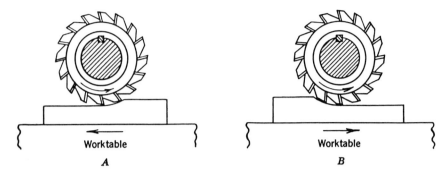

Figure 11 Methods of feeding work on milling machine. (*A*) Conventional or up milling. (*B*) Climb or down milling. (From Ref. 6 with permission.)

The major advantage of up milling is that tool wear is not affected by the surface condition of the workpiece. A smoother surface is possible if the cutting edges are sharp. Larger clamping forces are needed since the cutter tends to lift the workpiece from the machine table.

Since the teeth dig into the workpiece in down milling, a rigid setup is required. However, this downward motion does help to hold thin parts against the table and less chatter is experienced. When down milling materials with a hard surface, such as castings and hot-worked metals, the teeth will wear faster and can be damaged. Advantages of down milling include a reduced tendency to have tool marks on the part and a surface finish that is not affected by a BUE if one is created on the cutting edge.

VII. THREADING AND TAPPING PROCESSES

A screw thread is a ridge of uniform cross section in the form of a spiral on the surface of a cylinder or cone [5]. The process of cutting external or internal threads with a metal cutting tool is called threading. When internal threads are cut with tools called taps, the machining process is referred to as tapping. The terminology for a tap is shown in Fig. 12. Several thread forms have been standardized. Some of the more popular thread forms are as follows: 60° V-thread, unified thread, American National standard pipe thread, ISO metric thread, square thread, Acme thread, and buttress thread [11]. Various types of threads are shown in Fig. 13.

Threads can be cut on a lathe, but most threads are cut on production machines. Lathes are used when only a few threads are required for small production runs or when special thread forms or special workpieces are involved. Die-hard chasers can be used on lathes and in threading machines to increase production and simplify setups. The die head on a lathe is stationary and the work rotates. At the end of the cut, the chasers open so that they can be withdrawn

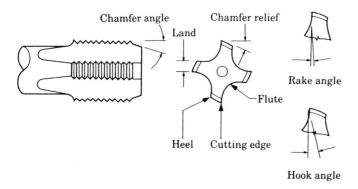

Figure 12 Terminology for a tap. (From Ref. 9 with permission.)

without damaging the workpiece. In a threading machine, the work is fed into rotating dies [6].

Special machines are designed specifically for production threading or tapping. A threading machine may have one or two spindles with a self-opening die head on each spindle, and a mechanism for feeding and clamping the work [3]. Automatic screw machines may have from four to eight spindles, arranged in a circle, each with an individual workpiece. A computer numerical control

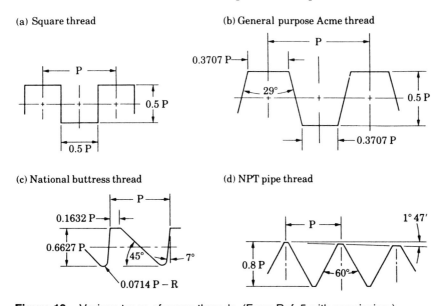

Figure 13 Various types of screw threads. (From Ref. 5 with permission.)

(CNC) tapping center provides spindle feeds in both directions, with the speed and feed more rapid during retraction.

In thread milling, a milling machine can be used to cut large threads which need to be accurate. Even though threads cut on a lathe ordinarily require more than one pass, thread milling can be used to cut a thread to full depth with only one pass of the multitooth cutter along the work. This process is shown in Fig. 14.

When die-head chasers are used to cut threads, high metal removal rates are expected and large forces develop due to the friction between the tool and workpiece. Since these frictional forces generate excessive amounts of heat, lubricity is an important characteristic of the cutting fluid [18]. Liquid film and metal film lubricants can be used to increase the lubricity of a coolant. Cutting oils have the ability to provide a liquid film between the tool and workpiece at extremely high temperatures. Additives such as sulfur and chlorine are used to form metal film lubricants for threading. Good thread-cutting fluids can serve a wide range of workpiece materials, but a single fluid cannot be perfect for all materials. If a single threading fluid is used in a shop, it may be necessary to reduce the cutting speed for difficult-to-machine materials.

The cutting fluids selected for threading operations must also remove the heat from the cutting zone to reduce die wear. Excessive temperatures can cause the material being threaded to weld to the die, producing a BUE. The BUE can lead to torn threads.

The best fluids for threading have the cooling capacity of soluble oils and synthetic coolants and the lubricity provided by fluids which have sulfur and chlorine additives. Metal film lubricants are formed when the cutting fluid absorbs the energy from the die and workpiece, allowing the chemical reactions to proceed to completion more rapidly.

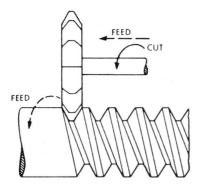

Figure 14 Thread milling with a revolving multitooth cutter. (From Ref. 7 with permission.)

In order to completely flood a cylindrical bar or pipe with coolant for threading, a distribution manifold is recommended. Most threading machines are supplied with a distribution manifold. Die hards used on other machine tools normally require the fabrication of special distribution manifolds from tubing or pipe.

The water-soluble coolants used in most machining operations may not be suitable for threading difficult-to-machine materials. It may be best to thread these materials on a separate machine with a special coolant. If this option is unavailable, the cutting speeds should be reduced as much as practical and the cutting edges of the die should be sharpened regularly.

Cutting fluids used in tapping should be kept as clean as possible since foreign objects, which get trapped between the tap and workpiece, could damage the new threads. A high coolant flow rate is recommended to remove chips. The best tool life occurs when a high-pressure coolant is used to flush out the chips.

VIII. BROACHING PROCESSES

Broaching can be used to machine either an internal or external surface. Slots or grooves, which are difficult to machine by other processes, can be broached by an elongated tool. The cutting tool, called a broach, consists of a series of spaced teeth that make progressively deeper cuts. This broaching action is shown in Fig. 15. Roughing, semifinishing, and finishing can be done in a single pass.

In most broaching machines, the broaching tool is pulled or pushed through or across the work which is stationary. In surface broaching, the work can also move across the tool. In continuous broaching, the work is moved continuously against stationary broaches [6].

Examples of features that can be broached in one operation are flat surfaces on internal combustion engine blocks and internal splines for holes in gear blanks. Other examples of shapes produced by broaching include external splines, contours, and finished holes.

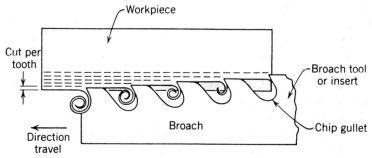

Figure 15 Broach tool and broaching action. (From Ref. 6 with permission.)

In addition to the advantages already mentioned, the following advantages have made broaching a popular process for high-production machining: (a) cycle time is short, (b) loading and unloading of parts are fast, (c) virtually any shape or contour can be broached, (d) dimensional accuracy for interchangeable parts can be achieved, and (e) surface finishes comparable to milling work can be produced [6].

Since most broaching operations are done in one pass, a cutting fluid can cool the cutting teeth after each engagement with the workpiece. However, since broaching is classified as the machining process with the highest-degree of severity, compounded oils are normally recommended [19].

IX. SUMMARY OF RECENT RESEARCH

Much of the metal cutting research performed to evaluate the performance of cutting fluids has involved drilling and turning processes. Additional processes that have been studied include milling, tapping, and broaching.

A. Drilling Research

The performance of cutting fluids can be evaluated by using drilling operations. A synthetic cutting fluid was evaluated by Leep [20]. A one-half replicate of a 2^5 factorial experiment, which included workpiece hardness, hole diameter, cutting speed, feed rate, and fluid concentration, was designed. One of the dependent variables used in this analysis was the average wear on the cutting edges of HSS drills. The main effects were included at the following levels: hardnesses of the plain carbon steel workpieces (160 and 235 HB, where HB is the Brinell hardness number), diameters of the drills (6.350 and 9.525 mm), cutting speeds of the drills (30.0 and 47.2 cm/s), feed rates of the drills (0.178 and 0.254 mm/rev), and concentrations of the cutting fluid (4.8 and 9.1 vol. %). Workpiece hardness, hole diameter, and cutting speed were found to be significant, at the 5% level, in affecting the tool wear. Feed rate and fluid concentration were not significant at the levels tested.

Cutting fluid concentrations, ranging from 2.5 to 5.5 wt. % were examined by Leep and Kelleher [21]. Drilling tests in a 4 by 2^3 experimental design were used to produce tool wear while using a synthetic cutting fluid at four levels of concentration. The commercially available synthetic cutting fluid was recommended for heavy machining operations. The following factors were included at the indicated levels: concentrations of the cutting fluid (2.5, 3.5, 4.5, and 5.5 wt. %), hardnesses of the plain carbon steel workpieces (146.5 and 273.6 HB), cutting speeds of the drills (25.4 and 38.1 cm/s), and diameters of the drills (6.350 and 9.525 mm). The feed rate was held constant at 0.178 mm/rev since it was not a significant factor in the study cited above. Also, lower levels of concentration than used in the previous study were examined. All four factors were found to be significant at the 5% level over the ranges tested. Interactions between hardness and speed, and also hardness and diameter, had significant effects on tool wear.

An experimentally formulated semisynthetic cutting fluid and a commercially available premium soluble oil were studied by Block [22]. The workpiece material was aluminum alloy 390, which contains 16 to 18% silicon and is used to produce cylinder blocks for automotive engines. The semisynthetic fluid concentrate consisted of mineral oil, corrosion inhibitors, biocides, a cleaner, and a water-soluble ester which was used as the major lubricant. Ingredients in the soluble oil included chlorinated paraffins, fatty acid soaps (amine), sulfurized triglycerides, petroleum sulfonate, and a biocide. The total extreme-pressure additive content was from 10 to 15% and the viscosity was approximately 650 s at 40°C using a Saybolt universal viscometer.

A 2^3 experimental design was used by Block to study the effects of cutting speed (at 24.4 and 50.8 cm/s), drill diameter (at 6.350 and 9.525 mm), and fluid concentration (at 4.0 and 8.0 wt. %) on tool wear. Blind holes were drilled 25.4 mm deep at a feed rate of 1.016 mm/rev. The results showed no appreciable difference between the cutting fluids for overall tool wear. For the experiment with the soluble oil, the interaction between cutting speed and drill diameter was the most significant source of variation. For the experiment with the semisynthetic cutting fluid, the most significant source of variation was the interaction between cutting speed and fluid concentration.

The machinability of aluminum 390 was also investigated by Leep and Sims [23]. The soluble oil used by Block was compared to a different semisynthetic cutting fluid, which used a proprietary water-soluble fatty ester of molecular weight between 200 and 500 as the major lubricant. The same experimental design, factors, levels, and cutting conditions used by Block were used in these experiments. However, the number of observations was doubled by adding a replicate for each set of conditions in order to increase the precision associated with estimating the population variance.

Wear on the drills while using the semisynthetic cutting fluid was approximately 3% less than that for the soluble oil. Concentration of the semisynthetic cutting fluid was the most significant factor in the experiment with this fluid. The most significant sources of variation for the soluble oil experiment were the cutting speed and the interaction between concentration and diameter.

Eldridge studied the effects of cutting conditions on aluminum 380 while using a commercially available soluble oil formulated for drilling aluminum [24]. Aluminum 380 contains 7.5 to 9.5% silicon and is used to make carburetors for automotive engines. Each test consisted of drilling 460 blind holes which were 25.4 mm deep. HSS drills with a diameter of 6.350 mm were used. The cutting fluid had a concentration of 5 vol. %. Tool wear was measured for two values each of cutting speed and feed rate. The speeds were 99 and 198 cm/s and the feeds were 0.406 and 0.812 mm/rev. When the cutting speed was doubled, the tool wear increased 49%, but when the feed rate was doubled, the tool wear increased only 6%.

Cutting fluids, which are formulated for machining aluminum alloys, normally contain additives which inhibit the rapid formation of an aluminum oxide film on the machined surface [25]. This hard film can increase tool wear on the cutting edges. Aluminum die casting alloys, which contain more than 7% silicon, can also be very abrasive to the cutting tools. In addition, the high coefficients of thermal conductivity for aluminum alloys require cutting fluids that rapidly dissipate heat that collects during machining.

Methods of cutting fluid application were studied by Rao in drilling tests [26]. A medium carbon steel was used as the workpiece material. Surface roughness was one measure used to compare the flood method with the mist method for a soluble oil. In the flood application, the top of the workpiece was covered with the cutting fluid. The method used in the mist application was to spray the cutting fluid into the machining area. Feed rates of 0.16, 0.22, and 0.31 mm/rev were used with HSS drills having a diameter of 19.05 mm. Very poor surface finish was observed over the entire range of feed rates tested when the mist was used. The flood application was superior in the lower feed range.

The cutting fluids studied by Block in a flood application were also evaluated by Leep on the basis of surface finish inside the holes [27]. The same experimental design was used. Variations in drill diameter, cutting speed, and the interaction between these factors were significant for both cutting fluids. Concentration of the semisynthetic cutting fluid also had a significant effect on the surface finish. The surface finish produced while using the semisynthetic cutting fluid was approximately the same as that with the soluble oil.

Results from the research by Leep were used by Leep, Halbleib, and Jiang to design an experiment specifically for the evaluation of the surface finish of holes drilled into aluminum 390 when a commercially available synthetic cutting fluid was used [28]. The experimental design was a 2^4 full factorial. In addition to the factors of cutting speed, drill diameter, and fluid concentration, feed rate and a second replicate were added to the design.

The factors were evaluated at the following levels: speed (63.5 and 127.0 cm/s), diameter (6.350 and 9.525 mm), concentration (3 amd 6 wt. %), and feed (0.305 and 0.610 mm/rev). Blind holes were drilled 25.4 mm deep. Feed rate, fluid concentration, and the interaction between feed rate and drill diameter were significant. Cutting speed and drill diameter were not significant within the ranges tested.

Thomas investigated the relationship between the surface finish of drilled holes and the machining parameters of feed rate, drill diameter, and fluid concentration [29]. A commercially available synthetic cutting fluid was used in this experiment to machine stainless steel 304. The design of a 2^3 factorial experiment included two replications and the main effects at the following levels: feed (0.051 and 0.102 mm/rev), diameter (6.350 and 9.525 mm), and concentration (3 and 6 wt. %). Drill diameter was the only factor significant at the 5% level.

The same cutting fluid was used by Spond to evaluate the surface finish of holes drilled into a medium carbon steel [30]. The main drilling experiment consisted of a 2^2 factorial design with two levels of feed rate (0.127 and 0.254 mm/rev), two levels of drill diameter (6.350 and 9.525 mm), and four replications. From the statistical analysis, the interaction between feed rate and drill diameter was significant at the 5% level even though the main effects were not. The mean total roughness parameter provided better models than the arithmetic average roughness parameter.

Eldridge also measured the surface finish of holes drilled into aluminum 380 while using a commercially available soluble oil [24]. The experimental details were the same as described above. A 16% increase in surface roughness was associated with increasing the cutting speed from 99 to 198 cm/s. Increasing the feed rate from 0.406 to 0.812 mm/rev caused the surface roughness to increase by 15%.

Rao also compared the flood method to the mist method by studying the variations of thrust force and cutting torque with cutting speed [26]. Cutting speeds ranged from 17 to 47 cm/s for 12.70 mm drills and 18 to 50 cm/s for 19.05 mm drills. The mist application of the cutting fluid usually produced the lower thrust force for most of the speeds tested and the feeds of 0.16, 0.22, and 0.31 mm/rev. For most of the speeds tested, the mist application also lowered the cutting torque. The reduction in cutting torque was more predominant at the highest feed and the higher speeds.

A commercially available synthetic cutting fluid was used by Leep and Peak in the development of models that predict tool wear from measured cutting forces [31]. These predictive models can be used in adaptive control systems which allow maximum feed rates while protecting the machine tool, parts, and cutting tools from damage in unattended machining operations. Drills with diameters of 6.350 and 9.525 mm were used to drill a titanium alloy (6Al4V) and a medium carbon steel (AISI 1045), respectively. The thrust force was significant at the 5% level in the models for both materials. Also, the cutting torque was significant at the 5% level in the model for the titanium alloy.

The concentration of cutting fluid was 5 wt. % for both models and the holes were drilled 25.4 mm deep to the shoulders of the drills. Since the titanium alloy was difficult to drill, a 135°, split-point, cobalt HSS drill was selected. The cutting speed and feed rate were 25.4 cm/s and 0.152 mm/rev, respectively. The medium carbon steel was drilled with a 118° HSS drill. A speed of 81.3 cm/s and a feed of 0.254 mm/rev were used with this medium carbon steel workpiece.

Herde developed production drilling models, similar to the models developed by Leep and Peak, but for a composite material [32]. The workpiece material was flooded with the same commercially available synthetic cutting fluid. Machining parameters considered included cutting speed, feed rate, and workpiece hardness. Blind holes, 6.350 mm in diameter, were drilled into an aluminum matrix composite containing 16 vol. % alumina particles. This 6061 aluminum alloy composite was hardened and tempered to a T6 condition. Carbide-tipped drills were employed.

In addition to varying the three machining parameters, the thrust force and cutting torque were measured by Herde. Six regression models were generated. Cutting torque was found to be significant in all models. The most general model pooled the data from three tests with different speed/feed combinations and variations in workpiece hardness.

Radioactive cutting tools were used by Jeremic and Lazic to study the tribological characteristics of semisynthetic cutting fluids [33]. This method provided the possibility for detecting subtle differences between cutting fluids which were chemically very similar. Relationships were developed between the tool wear measured with a microscope and the known tool radioactivity. The criterion for tool life of the 12.5 mm diameter HSS drills was 0.30 mm wear. When three semisynthetic cutting fluids were used, the life of the drills varied from 77 to 107 min while drilling a low-machinability steel which was heat-treated to a hardness of 270 HB.

Magnetic cutting fluids were studied by Podgorkov in drilling operations [34,35]. These fluids had either a polyethylene base or a mineral-oil base. A magnetic field was used to cause these ferromagnetic fluids to penetrate into the cutting zone. Measurements of drilling torque were used to evaluate the effectiveness of the cutting fluids. The effects of drill diameter, feed rate, and cutting speed upon drilling torque were investigated. For drill diameters ranging from 4 to 10 mm, the torque was reduced 20 to 25% when ferromagnetic fluids were used compared to dry cutting. When feed rates from 0.1 to 0.4 mm/rev were examined, torque reductions of 10 to 12% were observed. The reduction in torque was small for the cutting speeds examined.

The results obtained by Podgorkov indicated the effectiveness of using ferromagnetic fluids in drilling nonmagnetic structural materials such as aluminum, zinc, and titanium alloys, and stainless steel in special environments such as high altitude, high vacuum, or low gravity. However, high cost prevents them from being a feasible substitute for conventional cutting fluids.

Vasyshak and Soshko evaluated the effectiveness of polymer additions to cutting fluids used in drilling operations [36]. Each cutting fluid was prepared by dissolving a polymer with a molecular weight greater than 18,000 in an industrial oil, which was also used as the control fluid. The polymer concentration in the oil for all seven fluids was 2 wt. %. A high-temperature strength steel was machined with 6.3 mm diameter steel drills. The holes were drilled 15 mm deep at a cutting speed of 46.2 cm/s. The fluids were evaluated mainly on the basis of the thickness of the chip removed and the plunge time of the drill. These experiments showed that the high molecular SKD rubber and the type P-20 polyisobutylene were the most effective additions to the cutting oil.

The following conclusions were drawn from this review of the literature on drilling research.

1. Drilling can be used to evaluate the performance of cutting fluids.
2. Variations in workpiece hardness, cutting speed, drill diameter, and fluid concentration had significant effects on tool wear while drilling carbon steels.
3. Tool wear results from drilling aluminum 390 while using a semi-synthetic cutting fluid and a premium soluble oil were approximately the same.
4. Tool wear resulting from drilling aluminum 380 while using a soluble oil was very sensitive to cutting speed.
5. Cutting fluids formulated for machining aluminum alloys should contain additives that suppress the formation of aluminum oxide and should rapidly dissipate heat that collects during machining.
6. The flood application was superior to the mist method when the objective was to produce good surface finish at low feed rates.
7. Surface finish results from drilling aluminum 390 while using a semisynthetic cutting fluid and a premium soluble oil were approximately the same.
8. Variations in feed rate, fluid concentration, and the interaction between feed rate and drill diameter had significant effects on surface finish while drilling aluminum 390.
9. Surface finish resulting from drilling stainless steel 304 while using a synthetic cutting fluid was significantly affected by drill diameter.
10. The interaction between feed rate and drill diameter had a significant effect on the surface finish resulting from drilling a medium carbon steel while using a synthetic cutting fluid.
11. Surface roughness resulting from drilling aluminum 380 while using a soluble oil increased with increases in cutting speed and feed rate.
12. The mist application produced a lower thrust force and a lower cutting torque for most cutting speeds when compared to the flood application. Cutting torque was also reduced at higher feed rates and higher cutting speeds by using the mist.
13. Thrust force was significant in drilling models that predicted tool wear for a medium carbon steel workpiece and a titanium alloy

workpiece when a synthetic cutting fluid was used. Cutting torque was significant in the model for the titanium alloy.
14. Cutting torque was significant in drilling models that predicted tool wear for a composite material workpiece when a synthetic cutting fluid was used.
15. Radioactive drills can be used to study the tribological characteristics of cutting fluids.
16. Ferromagnetic cutting fluids were effective in drilling nonmagnetic structural materials in special environments, but high cost prevents them from being a feasible substitute for conventional cutting fluids.
17. A high molecular SKD rubber and a type P-20 polyisobutylene were effective additions to an industrial cutting oil for drilling into a high-temperature strength steel.

B. Turning Research

In turning operations where the cooling mechanism is important, the critical factors which influence the rate of heat transfer from the cutting tool to the coolant are the cutting fluid formulation, flow rate, and direction of application. Childs, Maekawa, and Maulik conducted both theoretical and experimental studies of the effects of coolant on the tool temperature in turning processes [4]. These researchers used finite element analysis to model the temperature distribution in the tool, tool holder, chip, and workpiece. The experimentation involved the turning of an annealed medium carbon steel workpiece. HSS tools were fed at a rate of 0.254 mm/rev and the cutting speeds were 55, 75, and 102 cm/s. A water-based coolant was used. The best match between theory and experimentation was obtained for tests at a cutting speed of 102 cm/s. Increasing the coolant flow rate from 0.25 l/min, which was directed onto the workpiece ahead of the tool, to 2.5 l/min, which was flooded directly onto the tool, caused the heat transfer coefficient from the tool to the coolant to increase by a factor of five. These coefficients are typical of those found in industry.

Cutting fluids should be formulated according to the requirements of the cutting tool materials. For example, synthetic cutting fluids are recommended when the cutting tools are made from tungsten-free carbides. Rutman et al. studied the effects of five cutting fluids on the wear resistance of tools tipped with tungsten-free carbides in the turning of plain carbon steel, 197 to 207 HB [37]. New organic compounds formed as a result of the chemical-mechanical transformations in synthetic cutting fluids improved the conditions in the contact zone between the tool and workpiece. These more active reactions made machining easier by raising the temperature in the contact zone due to the low thermal conductivity of the tungsten-free carbides. The most effective cutting fluids were a synthetic (3% concentration) that contained water soluble polyglycols and a 1% solution of a tribopolymer-forming additive and corrosion inhibitors. These fluids were compared to two water emulsions and an oil-based emulsion. The synthetic cutting fluids produced an appreciable improvement in cutting tool life and satisfactory surface finish.

Gartfel'der and Fedorov performed another research project in which the cutting fluid formulation influenced the wear resistance of a different cutting tool material [38]. These researchers studied the influence of 12 cutting fluids, from four groups, on the wear resistance of cubic boron nitride (CBN) tools during the turning of a hardened steel (50 HRC, where HRC is the Rockwell hardness number on the C scale). The most effective group of cutting fluids reduced the amount of oxygen in the cutting zone. Special cutting fluids should be developed for applications where CBN tools are used.

Several methods are available for evaluating the performance of cutting fluids in turning operations. De Chiffre employed three testing procedures to investigate the cooling properties of cutting fluids, the lubricity of cutting fluids, and the wear reduction as a result of using cutting fluids [39]. The cooling properties were studied while facing a medium carbon steel on a lathe. Cutting temperature was measured for cutting speeds ranging from 33 to 495 cm/s. Using a thermoelectric principle, an average contact temperature on the tool was measured by recording the electromotive force induced in the tool–workpiece circuit. The lubricity of cutting fluids was investigated in experiments conducted on an electrolytic copper. A restricted contact length of 0.4 mm was used. A 5 min tool-wear test was used to study the combined effects of cooling and lubrication for two cutting oils used to machine an austenitic stainless steel. Good reproducibility was obtained from the results of the tool-wear tests. All three testing procedures were quick and inexpensive.

Naerheim and Kendig used rapid electrochemical techniques to evaluate the effectiveness of cutting fluids in terms of cutting forces measured with a Kistler piezoelectric dynamometer [40]. Turning experiments were performed on a bar of 4340 steel. The reduction in cutting forces was related to the extent of absorption of the cutting fluid surfactant on the workpiece. The tools and chips from these experiments were analyzed by Naerheim, Smith, and Lan [41]. Chips produced when CCl_4 was used as a model cutting fluid were studied with a scanning Auger microscope. The researchers concluded that the cutting fluid reduced the adhesion forces between the tool and chip. They also concluded that the interaction between the cutting fluid and the chips reduced the cutting forces by aiding the plastic deformation (i.e., the Rebinder effect).

In addition to using radioactive cutting tools to evaluate cutting fluids for drilling operations, Jeremic and Lazic also studied turning operations [33]. Three soluble oils were included in their investigations. The criterion for the life of the carbide inserts with a titanium carbide coating was flank wear of 0.30 mm. Tool life varied from 133 to 169 min while turning a low-machinability steel which was heat-treated to a hardness of 270 HB. The researchers also studied a synthetic cutting fluid which contained a water-soluble concentrate. The concentrate consisted of EP additives and additives with high molecular weights. It contained no oil, fatty additives, or inorganic salts. As the cutting speed was increased from 217 to 317 cm/s, the tool life decreased from 66 to 37 min while turning a low-machinability steel which was heat-treated to a hardness of 320 HB.

In an effort to reduce the detrimental effect from electrochemical wear due to galvanic corrosion in aqueous cutting fluids, Kurimoto proposed galvanic-cathodic protection of cutting tools [42]. A relatively large zinc-coated area on the HSS cutting tool served as an auxiliary anode. The electrolytic cells included the tool–workpiece, the tool–auxiliary anode, and the workpiece–auxiliary anode. Tool-wear tests were performed while turning normalized steel bars. The three aqueous fluids tested were as follows: (a) a $NaNO_2$ solution in water at 0.1 wt. % concentration, (b) a synthetic cutting fluid at 1.2 vol. % concentration, and (c) a soluble oil at 4.8 vol. % concentration. With the conditions used in the experiment, a zinc-coated tool was very effective in reducing crater wear but rather ineffective in reducing flank wear. Galvanic-cathodic protection of cutting tools was effective when highly electrolytic water-based fluids were applied but was less effective with weaker electrolytes such as soluble oils.

When the surface integrity of machined surfaces is critical, the selection of a cutting fluid can be important. Jeelani and Ramakrishnan studied the surface damage produced during facing cuts, on a lathe, of titanium 6242 alloy which contained 6 wt. % aluminum and 4 wt. % zirconium [43]. Cutting speeds ranged from 10 to 80 cm/s and a highly chlorinated water-soluble oil was used. A scanning electron microscope made it practical to examine detailed variations in the machined surfaces. Application of the cutting fluid reduced the damage to the workpiece surface at all cutting speeds tested.

Sakuma and Fujita also studied the surface finish produced by a turning operation [44]. Carbide tools were used in finish turning of a carbon steel, and surface roughness was expressed in terms of grooving wear or boundary wear. A linear relation was found between the surface roughness and the depth of grooving wear, i.e., the deeper the groove, the rougher the surface. The cutting fluids used for the test included an emulsion-type fluid and four cutting oils. The emulsion concentrate was diluted with seven parts water and had the lowest viscosity. In fact, the authors commented that its viscosity was so low that it could not cover the cutting edge properly. Therefore, this cutting fluid could not suppress oxidation at the cutting edge. Because of this phenomenon, the water-based cutting fluid was the least effective in reducing grooving wear.

Additives are introduced into cutting fluids to maximize their effectiveness. Cutting fluids containing polymers which are soluble in water have been evaluated in turning operations for various steels. The steels tested by Soshko and Shkarapata had surface hardnesses ranging from 160 to 240 HB and were machined with a sintered carbide turning tool [45]. The workpiece materials included carbon and alloyed constructional steels and stainless steels. The cutting fluids included an experimental fluid containing polyvinyl chloride (PVC) latex and a commercial soluble oil with and without the addition of PVC latex. Flank wear of 0.3 mm was taken as the criterion for tool wear. It was concluded from the results of the tests that the relative cutting tool life increased with cutting speed when the experimental fluid was compared to the soluble oil without the addition. Cutting speeds ranged from 83 to 142 cm/s. When the same fluids were compared on the basis of surface finish, the improvement was noticed only at the low cutting speeds.

Makarov et al. also studied cutting fluids with a polymer content for turning steels [46]. The values of the surface hardness of the three steels ranged from 140 to 220 HB. Cemented carbide cutters were tested at speeds of 167, 181, and 210 cm/s. Wear criterion for these tools was 1 mm. The polymer used in the experimental fluid was a polyacrylic acid with iron. The other cutting fluids used in the study included fluids with an oil base and emulsions. The author concluded that the polymer-based cutting fluid had performance properties that enabled it to be considered as a replacement for oil-based fluids.

Synthetic cutting fluids with trioactive additives were studied by Lobantsova and Chulok [47]. These additives can form a protective polymer film during the machining operation, by frictional and chemical reactions, which reduces tool wear. The researchers studied the machining of a constructional carbon steel, a corrosion-resistant steel, and a difficult-to-machine alloy steel with HSS tools and cemented carbide inserts. The cutting fluids which were examined in turning processes contained the following trioactive additives: (a) cinnamon, (b) maleic, (c) fumaric, and (d) amide. Control cutting fluids included a soluble oil and two synthetic fluids. Criteria for evaluating the cutting fluids included both the durability of the tool and the thermal stability of the additive. The authors presented a method of predicting the efficiency of trioactive additives in cutting fluids.

Coolant systems which deliver the cutting fluid at a high velocity and a high flow rate are available for turning applications. These systems help to control chips and improve tool life. Uncontrolled chips can prohibit the implementation of an unattended machining operation, be dangerous to the operator, damage workpieces machined from expensive materials, and present a disposal problem. A coolant velocity of 100 m/s and flow rates of 15 to 20 l/min are possible with these systems. Ringler reported the results of tool-life tests for turning Ti6Al4V, a difficult-to-machine alloy [48]. A high-velocity coolant distribution system was used when the material was machined at a cutting speed of 140 cm/s and feed rates of 0.13, 0.25, and 0.38 mm/rev. In all cases, the tool life was doubled when compared to tests in which the same cutting conditions were used but the high-velocity coolant system was not used.

A high-pressure cutting fluid delivery system was demonstrated at Metcut Research Associates Inc. in 1990 [49]. A high-velocity jet of water-based cutting fluid was mixed with liquid carbon dioxide to provide chip control when machining advanced metals. These metals included grade 1010 steel, stainless steels (304 or 17-4PH), alloy steels (4340 at 300 HB), titanium (6Al4V), and Inconel 718. Additional benefits include increased tool life and increased metal removal rate due to the dramatic removal of heat from the machining zone. In some applications, surface integrity can also be improved. Surface integrity is extremely critical in the aerospace industry where parts are subjected to fatigue.

The following conclusions were drawn from this review of the literature on turning research:

1. Finite element analysis can be used to model the effects of a coolant on the temperature distribution in metal turning operations.
2. Cutting fluids used with tungsten-free carbide and CBN tools require special formulations.
3. Simple laboratory testing can be done to investigate the cooling properties and lubricity of cutting fluids, and the wear reduction as a result of using cutting fluids.
4. Rapid electrochemical techniques can be used to evaluate the effectiveness of cutting fluids.
5. The Rebinder effect can cause the interaction between the cutting fluid and the chips to reduce the cutting forces by aiding the plastic deformation. Adhesion forces between the cutting tool and the chips can also be reduced by the cutting fluid.
6. Radioactive carbide inserts can be used to study the tribological characteristics of cutting fluids.
7. Galvanic-cathodic protection in the form of a zinc-coated HSS tool was effective in reducing crater wear but not flank wear. Wear on the zinc-coated tool was reduced more by the water-based cutting fluids than the soluble oil.
8. Highly chlorinated water-soluble oils can be used to reduce surface damage while machining titanium alloys.
9. Cutting oils were found to be more effective in reducing grooving wear on the cutting edges of end-cutting tools in finish turning of steels than an emulsion concentrate diluted with water. The reduced grooving wear improved the surface roughness.
10. Cutting fluids containing PVC latex can improve the relative tool life at high cutting speeds while machining various steels compared to soluble oils without the PVC latex. The improvement in surface finish while using polymer-containing cutting fluids occurred only at the low cutting speeds.

Metal Cutting Processes

11. Cutting fluids containing polyacrylic acid with iron can be considered as a replacement for oil-based fluids when turning steel.
12. Cutting fluids containing cinnamon, maleic or fumaric acids, or an amide increased tool life compared to emulsified and/or synthetic fluids when turning constructional steels.
13. A high-velocity coolant system can control chips and improve tool life when turning difficult-to-machine alloys.
14. A high-velocity coolant system that uses liquid carbon dioxide can control chips, increase tool life, increase metal removal rate, and improve surface integrity when machining advanced metals.

C. Milling Research

Jeremic and Lazic also used radioactive end mills to evaluate cutting fluids [33]. The same three synthetic cutting fluids used in their drilling tests were used in these milling tests. The criterion for tool life of the 12 mm diameter HSS end mills with four cutting edges was 0.35 mm wear. Tool life varied from 81 to 99 min while milling a low-machinability steel which was heat-treated to a hardness of 270 HB.

Ivkovic [50] conducted some milling tests in the same metal cutting and tribology laboratory that Jeremic and Lazic used. The criterion for tool life was 0.40 mm wear. Tool life, when a semisynthetic cutting fluid was used, was 15% longer than when an emulsifying mineral oil-in-water was used. The optimal concentration for both cutting fluids was 6%.

Tests were performed by Makarov et al. in which a single tooth end mill, 90 mm in diameter and fitted with a HSS cutter, was used [46]. The criterion for tool life was 0.40 mm wear. When milling a steel with a surface hardness in the range from 140 to 190 HB, two cutting fluids with a 1% solution of a polymer containing a polyacrylic acid and iron were applied. A different molecular mass of the polymer was included in each cutting fluid. Results showed that an increase in the molecular mass of the polymer from 46×10^4 to 110×10^4 increased the tool life by 15%. In another test, an oil-based cutting fluid was compared to a 1% solution of the same polymer (with a molecular mass of 80×10^4) in water. The workpiece material was a steel with a hardness of 190 to 220 HB. A 22% increase in tool life was observed while machining with the cutting fluid containing the polymer compared to the oil-based cutting fluid.

Cutting fluids based on triboadditives were studied by Lobantsova and Chulok in face-milling operations, also [47]. The machining was performed with milling heads having mechanically secured, four-edged, HSS inserts. When the additive was a crotonic organic acid and the workpiece material was a constructional carbon steel, the tool life increased 10 to 100% when compared to two synthetic cutting fluids. The cutting conditions varied from 62 cm/s and

0.40 mm/rev to 98 cm/s and 0.15 mm/rev. In another test, a cutting fluid containing a malonic organic acid was applied while face milling an alloy steel. When compared to a soluble oil and a synthetic cutting fluid, tool life increased 50 to 80% for cutting speeds ranging from 67 to 100 cm/s and a feed rate of 0.40 mm/rev.

Poduraev et al. evaluated and ranked six cutting media used to mill a high-strength alloy steel with a hardness of 56 HRC [51]. The life of the milling cutters was evaluated over the speed range from 40 to 67 cm/s. For a given cutting speed, the cutting media were ranked in the following order with the best one first: (a) a finely dispersed powder of a low melting point metal (PO-2) in polyorganosiloxane resins (K-40), (b) K-40, (c) PO-2 in an oil-based cutting fluid (MR-1), (d) MR-1, (e) dry cutting, and (f) a 5% soluble oil.

Experimental investigations on the milling of key slots along long shafts were performed by Agadzhanyan and Bogdanenko [52]. The major objectives of their research were to determine the most effective method of supplying the cutting fluid, to determine the effect of heat treating the cutter, and to determine the optimal cutting conditions.

Tests were performed with circular cutters of HSS and HSS cutters which had been carbonitrided. The three methods for supplying the cutting fluid were flooding, spraying, and a jet delivery to the cutting zone under high pressure. Carbonitriding the cutters and using jet delivery of the cutting fluid increased tool life by 50%. Slots were machined into shafts of steel and monel alloy. The cutting fluids were a sulfurized oil, a soluble oil, and an industrial oil. The following machining conditions were used: milling speed (62.5, 73.3, 83.3, and 104.2 cm/s), cutter feed (3.8 and 5.0 mm/s), and cut depth (0.9 and 2.2 mm).

Agadzhanyan and Bogdanenko defined tool life as 0.70 mm wear on the flank of the cutter teeth. Even though they were able to extend the tool life of HSS cutters to 20 to 25 min, the productivity was not high enough to meet industry demands. Additional research was performed with cutters having sintered carbide tips and more effective cutting fluids. Field testing demonstrated that under these conditions, tool life could be increased by a factor of 12 to 15, or to 300 min.

The following conclusions were drawn from this review of the literature on milling research:

1. Radioactive end-milling cutters can be used to study the tribological characteristics of cutting fluids.
2. A semisynthetic cutting fluid can increase tool life in milling operations compared to an emulsifying mineral oil in water.
3. Cutting fluids containing polyacrylic acid with iron increased tool life by 15 to 20% over oil-based fluids when milling steel.

4. Cutting fluids containing crotonic or malonic organic acids increased tool life compared to emulsified and/or synthetic fluids when milling constructional steels.
5. A finely dispersed powder of a low melting point metal in polyorganosiloxane resins was found to be better than five other cutting media for milling a high-strength alloy steel.
6. The life of slot-milling cutters can be increased by using a high-pressure jet to deliver the cutting fluid to the cutting zone.

D. Tapping Research

Podgorkov also studied magnetic cutting fluids in tapping operations [34]. Cutting edges of the taps were magnetized in order to attract the cutting fluid. The dependence of torque upon cutting speed was investigated. For a given tap size, the depth of cut was specified by the thread profile and the feed rate was determined by its pitch. At a cutting speed of 2.5 cm/s, the torque required to tap holes in a titanium alloy was reduced 25 to 30% compared to cutting dry. However, at a speed of 10.0 cm/s, the torque was reduced by only 10 to 12%.

A table-mounted dynamometer was used by Hartley to study the effects of cutting fluids in tapping operations [53]. When a neat oil was compared to a heavy duty cutting fluid with EP additives, the cutting torque was reduced 27% by the EP cutting fluid while tapping an alloy steel. Hartley cautioned against using only torque to evaluate cutting fluids in holemaking operations since chips caught in the flutes of cutting tools can provide misleading results.

Childs and Hartley identified two modes of tap failure with their experiments: too high a cutting force and jamming of the chips in worn corners of the cutting edges of the tap [54]. The researchers found that a mineral oil containing an EP additive (sulfur) outperformed a solid-film lubricant.

The following conclusions were drawn from this review of the literature on tapping research:

1. The cutting torque required to tap holes in a titanium alloy was reduced 10 to 30% by using magnetic cutting fluids instead of cutting dry.
2. The cutting torque required to tap holes in an annealed alloy steel was reduced 27% when a heavy duty cutting fluid with EP additives was compared to a neat oil.
3. A mineral oil containing an EP additive (sulfur) outperformed a solid-film lubricant when holes in an annealed alloy steel were tapped.

E. Broaching Research

A study was conducted by Schmenk to determine if the horizontal broaching machines used in the automotive industry to broach cast iron engine components could also be used to broach aluminum engine components [55]. These components included cylinder heads and crankcases.

Both hard and soft aluminum alloys were studied. The hard aluminum alloys consisted of aluminum 319 and aluminum 390 in the T6 heat-treat condition. These alloys had a hardness greater than 95 HB for a 500 kg load. The softer alloys involved aluminum 319, aluminum 355, and aluminum 380 in the T5 condition. The hardness for these alloys ranged from 70 to 80 HB.

Broaching was simulated by using a large-diameter, single-tooth fly cutter in a vertical-milling machine. The cuts were relatively straight across the workpiece since the diameter of the fly cutter was much larger than the width of the workpiece in the direction of the cut. The tool holders for the inserts were very similar to stand-up broaching tool holders.

Many tests were conducted using a heavy duty synthetic cutting fluid at a concentration of 4.8 vol. %. Some tests were performed using a wide variety of water-based coolants. Typical cutting speeds ranged from 76 to 169 cm/s.

Each broaching tool was expected to machine approximately 5000 m. For a crankcase or manifold which required a broaching cut approximately 500 mm long, this tool-life criterion represented broaching 10,000 workpieces.

Aluminum 390, with a hardness of 100 HB, was machined at 76 cm/s. The flank wear after machining the equivalent of 10,000 workpieces using carbide inserts was less than 0.10 mm. This tool life was estimated to be approximately three times better than the tool life resulting from broaching cast iron under similar conditions. Similar tests were initiated on aluminum 319 in the T5 condition, but were terminated when no tool wear could be observed after machining the equivalent of 2460 parts from the softer alloy. For carbide tools with rake angles from –5 to 15° and shear angles from 0 to 15°, tool geometry did not have an appreciable effect on wear.

The machining of hard and soft aluminum alloys was evaluated on the basis of cutting forces, part deformation, edge breakout, tool wear, surface finish, tool geometry, and coolant selection. The following conclusions were drawn from this study:

1. Hard aluminum alloys can be broached using the same tooling and cutting conditions used to broach cast iron engine components.
2. The maximum broaching speeds will maximize the metal removal rate and improve surface finish when machining soft aluminum alloys.
3. Positive tool rake angles will outperform negative rake angles.
4. The application of a coolant will improve surface finish and chip removal.

REFERENCES

1. M. E. Merchant, *J. Appl. Phys. 16*(5): 267–275 (1945).
2. M. E. Merchant, *J. Appl. Phys. 16*(6): 318–324 (1945).
3. E. P. DeGarmo, J. T. Black, and R. A. Kohser, *Materials and Processes in Manufacturing*, 7th ed., Macmillan, New York (1988).
4. T. H. C. Childs, K. Maekawa, and P. Maulik, *Mater. Sci. Technol. 4*: 1006–1019 (1988).
5. S. Kalpakjian, *Manufacturing Engineering and Technology*, Addison–Wesley, Reading, MA (1989).
6. B. H. Amstead, P. F. Ostwald, and M. L. Begeman, *Manufacturing Processes*, 8th ed., Wiley, New York (1987).
7. H. D. Moore and D. R. Kibbey, *Manufacturing: Materials and Processes*, 3rd ed., Grid Publishing, Columbus, OH (1982).
8. B. W. Niebel, A. B. Draper, and R. A. Wysk, *Modern Manufacturing Process Enginerring*, McGraw–Hill, New York (1989).
9. S. Kalpakjian, *Manufacturing Processes for Engineering Materials*, 2nd ed., Addison–Wesley, Reading, MA (1991).
10. T. Smith, Y. Naerheim, and M. S. Lan, *Tribol. Int. 21*: 239–247 (1988).
11. J. E. Neely and R. R. Kibbe, *Modern Materials and Manufacturing Processes*, Wiley, New York (1987).
12. V. Latinovic and M. O. M. Osman, *Int. J. Prod. Res. 24*: 1319–1329 (1986).
13. S. Azad and S. Chandrashekhar, *Mech. Eng. 107*: 60–68 (1985).
14. M. Albert, *Mod. Mach. Shop 56*: 62–67 (1984).
15. C. Trost, *Cutting Tool Eng. 37*: 32–34 (1985).
16. T. Geiser, *Ind. Prod. Eng. 9*: 46, 49–51 (1985).
17. S. Kalpakjian, *Manufacturing Processes for Engineering Materials*, Addison–Wesley, Reading, MA (1984).
18. B. A. Washington, *Cutting Tool Eng. 37*: 37–38 (1985).
19. E. S. Nachtman and S. Kalpakjian, *Lubricants and Lubrication in Metalworking Operations*, Marcel Dekker, New York (1985).
20. H. R. Leep, *Lubr. Eng. 37*: 715–721 (1981).
21. H. R. Leep and S. J. Kelleher, *Lubr. Eng. 46*: 111–115 (1990).
22. R. A. Block, "Effects of cutting fluids on drilling an aluminum-silicon alloy," M. Eng. thesis, Univ. of Louisville, Louisville, KY (1990).
23. H. R. Leep and R. W. Sims, *J. Synth. Lubr. 4*: 283–305 (1987–88).
24. T. W. Eldridge, "Effects of cutting conditions on the drilling of aluminum alloy 380," M. Eng. thesis, Univ. of Louisville, Louisville, KY (1989).
25. J. Franklin, *Die Cast. Eng. 29*: 48, 50 (1985).
26. P. N. Rao, "A comparative study of methods of cutting fluid application in turning and drilling," Proceedings of 3rd International Colloquium "Lubrication in Metal Working," Esslingen, West Germany, pp. 101.1–101.8 (1982).
27. H. R. Leep, *J. Synth. Lubr. 6*: 325–338 (1990).
28. H. R. Leep, E. D. Halbleib, and Z. Jiang, *Int. J. Prod. Res. 29*: 391–400 (1991).
29. R. J. Thomas, "Evaluation of surface finish of holes drilled into stainless steel," M. Eng. thesis, Univ. of Louisville, Louisville, KY (1990).

30. K. S. Spond, "Surface finish model for holes drilled into medium carbon steel," M. Eng. thesis, Univ. of Louisville, Louisville, KY (1991).
31. H. R. Leep and M. A. Peak, "Drilling models for a synthetic cutting fluid," Proceedings of 8th International Colloquium "Tribology 2000," Esslingen, Germany, pp. 23.4-1–23.4-3 (1992).
32. D. L. Herde, "Production drilling models for a hardened alumina-reinforced composite material," M. Eng. thesis, Univ. of Louisville, Louisville, KY (1991).
33. B. Jeremic and M. Lazic, "Tribological characteristics of cutting fluids (emulsions and solutions)—Applications of thin layer activation," Proceedings of 3rd International Colloquium "Lubrication in Metal Working," Esslingen, West Germany, pp. 77.1–77.11 (1982).
34. V. V. Podgorkov, *Sov. Eng. Res.* 4: 41–42 (1984).
35. V. V. Podgorkov, *Sov. J. Frict. Wear 9*: 78–82 (1988).
36. M. O. Vasyshak and A. I. Soshko, *Sov. Mater. Sci.* 22: 339–340 (1986).
37. P. A. Rutman et al., *Sov. Eng. Res.* 3: 90–91 (1983).
38. V. A. Gartfel'der and V. M. Fedorov, *Sov. J. Superhard Mater.* 6: 70–72 (1984).
39. L. De Chiffre, "Laboratory testing of cutting fluid performance," Proceedings of 3rd International Colloquium "Lubrication in Metal Working," Esslingen, West Germany, pp. 74.1–74.5 (1982).
40. Y. Naerheim and M. Kendig, *Wear 114*: 51–57 (1987).
41. Y. Naerheim, T. Smith, and M. S. Lan, *J. Tribol. Trans. ASME 108*: 364–367 (1986).
42. T. Kurimoto, *Wear 127*: 241–251 (1988).
43. S. Jeelani and K. Ramakrishnan, *J. Mater. Sci.* 20: 3245–3252 (1985).
44. K. Sakuma and T. Fujita, "Characteristic of grooving wear on end-cutting-edge and effect of cutting fluids in finish turning of steels," Proceedings of 3rd International Colloquium "Lubrication in Metal Working," Esslingen, West Germany, pp. 98.1–98.5 (1982).
45. V. A. Soshko and Y. E. Shkarapata, *Sov. Mater. Sci.* 22: 539–540 (1986).
46. R. V. Makarov et al., *Soc. Eng. Res.* 6: 35–36 (1986).
47. V. S. Lobantsova and A. I. Chulok, *Sov. Eng. Res.* 4: 69–71 (1984).
48. A. G. Ringler, *Cutting Tool Eng.* 40: 26, 28 (1988).
49. B. Weiss, *Metalworking News*, p. 10 (March 6, 1990).
50. B. Ivkovic, "Possible approaches for measurements of tribological characteristics of cutting fluids," Proceedings of 3rd International Colloquium "Lubrication in Metal Working," Esslingen, West Germany, pp. 79.1–79.6 (1982).
51. V. N. Poduraev et al., *Sov. Eng. Res.* 6: 29–33 (1987).
52. R. A. Agadzhanyan and Y. G. Bogdanenko, *Chem. Pet. Eng.* 22: 400–401 (1986).
53. D. Hartley, "Observations of lubricant effects on chip interference in tapping operations," Proceedings of 3rd International Colloquium "Lubrication in Metal Working," Esslingen, West Germany, pp. 97.1–97.7 (1982).
54. T. H. C. Childs and D. Hartley, *J. Lubr. Technol. Trans. ASME 105*: 507–513 (1983).
55. M. J. Schmenk, "Broaching of aluminum automotive engine components," Paper 800488, SAE International Congress & Exposition, Detroit (1980).

4
Performance of Metalworking Fluids in a Grinding System

CORNELIS A. SMITS
Cincinnati Milacron
Cincinnati, Ohio

I. INTRODUCTION

Metalworking fluids have been applied for many reasons to enhance grinding as a metal removal operation for precision components. The importance of metalworking fluids and proper fluid application is proved every day in shops where precision components are produced.

Various publications on the performance of metalworking fluids, their application, and their secondary functions have increased the knowledge of fluid applications in the industry. The wide range of grinding operations and specific machine tools involved, as well as the environmental and health issues, creates a very complex technology with many constraints. Other considerations include the maintenance and ongoing contamination of the metalworking fluid by grinding chips, grinding wheel particles, machine tool and other oil leaks, make-up water minerals, elevated temperatures caused by grinding energy, and other possible pollutants which can cause a significant change in metalworking fluid behavior. Also, the carry-off of certain chemicals which stick to ground workpieces and metal fines will cause changes by depletion.

One can list the variables that are of importance under machine tool, grinding wheel, or metalworking fluid-related variables. The machine tool-related variables change over time due to wear of machine tool elements. However, this change is very slow and is measured in years instead of hours. The grinding wheel variables will be constant depending on duplication of manufacturing and the availability of facilities for maintaining a constant circumferential speed over the wear range of the grinding wheel. The metalworking fluid is the most dynamic element in the grinding system as it is continuously under change due to carry-off and the input of pollutants as mentioned earlier.

All of these conditions must be considered when selecting, applying, and maintaining metalworking fluids. In this context, the variables affecting performance will be discussed and the proper engineering criteria defined in order to choose the best selection of metalworking fluids and to correctly design its application.

II. METALWORKING FLUIDS AS AN ELEMENT OF THE TOTAL GRINDING SYSTEM

It is essential to know the grinding system that will be served by the metalworking fluid. This system includes the conditions of the workpieces before beginning the grinding operations. Figure 1 shows the most essential workpiece variables. The workpiece dimensions and static stiffness will define the grinding contact area with the grinding wheel and level of grinding force that can be applied. (Workpiece stiffness can be a significant constraint.) Also, the dimensions and shape of the part, combined with its thermal conductivity, will define the heat sensitivity of the workpiece—a constraint for maximum energy input.

Grinding is an energy inefficient method of shaping components and it is obvious that the thermal balance in the grinding zone is of significant importance. The thermal conductivity and diffusivity of the workpiece material has a significant impact on the grinding results. The flow of heat from the contact zone of workpiece and grinding wheel is very important in order to limit the thermal damage to the workpiece material. Application of the metalworking fluid can have a significant influence on this heat flow. The fluid speed leaving the nozzle, the flow rate, and the direction of the flow (nozzle position), are important factors. Other factors such as grinding wheel composition, wheel and workpiece speed, and the amount of energy input will influence the heat flow.

The metalworking fluid will carry away most (96%) of the input energy by its contact with the workpiece, chips, and grinding wheel. The energy input will end up in the metalworking fluid where it will be transferred to the surroundings by evaporation, convection, or in a forced manner, by a chiller. Evaporation is the

Metalworking Fluids in Grinding System

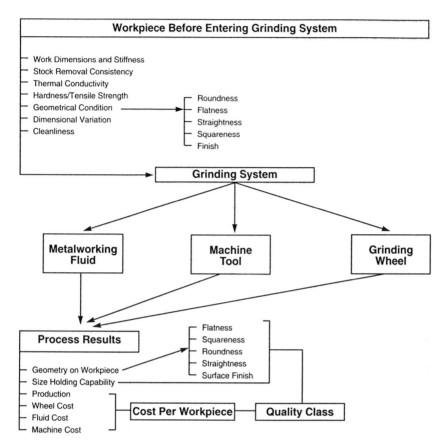

Figure 1 Workpiece variables.

dominating factor in obtaining a balance between energy input and output. The metalworking fluid will warm up until a temperature is reached that balances energy input and output to the ambient through convection and evaporation. With a very high input of energy, a high make-up rate can be expected which causes a possible high rate of contamination if nontreated water is used.

Another workpiece material characteristic is its hardness (R_c). When hardness increases, higher forces are required to penetrate abrasive grains into the material surface. This will result in higher grinding forces and the normal force (F_n) will increase. Figure 2 shows that as you approach 50–60 R_c it is very difficult for the grinding abrasive to penetrate into the surface of the workpiece. This results in very thin chips and considerable machine deflection, particularly for weak grinding systems such as internal grinding. Some of this behavior is

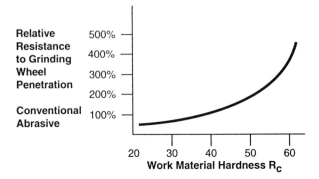

Figure 2 Resistance to grinding wheel penetration increases with increased work material hardness. As you approach 50–60 R_c, it is very difficult for the grinding abrasive to penetrate into the surface of the workpiece. This results in very thin chips and considerable machine deflection, particularly for weak grinding systems such as internal grinding. Some of this behavior is due to the fact that the wheel itself flattens out in the contact area—sort of like the way a tire flattens against the road due to the weight of the car.

due to the fact that the wheel itself flattens out in the contact areas, similar to the way a tire flattens against the road due to the weight of the car.

The normal force F_n is also dependent upon the lubricant capability of the metalworking fluid, the grinding wheel grade, the truing technology, and the aggressiveness of the cut, see Fig. 3. The stock removal consistency, geometry, and dimensional variations are mechanically related properties that cannot be influenced by metalworking fluids.

Cleanliness of workpieces entering the system can affect fluid performance. Carry-over of oil and grease from other operations will be an influence on the grinding operation and present an ongoing contamination of the fluid.

Figure 4 shows the most important variables that affect manufacturing cost and quality class. The consistency of these listed variables is very important. Some will remain constant when implemented, others change over time. The metalworking fluid represents an essential element of these variables.

Grinding wheels depend on duplication capability when the optimum composition (grade) has been obtained. The truing tool and truing conditions, however, can change the cutting action of the grinding wheel which can lead to significant performance changes. Consistency in truing technology is essential.

III. PERFORMANCE OF METALWORKING FLUIDS

To express performance of a metalworking fluid, a wide range of criteria are applied. The criteria can be grouped as shown in Fig. 5.

Factors Affecting Normal Force

- Workpiece Hardness
- Bond and Grain Volume of Wheel
- Truing Technology
- Grinding Fluid
- Aggressiveness of Cut

Importance of Normal Force

- The Normal Force Acting Between Wheel and Work Causes Them to Spring Apart Reducing the Intended Stock Removal

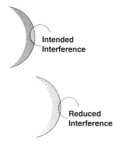

Figure 3 Factors affecting normal force.

The environmental criteria will not be discussed in this section. Some attention will be given to the criteria for maintainability of the metalworking fluid but only where an effect on the grinding operation is measurable.

The overall performance of a metalworking fluid is a result of the criteria listed. Most items listed under the environmental and maintainability categories are well known and areas where attention has been focused. However, the reason for the application of a metalworking fluid is its ability to aid the grinding performance, which can be expressed by:

Productivity
Tool life
Energy consumption
Quality

These are the four significant process parameters that we have defined in cylindrical grinding operations.

Optimizing the operation must be the ultimate objective in any study of a metal cutting process. Optimization requires first a thorough understanding of

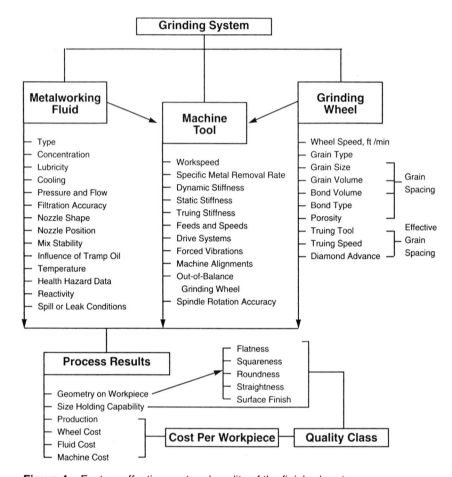

Figure 4 Factors affecting cost and quality of the finished part.

the interrelationships of the significant variables; then these variables must be measured to find their quantitative influence on quality and cost.

A. Productivity

The productivity of a grinding process is expressed by the specific metal removal rate, Q', defined as the volume of metal removed per unit of time per unit of effective wheel width. Its units are in inches squared per minute (in.²/min). This is the most important of all parameters since it determines production rate. Q' may be determined as follows:

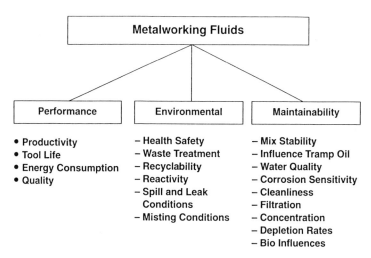

Figure 5 Metalworking fluid selection criteria.

For cylindrical grinding (internal or external):

$$Q' = \frac{\pi}{2} \frac{DSL}{WT} = \frac{\text{in.}^2}{\text{min}}$$

For surface grinding:

$$Q' = A \times V_w = \frac{\text{in.}^2}{\text{min}}$$

where

D = diameter or bore of workpiece (in.)
S = stock removal on diameter (in.)
L = work length (in.)
T = grinding time (min)
W = effective wheel width (in.)
A = down feed increment (in.)
V_w = table speed (in./min)

These equations represent the cubic inches of metal removed from a workpiece in one minute by one inch of usable wheel width.

The formula for cylindrical grinding has been derived as follows:
The volume of metal removed per minute is:

$$V_m = \frac{\pi(D_1^2 - D_2^2)L}{4T}$$

where

D_1 = work diameter before grinding (in.)
D_2 = work diameter after grinding (in.)
L = length of workpiece (in.)
T = grinding time (min)

Then

$$Q' = \frac{V_m}{W} = \frac{\text{in.}^3/\text{min}}{\text{in.}}$$

$$Q' = \frac{\pi(D_1^2 - D_2^2)L}{4WT} = \frac{\text{in.}^3/\text{min}}{\text{in.}}$$

where W = effective wheel width (in.).
Since $(D_1^2 - D_2^2)$ can be written $(D_1 + D_2)(D_1 - D_2)$ and substituted, then:

$$Q' = \frac{\pi(D_1 + D_2) \times (D_1 - D_2)L}{4WT}$$

Since $(D_1 + D_2)/2$ = average diameter, or D; and $(D_1 - D_2)$ = stock on the diameter, or S; then:

$$Q' = \frac{\pi DSL}{2WT} = \frac{\text{in.}^3/\text{min}}{\text{in.}}$$

This means that Q' equals the cubic inches of metal removed from a workpiece in one minute by one inch of effective wheel width.

In plunge grinding, where the workpiece length L always equals the effective width W, the formula is:

$$Q' = \frac{\pi DS}{2T}$$

B. Tool Life

Tool life in grinding is expressed as a grinding ratio (G) defined as the volume of material removed with one volume unit of grinding wheel. This is an important parameter because wheel wear determines wheel costs in heavy metal removal operations, and it greatly affects quality when finish grinding. The expression for the G-ratio is:

$$G = \frac{DSLP}{2qdW} = \frac{\text{metal removal}}{\text{wheel wear}} = \frac{\text{in.}^3}{\text{in.}^3}$$

where

q = radial wheel wear (in.)
$d = (d_1 + d_2)/2$ = average wheel diameter
P = number of workpieces ground
L = length of one part (for plunge infeed, L and W cancel)

What this expression really describes is the volume of metal removal for each volumetric unit of grinding wheel worn or dressed away.

C. Energy Consumption

The energy consumption in grinding is described by the specific energy (U) defined as the horsepower required to remove one unit volume of material per unit of time. The importance of specific energy lies primarily in its role as a process limitation or evaluation, since it expresses the energy required to perform the metal removal operation and must be known and controlled in order to stay within the power available. It is also a very important parameter in predicting the behavior of nonrigid parts under grind. The expression for specific energy is:

$$U = \frac{N}{Q'W}$$

where

N = horsepower (hp, kW, or Y)
Q' = specific metal removal rate (in.^{2}/min)
W = effective wheel width

This represents the power required to remove one cubic inch of metal in one minute.

D. Quality

The quality of a part is often measured as surface finish (f). This is one of the most important parameters in a finish grinding operation, and it is very closely related to specific metal removal rate and G-ratio. It can be described by $f = R_a$, where R_a is the average surface roughness value in use today throughout the world. The following definitions apply:

R_a, arithmetic mean of departures of roughness profile from the mean line;
R_q, root mean square (RMS) parameter corresponding to R_a;
R_z, ISO ten-point numerically average height difference between the five highest peaks and the five lowest valleys within the measuring length.

IV. INTERRELATIONSHIP OF GRINDING PARAMETERS AND THE INFLUENCE OF METALWORKING FLUIDS

In 1907, F. W. Taylor reported that in metal cutting both tool wear and specific cutting force were exponentially related to cutting speed [1]. About 50 years later, after a large mass of grinding data had been collected and recorded, it was discovered that the G-ratio and specific metal removal rate relationship for grinding also holds true in the Taylor equations for metal cutting processes. This application of the Taylor formulas to grinding has undoubtedly been the most important breakthrough in the quantitative understanding of the process. The formulas for G-ratio and specific energy as related to specific metal removal rate are shown below.

Productivity and tool life relationship:

$$Q'^{n_1}G = C_1$$

where

G = grinding ratio
n_1 = grinding ratio exponent
C_1 = grinding ratio at $Q' = 1$ (see Fig. 6)

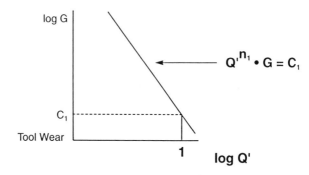

C_1 = G Ration at Q' = 1
C_1 = Influenced By: Wheel Grading
| Metalworking Fluid |

n_1 = Influenced By: Workmaterial
| Metalworking Fluid |

Type of Abrasive Material (Conventional)

Figure 6 Productivity vs. tool life.

Metalworking Fluids in Grinding System

Productivity and energy consumption relationship:

$$Q'^{n_2} U = C_2$$

where

U = specific energy
n_2 = specific energy exponent
C_2 = specific energy at $Q' = 1$ (see Fig. 7)

An additional parameter that fits the exponential equations is surface finish (f). Productivity and quality relationship:

$$Q'^{n_3} f = C_3$$

where n_3 = surface finish exponent and C_3 = surface finish at $Q' = 1$. This parameter is tied very closely to G and Q' and varies inversely to G and Q' as seen in Fig. 8.

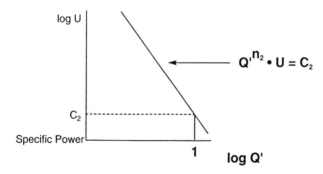

C_2 = U Value at Q' = 1
C_2 = Influenced By: Wheel Grading
  | Metalworking Fluid |
n_2 = Influenced By: Workmaterial
  | Metalworking Fluid |
  Type of Abrasive Material (Conventional)
  Grain Size

Figure 7 Productivity vs. energy consumption.

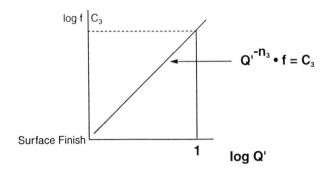

C_3 = Surface Finish at $Q' = 1$
C_3 = Influenced By: Grain Size
  Metalworking Fluid
  Workmaterial Hardness

n_3 = Influenced By: Hardness-Structure Wheel
  Workmaterial
  Metalworking Fluid

Figure 8 Productivity vs. quality.

A. Grinding Performance Diagram

One of the most important qualities of the Taylor formulas as working instruments for optimizing the process is the simplicity and usefulness of their graphic representation. When a certain material is ground with a certain wheel at several different specific metal removal rates, and the resulting G-ratio and specific power data points are recorded on log-graph paper, the data lines drawn through the points are straight. The slope of the lines then represents the exponents (see Figs. 6–8).

To optimize the process, we try to find the wheel grading and cutting fluid combination that gives us the highest C_1 value (or G-ratio at $Q' = 1$) and the lowest possible exponent, n_1. This would mean the least possible influence of increased metal removal rate on the G-ratio. At the same time, we would like to have the lowest possible C_2 value (specific power at $Q' = 1$) and the highest possible exponent—or steepest slope of U—so that the increased metal removal rate would require the least possible increase in horsepower.

The f or finish line is always sloped in the opposite direction to both G and U, because finish values become larger as Q' is increased.

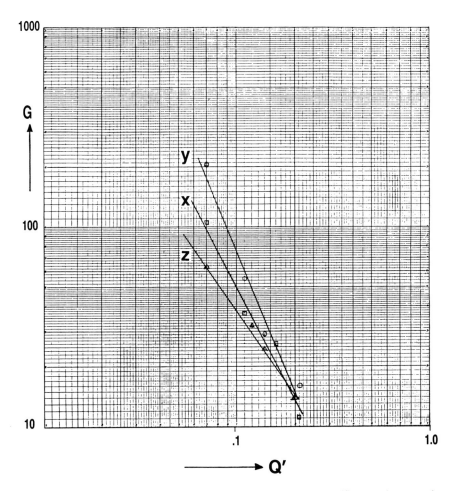

Figure 9 Wheel life with three grinding fluids at various specific metal removal rates.

The development of grinding wheels and metalworking fluids in recent years has shown considerable influence on the values for C_1 and n_1 as well as on C_2, n_2 and C_3, n_3. CBN grinding technology has shown very flat G-ratio lines and also very flat specific power lines. The values for C_1 are 100 to 1000 times higher while the values for C_2 are 50–75% of the values for conventional abrasives.

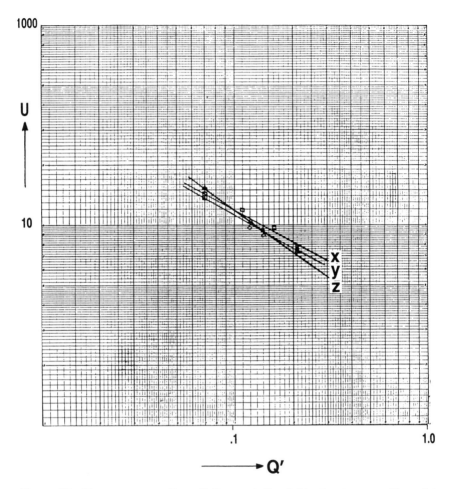

Figure 10 Energy consumption with three grinding fluids at various specific metal removal rates.

B. The Influence of Grinding Wheels on Grinding Performance

The influence of a grinding wheel on both wheel life and horsepower consumption has been measured in many applications and research activities. Changing grain type, grain size, grain volume, bond type, and hardness of a grinding wheel will have an influence on the G-ratio line as well as the specific energy line and consequently on the surface finish line. The changes can be expressed in the values for the constants and the exponents.

The key is to select wheel gradings and to generate a wheel manufacturing process that produce the following:

High value of C_1 and a flat G-ratio line or a small value for n_1 which results in high productivity with controlled wheel wear. In general, a CBN grain will produce high values for C_1 and very low values for n_1.

Low value of C_2 and a value for n_2 of near 1. This means an increase in specific metal removal rate that does not result in an excessive increase of horsepower requirements and consequential grinding forces.

Low value for C_3 (good surface finish) and a low value for n_3. This means that an increase of the specific metal removal rate only has a small to moderate effect on surface finish.

Because of the many variables possible on grinding wheel compositions and manufacturing, an enormous data bank is needed to predict or analyze the influence of the grinding wheel. Such a data bank is essential, however, to generate a successful grinding system.

C. The Influence of a Metalworking Fluid on Grinding Performance

To measure the performance of a metalworking fluid, an empirical test is required. From the test data over a range of specific metal removal rates, the values for C_1 and the exponent n_1 as well as $C_2 n_2$ and $C_3 n_3$ can be defined. Figure 9 shows the G-ratio lines obtained with various cutting fluids x, y, and z.

To compare performance based on G-ratio at only one value of Q' is very misleading, as can be seen from Fig. 9. We can observe significant changes in value for C_1 and n_1. Many times performance has been measured on the basis of energy consumption. Figure 10 shows the relationship between U and Q' for the three fluids tested in Fig. 9. Here again, we see significant changes in values for C_2 and n_2. For a true performance factor we have to consider both tool life and energy consumption. The ideal performance would produce a very high G-ratio with a very low-energy requirement.

The performance factor (efficiency) can be expressed as:

$$E = \frac{G}{U} \qquad Q'^{n_1}G = C_1, \qquad Q'^{n_2}U = C_2$$

$$E = \frac{C_1}{C_2} Q'^{(n_2-n_1)} \qquad \frac{C_1}{C_2} = C_E$$

$$E = \frac{C_E}{Q'^{(n_1-n_2)}}$$

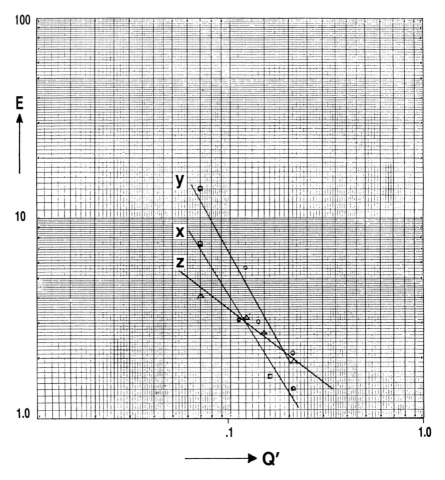

Figure 11 Grinding efficiency with three fluids at various specific metal removal rates.

where E = volume of material removed with one volume unit of grinding wheel per unit of specific energy.

The data from Figs. 9 and 10 are combined in Fig. 11 using the performance factor $E = G/U$. Now it can be clearly seen that product y outperforms products x and z up to a Q' value of 0.2 in.^{3}/min/in. Over a Q' value of 0.2, product z outperforms both y and x. For performance in grinding operations product x can be considered as an underperforming product relative to product y, but relative to product z below $Q' = 0.12$ it outperformed z.

Using this method, fluid performance can be evaluated on the basis of two criteria:
1. The level of the E value.
2. The rate of change of the E value as a function of Q'.

The ideal fluid has a very high efficiency (E value) and increases with increasing specific metal removal rate (Q'). However, this is very unlikely. Usually the E value decreases with an increase of the specific metal removal rate Q' (see Fig. 12).

Under ideal conditions, the efficiency is sometimes nearly consistent and independent of Q'. This has been the case with a combination of superabrasives and straight oil applications. In these cases, n_2 is approximately equal to n_1 which means that the slopes of the G-ratio and U value were near, or equal to one ($n_2 = n_1 = 1$). Note the following statements:

$n_2 = n_1$ Obtained with superabrasives and special metalworking fluid

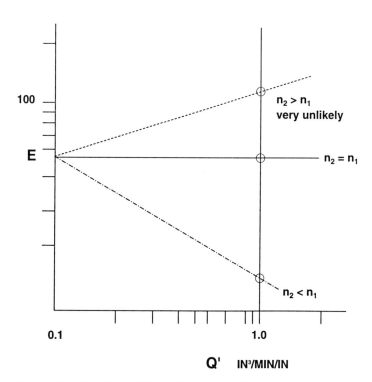

Figure 12   Grinding efficiency, three scenarios.

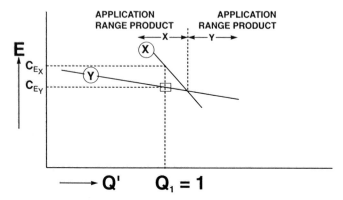

Figure 13 Selecting the proper fluid for the application. (The higher the C_E value the better the overall performance.)

or straight oil. On ballbearing steel material some synthetic abrasives are near this condition.

$n_2 < n_1$ Most applications are in this range of reduced efficiency at increasing specific metal removal rates.

See Fig. 12 for a graphical representation.

To compare metalworking fluid performance, the E value at a specific metal removal rate $Q' = 1$ can be used. The most favorable application range can be selected as shown in Fig. 13. Product X shows higher E values up to the intersection of the E value line for product y. As Q' values increase beyond the intersection point, product y has the best performance.

The primary function of the grinding performance factor E is in its expression of performance level as a function of the specific metal removal rate Q'. It can predict the best application range for metalworking fluids as far as grinding performance is concerned. Other factors related to the environment and to maintenance of the fluid must also be considered for the selection of fluids to be used in a specific metal removal rate range.

V. METALWORKING FLUID APPLICATION IN THE GRINDING ZONE

The application of metalworking fluids to the grinding zone has been discussed in many publications, but is still a "many times ignored" aspect. The application of the fluid to the grinding zone has to serve four basic functions: lubrication, cooling, chip removal, and workpiece surface protection. Lately, a study on flow mechanics resulted in a comprehensive flow model [2].

Various other models for lubrication and flow have previously been presented. Hahn [3] discusses the application of lubrication between grain and work surface. A model developed by Powell [4] discussed the fluid flow through porous wheels. The local lubrication between grain and work surface, however, has not been modeled to our knowledge. All these studies point to the importance of flow rate, fluid speed entering the flow gap, fluid nozzle position, and grinding wheel contact with the workpiece. For our purposes we will focus on the practical area of fluid application.

Various fluid nozzle design concepts have been applied on grinding operations. The function of a coolant nozzle is to supply fluid to the cut zone and to clean the grinding wheel. There are several problems in accomplishing the task. One of the principal problems is the air barrier generated by a rotating porous grinding wheel. The grinding wheel acts like an impeller in a centrifugal pump. Air is drawn in from the sides and forced out around the circumference. As wheels get wider, there is less of a problem with air.

It is possible to classify the coolant nozzles into the following categories:

Dribble
Acceleration zone
   Bourgoin Fluid Inducer
   Wedge
   Combined
Fire hose
Jet (medium pressure and super jet)
Wrap around

A. The Dribble Nozzle

The dribble nozzle is commonly used in most applications. About all that is accomplished is that some lubricant is deposited on the part's surface and some splash is sucked into the wheel. This does help in lubricating the cut zone. The major problem with this nozzle is that coolant cannot penetrate the air barriers and the fluid is not delivered efficiently into the cut zone. This can result in dry grinding despite the application of fluid.

B. Acceleration Zone Nozzle

The acceleration zone nozzle attempts to use the existing fluid systems to get the fluid up to the wheel speed, to cool the grinding wheel surface, and to wet the cutting grains. Several types exist.

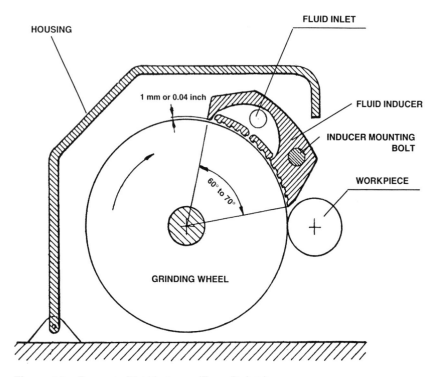

Figure 14 Bourgoin Fluid Inducer. (From Ref. 5.)

1. Bourgoin Fluid Inducer

A nozzle (Fig. 14) was developed by the French Research Center for Mechanical Industries (CETIM). Information on this device was published in 1976 and was assigned a patent [5]. The nozzle requirements were:

1. 60 to 70° of wrap around the wheel were required to accelerate the fluid.
2. The gap between the nozzle had to be held at 1 mm (0.04 in.).
3. Serrations inside the nozzle body to keep the liquid from separating tangentially.

The published results show evidence of improved wheel cleaning and reduced wheel wear. The device is no doubt much better than a dribble nozzle. The device is complex in that it must be adjusted for wheel curvature change as the wheel wears.

The true value in wheel cleaning was probably the wetting of the cutting

Metalworking Fluids in Grinding System

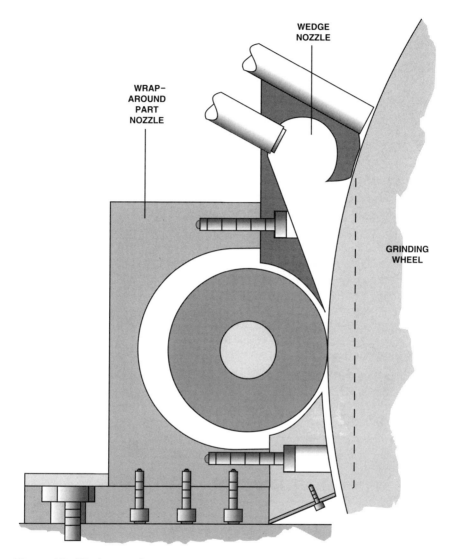

Figure 15 Wedge nozzle.

grits which led to the prevention of loading and smearing of metal on the wheel's surface.

2. Wedge Type

The wedge type nozzle is illustrated in Fig. 15. The incompressible metalworking fluid is forced into a grinding wheel and consequently forced up to wheel speed.

Complete development of this nozzle was never finished. The amount of wrap around the wheel no doubt should be the 60 to 70° worked out by the French. The nozzle wrap shown here was between 20 and 30°. The gap was controlled by maintaining a 30 to 60 psi nozzle pressure.

The sides of the wheel were covered with "horse blinder"-looking pieces. This forced the fluid alongside the wheel surface so it could be sucked into the wheel. This fluid would emerge out of the wheel's face, putting fluid in the cut zone.

This nozzle was very effective and data collected for ultrahigh speed (speeds above 16,000 sfpm) illustrated this fact.

3. Combined Inducer-Wedge

Probably the second most common nozzle to appear on the production floor is the combination nozzle. This nozzle is illustrated in Fig. 16 and its configuration is somewhat empirical. This nozzle has a gap between the wheel and surface of about 0.04 in. for a certain number of degrees of wrap, usually a minimum of 10 to 20°. There is the problem with wheel curvature between new and worn wheels in maintaining the small gap to force the acceleration of the coolant. This nozzle will wet the cutting grains, remove heat, and lubricate the cutting zone. It is a fairly effective nozzle.

C. Fire Hose Nozzle

The idea behind this nozzle is to force the fluid to slightly above the grinding wheel speed, thus having the fluid present in the cut zone. One of the first applications of this nozzle was on the Micro Centric grinding machines for bearing grinding. It did not become very popular due to the splash and mist and it required an enclosed machine. A recent version of this nozzle is shown in Fig. 17. Since liquid metalworking fluids are incompressible, the nozzle opening area can be calculated by the formula $Q = AV$, where Q is the fluid flow, V is the desired speed of the fluid, and A is the area of the nozzle opening. This type of nozzle again must overcome the air pressure present on the wheel's surface. An air scraper can do this job. However, some wheels do not have an air barrier because the wheel has no porosity (such as very fine grit resin wheels and hot-pressed resin wheels).

The benefits of this nozzle are that it wets the grinding wheel and lubricates the cut zone. It does not remove heat as effectively, because the coolant is not forced around the wheel surface for as long as the acceleration zone nozzle, for example. The fire hose nozzle is an effective nozzle, however, and worthy of consideration for many applications.

D. Jet Nozzle

There are two jet nozzles applied in industry. The high-pressure version has been in use for about 30 years, and the medium-pressure type was introduced in the early 1980s.

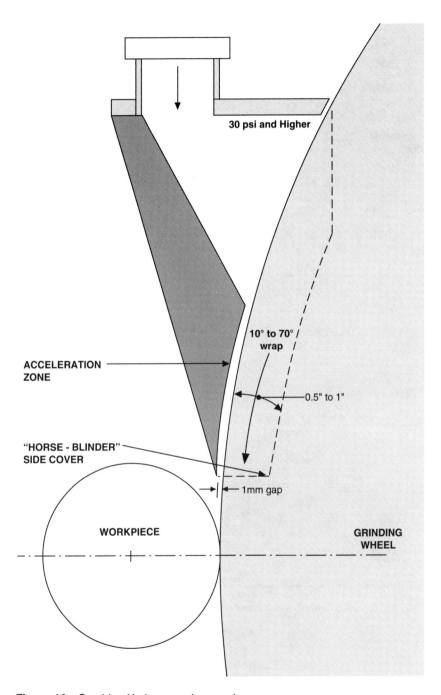

Figure 16 Combined inducer-wedge nozzle.

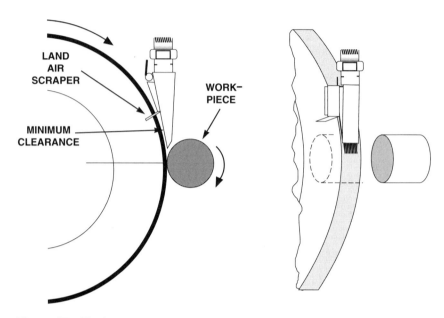

Figure 17 Fire hose nozzle.

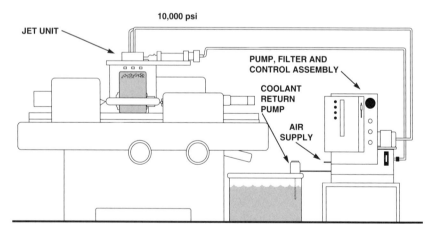

Figure 18 Jet wheel cleaning speeds grinding of high alloy materials.

Sheffield introduced their super jet wheel cleaner in the early 1960s, in conjunction with abrasive machining and crush dressing. The system is illustrated in Fig. 18. A number of these units are in use today. High-pressure coolant (10,000 psi) is forced against the wheel at about an 1/8 in. gap and moved back and forth across the wheel surface. The blast dislodges metal and swarf embedded in the wheel. Also, the fluid wets the grain surfaces. The jets are so powerful they easily overcome any air barrier.

One problem with the system is that the jet nozzles have very small openings and clog easily, but also wear fast, becoming ineffective. A good filter system is necessary, usually with a back flushing capability. The mist generated by this system usually requires an enclosure.

This system is effective in cleaning a wheel and wetting the grains on the wheel surface. The system does very little for cooling the wheel.

E. Wrap-Around Nozzle

A wrap-around nozzle is illustrated in Fig. 15. In the case shown, the part is surrounded by metalworking fluid essentially submerged in the cutting zone. This both wets the part and extracts heat. It does little for wetting or cooling the grinding wheel. This scheme is of limited value, but is probably as good as the dribble nozzle systems discussed earlier.

A similar scheme could be used on the wheel side. This has not been tried to the best of my knowledge. It may just act like a water brake and add heat to the wheel surface. The need is to extract heat from the wheel over a distance, especially for superabrasive grits.

F. Fluid Application for Superabrasive Grinding

A very interesting property of cubic boron nitride (CBN) is its thermal conductivity. A simple finite element model was presented by Glenn Johnson of General Electric [6] and is shown in Fig. 19.

The grain is simulated by a cross and the metal surface by a square. The initial metal temperature was assigned the value believed to be the interface temperature when a chip is being formed. The temperature of the CBN grit was set at room temperature. Only conduction between the two elements was permitted. The contact time between the two was 80 microseconds. (This simulated a grinding condition.) Note the big difference between CBN and aluminum oxide grain temperature in this short period of time.

This suggests several things:
1. Heat extraction into a superabrasive wheel is much more significant than with conventional abrasive.

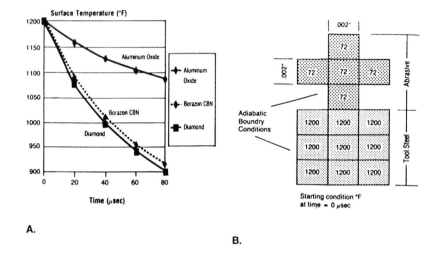

A.

B.

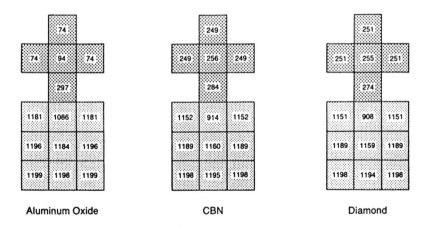

Aluminum Oxide   CBN   Diamond

C.

Figure 19 Finite element temperature analysis of abrasive grain and workpiece during grinding. A. Thermal model using finite elemental analysis on effect abrasive thermal properties have on cooling the surface in the pass of one abrasive grain during the formation of a chip. B. Finite elemental analysis model of heat transfer between workpiece and abrasive. C. Comparison of temperature (°F) distribution after 80 μs of contact between workpiece and abrasive for three types of abrasive. (From Ref. 6.)

2. Cooling a superabrasive grain as soon as it has completed its chip making task is extremely important.
3. Thermal damage of a workpiece is less likely with superabrasives than with a conventional wheel.
4. Heat extraction from resin-bonded superabrasive is very critical for good G-ratios.

Schemes like that in Fig. 20 have credibility [7]. Note the cooling high-pressure nozzle for the exiting wheel surface to drop the CBN grain temperature, the high-pressure jet nozzle to clean the film or load from the wheel, and the main supply for wetting the part and wheel before entering the cut zone.

VI. SELECTION OF FILTRATION SYSTEMS

Various parameters need to be defined for selecting the proper filtration system. The advantages and disadvantages of various systems will not be discussed here. In many publications this has been done in depth. An important element that is generally unrecognized will be reviewed in this chapter: the chip geometry and its effect on filtration systems.

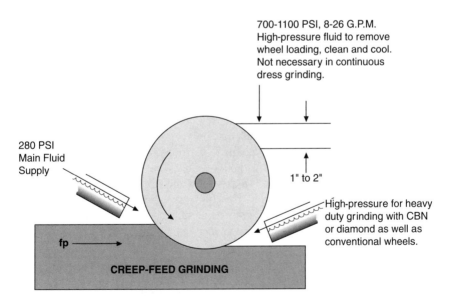

Figure 20   Use of high-pressure fluid supply in precision CBN grinding.

The chips generated by the grinding process are defined by the following grinding parameters.

Specific metal removal rate Q', in.^{3}/min/in.
Wheel speed V_s, in./min
Work speed V_w, in./min
Equivalent diameter De, in.

$$De = \frac{D_{\text{Workpiece}} \times D_{\text{Grinding Wheel}}}{D_{\text{Workpiece}} \pm D_{\text{Grinding Wheel}}}$$

Use + for OD grinding, use − for ID grinding, and for surface grinding use $De = D_{\text{Grinding wheel}}$.

The dominating parameters for chip thickness (dt) and chip length (L) are Q' and De. Wheel speed is of importance for the chip thickness, but has no influence on chip length. Work speed affects both chip length and thickness. Figure 21 shows the influence of Q' and De on chip length (L) and thickness (dt). The relation for the average undeformed chip thickness is as follows:

$$dt = Cx \sqrt[2+2a]{\frac{Q'V_w}{DeV_s^2}} \qquad L = \sqrt{\frac{DeQ'}{V_w}}$$

where

Cx depends on wheel grade (grain volume) (in.)
a depends on grit size in the grinding wheel
V_w = work speed, in./min
V_s = wheel speed, in./min

It is essential to understand that for most external grinding operations with work diameters below 2 in., the chips will be short and thick, while for larger ODs the chips become longer and thinner. For surface grinding operations long wire-type chips are formed. This means that cake-type filters might not be applicable at low De and Q' values. Thus, for external grinding on finishing conditions ($Q' < 0.2$ and $De < 1.0$) cake-type filters will not function well. In this area cartridge filtration or speed-related methods (cyclones, centrifugal) will be preferred.

Up to now, all discussions on chip form and size have concentrated on the average chip thickness. However, the chip size varies in thickness and length depending on the grinding wheel grain depth of penetration. At any time, in any grinding operation, there will be a variance in chip dimensions and very small chips will be generated.

Figure 22 shows a sampling of chips with various values of Q' for a hardened, as well as soft, 1045 steel material. The value of De was 1.0 in. during this test. The dimension on the horizontal axis shows the chip size. The vertical axis shows the number of chips in the sample taken during the grinding operation.

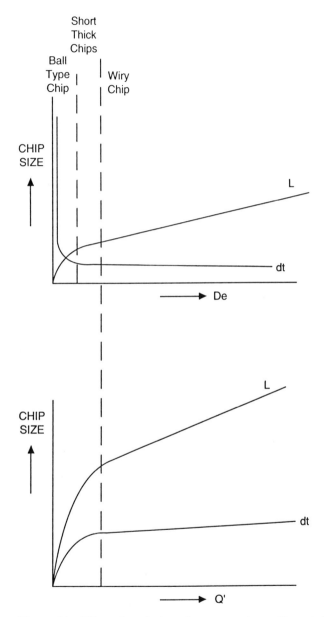

Figure 21 Effect of equivalent diameter and specific metal removal rate on chip size.

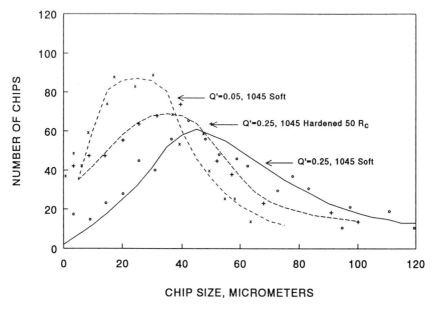

Figure 22 Chip size distribution for three grinding conditions.

The sample was collected directly after the grinding by means of a magnetic block, one inch below the grinding zone, attached to the workrest blade on a centerless grinder.

In Fig. 22, one can observe that larger as well as smaller size chips are formed. It can be observed that when Q' reduces to typical finish grinding conditions, the content of small size chips increases and the average chip size changes modestly. It is interesting to see that the higher Q' values will produce on average, larger chips, but also produce a wider spread of chip size. Chips of 5 μm and smaller were still found and the volume was still significant.

We can conclude that at the major grinding operations, smaller chips will be produced at a level of 5 μm or smaller. Filtration systems are difficult to design to filter these chips out of the fluid. Slowly, but surely, contamination will occur as a result of the build-up of small chips which remain in circulation. Filtration accuracy will dictate what size and what volume of chips will recirculate.

All the chips generally had a form varying from ball shape up to a wire type, with a 16:1 length-to-thickness ratio for the larger chips. However, these chips were mainly in the form of a curl, see Fig. 23. With larger values for De it is expected that the typical large chip shape will dominate, while for small De values ($De \leq 1.0$) the small chip form will be dominant.

For selection of a proper filtration system it is essential to have an

Typical Small Chip

 Typical Large Chip

Figure 23 Typical chip shapes.

understanding of the chip form and size range, as well as volume of grinding chips, grinding wheel wear particles, and other foreign materials which will pollute the metalworking fluids. Other important factors need to be considered, however, and this will be covered by other authors. The purpose of this contribution was to focus on chip geometry as a function of grinding conditions.

VII. KEEPING THE METALWORKING FLUID COOL

At high horsepower cuts, the temperature of the metalworking fluid will rise to a level that is no longer acceptable for work diameter tolerances, burning, etc. The main problem is how to dissipate the heat developed by the grinding process. Several methods can be used to realize this dissipation of heat. The methods used are:

1. Evaporation and convection
2. Cooling by air condensers
3. Cooling by forced evaporation
4. Cooling by refrigerating systems

A. Evaporation and Convection

If no forced cooling is available, the input energy must be transferred to the surroundings by evaporation and convection. Evaporation is usually dominant due to generally poor conditions for convection. Convection becomes significant only when a larger difference between fluid and ambient temperatures exists. In general, temperatures of metalworking fluids are 5 to 30°F (3 to 17°C) over ambient and the convection surface is a limited area. In this range, 80% of the input energy will be transferred by evaporation when ventilation of the ambient

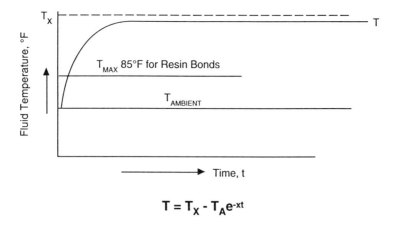

$$T = T_X - T_A e^{-xt}$$

Figure 24 Generalized grinding fluid warm-up curve over time. Increased coolant temperature affects dimensional accuracy of the workpiece, performance (apparent hardness) of the wheel, and the capacity to remove heat from the cutting zone. The steady-state temperature (T_x) of the fluid depends on energy input to the system. Convection and evaporation to ambient, as well as intentional chilling of the fluid supply, can reduce fluid temperature. T_x is generally 15 to 20°F over ambient. Approximately 80% of the grinding power contributes to a rise in temperature.

air exists. In a totally enclosed environment, the evaporation is low due to saturation of the air with water, and convection will take over. In such cases, very high metalworking fluid temperatures can be expected.

Most of the evaporation, however, will take place when the fluid is applied in the cutting zone. For an estimate of the volume of water evaporated to balance the input energy, the following formula can be used:

$$V_e = \frac{Q}{2300} = \text{pounds of water per hour}$$

where Q = energy input, BTU/h.

The warm-up curve of a fluid system depends on volume of fluid in the system, convection, ambient temperature, and the humidity of the surrounding air. Only an empirical temperature–time plot can provide the information needed.

A warm-up curve is shown conceptually in Fig. 24, while in Fig. 25 an empirically measured warm-up curve is shown. In this example, the peak fluid temperature was up to 40.7°C (105°F) while the ambient temperature was 21.6°C (70°F). In this test the peak temperature was never reached due to termination of the energy input after 5 h. The evaporation will be maximum if this peak temperature has been reached.

Metalworking Fluids in Grinding System

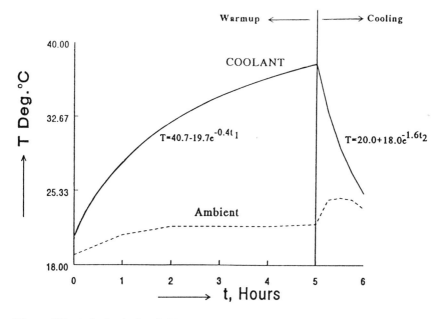

Figure 25 Actual grinding fluid warm-up curve.

B. Cooling by Air Condensers

Air coolers are all based on heat convection by blowing air through a condenser. For efficient convection a good temperature differential between ambient and fluid is needed. With a 15°F (8°C) temperature difference between fluid and ambient, a cooling surface of nearly 2000 square yards (1670 square meters) is needed. This method is basically not applied because of its inefficiency at the required small difference between fluid and ambient temperatures.

C. Cooling by Forced Evaporation

Another method is the evaporation system of a cooling tower. The dimensions are much smaller than for air coolers, but other disadvantages are introduced with this method. The cooling capacity of a cooling tower and the outlet coolant temperature depend on the ambient temperature and the humidity. The humidity under the conditions that the cooling tower is applied can be close to 90% if the cooling tower is installed near the machine in a badly ventilated area. One of the most important cost factors is the loss of concentrate together with the spray mist. Therefore, applying a direct circulation of a metalworking fluid through the cooling tower is not recommended. The best application of a cooling tower is to install the tower outside so that the relative humidity will always be the lowest possible and to use a heat exchanger combined with the cooling tower.

The required capacity of the cooling tower can be calculated with the following formula:

$$Q = \mu N \times 3406 = \text{BTU/h}$$

where N = energy input, kW·h and μ = power efficiency.

μ depends on many variables but is generally between 0.8 and 0.90 depending on machine type and power. The loss counts for bearings and drive system consumed for the grinding operation.

If a filtering system is used, the energy input of the pumps must also be considered. This means that the total capacity should be:

$$Q = 3406[\mu \times N + (1-f) \times Nf] = \text{BTU/h}$$

where f = energy efficiency of the filtering system and Nf = power input by the filtering system.

For a cyclone filter $f = 0.25 - 0.4$ (depending on cyclone type, piping, pressure, and differential over the cyclones). For nonspeed filtration systems (cake-type filters), $f = 0.65 - 0.7$.

If the cooling tower has this capacity the coolant temperature then depends on humidity and ambient temperatures. For resinoid-bonded wheels it is essential to hold the coolant supply temperature below 28°C (81°F). See Sec. E.

D. Cooling by Refrigeration and Heat Exchangers

For calculating the required capacity, the same formula can be used as for the cooling tower, thus: $Q = 3406[\mu \times N + (1-f) Nf] = \text{BTU/h}$

A refrigerating system is not as sensitive to ambient temperature, but the ambient temperature cannot be ignored. This means the cooling capacity must be available at the temperature that occurs in summer. In some cases, the energy from the grinding system can be utilized for purposes such as heating water or cleaner fluids or other heating purposes. In cases of larger systems with very high-energy input from grinding and machining operations, the savings could be considerable. For energy-saving purposes this concept should be seriously studied. Only by refrigerating systems can the energy be saved for other purposes.

E. Effect of Fluid Temperature on the Grinding Parameters with Resinoid-Bonded Wheels

A lot of work has been done on laboratory scales to determine the influence of the fluid and fluid temperature on the static hardness of the resinoid-bonded wheels. The actual effect on grinding performance was unknown. One of the applications where resinoid-bonded wheels are used is bar grinding. With the available equipment, data have been produced under production conditions. For

Metalworking Fluids in Grinding System 133

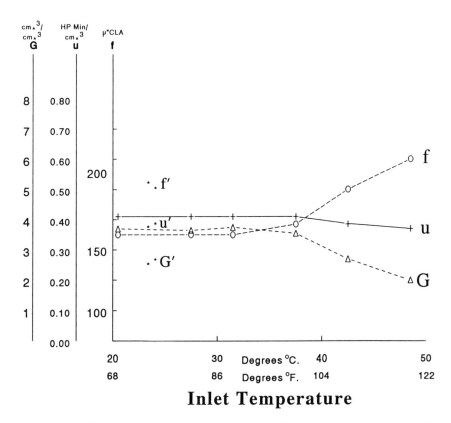

Figure 26 Effect of fluid temperature on the grinding parameters with resinoid-bonded wheels.

these tests 50CrV4 steel has been used. At $Q' = 0.56$ in.^{3}/min/in. the temperature of the coolant has been varied from 20 to 48.5°C (68 to 118°F). The grinding ratio (G) and specific energy (U) have been measured, as well as the surface finish.

For each test, the wheel was under the specific temperature condition with grinding for one hour. Then after this conditioning time a G-ratio measurement was done over 10 min grinding time and repeated three times. This measuring method was done for all tests except for the first test at 20°C, due to problems with the available cooling capacity. The data are shown in Fig. 26. After the test from 20–48.5°C fluid temperature, the wheel was brought back down to 20°C, and the grinding parameters were again measured. This test was then repeated again after a wear of 0.2 in. on the radius from the wheel. The data shows that the depth of influence is more than 0.2 in. into the resinoid wheel. (See data points marked *.)

Figure 26 shows the grinding ratio (G), specific energy (U), and surface finish (f), as a function of the fluid inlet temperature. Up to 32°C (90°F) no effect could be determined, but over 32°C fluid temperature, the G-ratio decreased rapidly with the increase of fluid inlet temperature. The surface finish grew rapidly worse above a 32°C fluid temperature. The close relationship between G-ratio, surface finish f, and the specific power U is quite clear.

All tests were done with a 21SC46Q9B2 (46 grit) grinding wheel. The effect of grain size on the temperature sensitivity is unknown but laboratory research showed coarser wheels to be less sensitive to temperature than finer wheels. Therefore, it might be considered that the critical temperature for coarser resinoid-bonded wheels will be higher. This means that a coolant temperature of 32°C (90°F) might be the acceptable maximum for 36–60 grit size wheels.

The last test data showed that the depth of penetration was more than 0.2 in. into the resinoid bonded wheel. (The data are shown with G, U, and f indexes.) This means that when a resin-bonded grinding wheel is under a temperature condition of over 32°C, the wheel hardness will change. Under more favorable conditions—temperatures less than 32°C—the previous high-temperature conditions will determine the results (at least over 0.2 in. of depth into the wheel).

REFERENCES

1. F. W. Taylor, *On the Art of Cutting Metals*, Society of Mechanical Engineers, New York (1907).
2. M. R. Schumack, J. Chung, W. W. Schultz, and E. Kannatey-Asibu, Analysis of fluid flow under a grinding wheel, *J. Eng. Ind. 113*: 190–197 (1991).
3. R. S. Hahn, Some observations on wear and lubrication of grinding wheels. in *Friction and Lubrication in Metal Processing*, F. F. Ling, R. L. Whitely, P. M. Ku, and M. Peterson, eds., ASME (1966).
4. J. W. Powell, The application of grinding fluid in creep feed grinding, Ph.D. thesis, Univ. of Bristol (1979).
5. B. Bourgoin, Centre Technique des Industries Mecaniques (France), Sprinkling device for grinding wheels, U.S. Patent 4,176,500 (1979).
6. G. A. Johnson, Beneficial compressive residual stress resulting from CBN grinding, Society of Manufacturing Engineers, MR86-625 (1986).
7. Y. Satow, Use of high pressure coolant supply in precision CBN grinding, Society of Manufacturing Engineers, MR86-643 (1986).

5
Metalforming Applications

KEVIN H. TUCKER
Cincinnati Milacron
Cincinnati, Ohio

I. INTRODUCTION

The working of metal into useful objects has been suggested as the oldest "technological" occupation known to mankind: metal has been formed into usable utensils, weapons, and tools for over 7000 years [1]. It would be hard to imagine the social metamorphosis that would have occurred had it not been for the increases in metalforming technologies throughout history. It is equally difficult to imagine how metalforming would have advanced without the aid of lubricants. Only recently has there been any historic record of lubricant use in metalworking.

In *Tribology in Metalworking*, John Schey describes the historical evolution of metalforming [1]. Most historians believe that forging was the first metalforming operation. Malleable metals, such as "native gold, silver, and copper were hammered into thin sheets, and then shaped into jewelry and household utensils as early as 5000 B.C.," probably without lubrication. The making of a coin by driving metal into a die with several punch strokes was recorded in the 7th century B.C., again with no reported lubricant addition, other than possibly from "greasy fingers." However, materials used as lubricants were in evidence as early as the 5th century

B.C. when Herodotus wrote about the extraction of light oil from petroleum. The manufacture of soap was recorded in 60 A.D., so a variety of lubricants would have been available throughout the history of metalworking. An 18th-century practice of forming rifle parts using a lubricant mixture of sawdust and oil was still used as recently as 50 years ago.

Metalforming lubricants are as varied as the many operations in which they are used. Animal fats were possibly the first lubricants used in primitive operations, as they were readily available from the rendering of animal carcasses. Stampers quickly learned that a coating of lard or tallow allowed the metal to be formed more easily, with more deformation, and with less tool wear [1].

Other metalforming processes, such as wire drawing and rolling, have been dated to the Dark Ages where lubricants such as lard oils or beeswax were used with success. One final historical note: The first sketch of a rolling mill is credited to Leonardo da Vinci [1]!

II. FACTORS AFFECTING FLUID REQUIREMENTS
A. Types of Metalforming Fluids
1. Oils

Petroleum mineral oil is probably the most widely used lubricant for metalforming. In light duty stamping, blanking, and coining operations, mineral oil provides the necessary separation of tool from metal. Many drawing operations can be performed with no lubricant application, relying solely on the ductility of the metal and part geometry. However, the application of mineral oil often enhances the speed at which parts can be produced. In order to optimize part–tool separation, the viscosity of the mineral oil can be varied to match draw severity.

Much work has been completed regarding the structure of mineral oil and its effect on performance. John Schey documented some of these effects [1]. Paraffinic oils are better suited for some operations due to their relatively high-viscosity index values. Paraffinic oils are less likely to be oxidized than naphthenic oils. Antioxidants, used to prevent oxidation, are more effective in paraffinic oils than naphthenic oils. Naphthenic oils though, are more easily emulsified and have more affinity for metal surfaces. However, most mineral oils contain some distribution of both straight and branched-chain hydrocarbons.

Synthetic oils are also of significance. Polybutene has been used as a synthetic rolling oil with good results for some time. Obviously, chain length affects viscosity, melting points, and suitability for specific operations. Cost is much higher than with mineral oils.

While viscosity is often the determining factor in performance, additives designed for specific applications are often included in straight oil formulations.

Fats for rolling operations, extreme-pressure (EP) additives, and even solids such as molybdenum disulfide and talc, are often used as lubricants.

2. Soluble Oils

Fluids based on mineral or synthetic oils which contain emulsifiers that allow for the dilution of the product into water are called soluble oils. They are sometimes sold in their diluted form and referred to as *preformed emulsions*. These products are generally mixed at dilutions of 10 to 50% in water. They may be formulated with fats or fatty acids for light duty applications, or may contain EP additives for severe forming. Seldom are solid lubricants used because they are difficult to suspend in an emulsion.

3. Semisynthetics

Fluids containing a lesser amount of mineral oil, usually under 30% of the total concentrate volume, are called semisynthetics or *chemical emulsions*. Compared to the volume of oil used in a soluble oil fluid, the mineral oil content in a semisynthetic is much lower. The mineral oil may have been replaced by hydrocarbons such as glycols and esters used as oil substitutes, by emulsifiers such as soaps and amides, and even by water. Generally, semisynthetics will mix into water more easily than soluble oils, but other than subtle differences in mix appearance and use–dilution ratios, they are similar to soluble oils.

4. Synthetics

Synthetic fluids are generally of two types. One is the "solution" group in which water serves as the carrier. In this class, also referred to as *water-based* synthetics, all additives are either water soluble or are reacted with some other component to be water soluble. Generally these fluids will range in mix clarity from clear to cloudy. A second type is a synthetic fluid that contains no mineral oil, but uses a hydrocarbon, such as a polybutene or glycol, as an "oily" replacement for mineral oil. The mixes of these fluids will generally range from hazy to milky.

B. Lubricant Additives Used in Metalforming Fluids

With the exception of solid lubricants or "pigment" materials, the additives used in metalforming fluids are very similar to those used in the metalworking industry as a whole.

Fats derived from animals and vegetables are very effective boundary lubricants. Tallow, lard, and wool grease are extremely good boundary lubricants. Oil from the sperm whale, which had many uses, was such a good lubricant that the species became endangered. Its use in lubricants today is prohibited. Vegetable oils such as coconut, rapeseed, and tall oils are good sources of boundary lubrication. Fatty acids, esters, waxes, and alcohols derived from natural materials are also frequently used lubricants.

Soaps formed from a reaction between a metal hydroxide and fatty material are excellent lubricants. For instance, in the working of aluminum, aluminum soaps are often formed as reaction by-products of the operation. These aluminum soaps will lower the coefficient of friction. However, they are not water soluble. While excellent lubricants, aluminum soaps are often difficult to remove from the tooling and can actually cause increased wear of the punch and die.

Without a doubt, chlorine, phosphorus, and sulfur have been the most frequently used extreme-pressure additives. Of these, chlorine is most commonly used because it addresses the widest variety of operations. It is often used in conjunction with the other two, such as in sulfochlorinated additives.

Solid lubricants are suspended into pastes or thick emulsions. Graphite, mica, talc, glass, and molybdenum disulfide are just a few examples. This type of additive is generally reserved for the most severe of drawing operations.

C. Physical Properties

One of the biggest physical differences between metalforming fluids and other metalworking fluids is the physical appearance of the fluids. A typical machining or grinding fluid will seldom be diluted in water at a dilution stronger than 10%. A stamping fluid will rarely be diluted at less than 10%. A machining and grinding fluid will have a viscosity near that of water, since water is its principal component. A stamping fluid often resembles paint in its ability to "coat" metal with a viscous covering.

The physical appearance of stamping and drawing fluids varies with the type of lubricant used. Straight oil stamping fluids range from very low viscosity, as in vanishing oils, to very high-viscosity "honey oils," so named because of its similar viscosity to honey. Pastes, which may be thick oil-in-water emulsions, or water-in-oil invert emulsions, are often used. These are generally used for very severe drawing operations in which the lubricant is painted onto the surface to ensure maximum carry-through of fluid through the die. Suspensions generally look like milky emulsions, although they may take on the appearance of the finely dispersed pigment material used as the lubricant, as in the case of graphite. Stamping fluids, especially those used for general purpose stampings, may also be thin, milky emulsions. Finally, new water-based stamping fluids have the viscosity of water.

D. Lubricant Functions

1. Separation

Unlike metal removal fluids that have the primary responsibility of cooling, the primary function of a metalforming fluid is to separate the part from the tooling, preventing metal-to-metal contact. To do this, the fluid must provide a barrier, either physical or chemical, to prevent the contact of punch or die to metal. The main objective for providing this separation is the protection of the tooling. The cost of the tooling in most forming operations is quite significant when compared to all other components in the process. A lubricant that improves tool life can pay for itself through reduced tooling changes.

2. Lubrication

Along with part–die separation, a metalforming fluid must provide lubrication sufficient to make the part. Good lubrication can be defined as the reduction of friction at the part–die interface. Lowering friction results in lower energy required to make a part. Further, lower friction results in less drag on the metal, which yields a more uniform flow of metal through the forming dies. Lower friction also benefits the operation by reducing the amount of wear, which in turn reduces the amount of metal fines and debris. With fewer metal fines in the punch zone, metal pickup or deposition on the punch is less likely. Although not as important as in cutting and grinding fluids, a lubricant may also provide cooling to the part and the tooling. In some metalforming applications, however, heat is necessary.

3. Corrosion Control

Another function of lubricants is to prevent corrosion. While a grinding or machining fluid may be expected to control corrosion for one or two days, stamping fluids are often required to provide corrosion protection for several months while parts are held in storage. Specialized testing equipment such as acid–atmosphere chambers, fog chambers, or condensing humidity chambers, are often used to predict fluid capabilities under a variety of storage conditions. A corrosion test involving a hydrochloric acid atmosphere is used in the automotive industry. Salt spray, ultraviolet light, condensing humidity, and elevated temperature chambers are all used to determine corrosion control of metalforming fluids.

Besides controlling corrosion in the finished part, a good fluid must also be safe for equipment. The fluid must prevent attack of metal ways, slides, and guide posts. Of even more importance, the fluid must be compatible with tooling material. Carbide tooling made with nickel or cobalt binders is often used for high-speed production punches and dies. The fluid must be formulated so that these binders are not leached from the metal matrix, which would result in punch or die degradation.

4. Cleanliness

After the part is formed, a fluid must be easily removed. An objective of a clean fluid should be the prevention of buildup of metal debris or *fines* in the forming die. Likewise, there should be adequate cleanliness to prevent residue deposition on the punch. The fluid should also be compatible with the selected cleaning equipment, whether it be vapor degreasing or alkaline wash. Along with part cleanliness, the fluid must also contribute to a clean working environment. Floors, walls, presses, and even operators need to be free of fluid residues.

5. Other Requirements

There are other aspects to a metalforming fluid that determine its acceptability for a specific operation. Operator acceptance factors such as ease of mixing, application method, and operator health and safety issues are of utmost importance. Press operators using reasonable safety practices must be able to work with the fluids, without the risk of dermatitis, respiratory distress, or other health problems. Fluid appearance, product odor, and rancidity control if the fluid is recirculated are aesthetic factors that need to be considered.

E. Lubrication Requirements

1. Boundary Lubrication

Boundary lubrication is defined as "a condition of lubrication in which the friction between two surfaces in relative motion is determined by the properties of the surfaces, and by the properties of the lubricant other than viscosity" [2]. Boundary lubricants can be defined as thin organic films that are physically adsorbed on the metal surface. There are several theories on how boundary lubrication occurs. One explanation is that the polar molecules of the boundary lubricant are attracted to the metal surface. Under practical conditions, layers of the lubricant are formed. As sliding friction occurs, some molecules are removed, but others take their place. This boundary film prevents metal-to-metal contact. If the film is broken, and metal-to-metal contact is made, part failure can occur. In practice, the majority of lubrication occurs as boundary lubrication [2].

2. Extreme-Pressure Lubrication

Another type of boundary lubrication is classified as extreme-pressure (EP) lubrication. Extreme-pressure lubricants are those lubricants that will react under increased temperatures and undergo a chemical reaction with the metal surface. In metalforming lubricants, chlorine, sulfur, and phosphorus have been the traditional EP lubricants. It was often believed that these compounds simply reacted with metals, to form chlorides, sulfides, and phosphides. However, it has also been shown that reaction species other than simple metallic salts are formed between the metal surface and the chemical lubricant. EP lubricants also work synergistically with other lubricant regimes. They appear to adsorb onto the metal surface as do organic-film, boundary lubricants [3].

Regardless of the mechanism, EP lubricants perform by chemically reacting with the metal substrate. Lubrication occurs through the sloughing off of the chemically reacted film through contact. Metal-to-metal contact is prevented by the chemical film. As removal occurs, regeneration of the film must occur through the presence of more lubricant, or failure occurs.

Even among tribologists there is some disagreement as to the activation temperature of extreme-pressure additives. Some lubrication engineers attribute the efficacy of extreme-pressure additives to the melting points of the iron salts of chlorine, phosphorus, and sulfur. If this is the case, chlorine would be the earliest to be activated, followed by phosphorus and sulfur [4]. Other guidelines for the necessary reaction of extreme-pressure additives have been established based on the observed process temperatures. Following this practical approach, phosphorus is effective in operations where temperatures do not exceed 400°F (205°C). For operations generating temperatures above 400°F (205°C), chlorine is selected. Chlorine will maintain effectiveness up to 1100°F (700°C). Sulfur is used in severe operations where temperatures exceed 1100°F (700°C), and will be effective up to 1800°F (960°C). This compares to fatty acid soaps which lose most of their effectiveness at 210°F [5,6].

3. Hydrodynamic Lubrication

Hydrodynamic lubrication can be described as a "system of lubrication in which the shape and relative motion of the sliding surfaces cause the formation of a fluid film having sufficient pressure to separate the surfaces" [7]. The viscosity of the lubricant system plays a significant role in determining the capability of the hydrodynamic lubricant film. In fact, some refer to this lubrication regime as "pressure–viscosity" lubrication [8].

4. Solid-Film Lubrication

Solid-film lubrication involves the use of solid lubricants for part–die separation. Ideally, the solids serve as miniature ball bearings on which the sliding surfaces ride during part formation. This prevents metal-to-metal contact. However, the solids will sometimes attach onto the surface of the tooling or the part. Removal of solid films from finished parts can be a major problem. Examples of this type of lubricant are mica, talc, graphite, and polytetrafluoroethylene (PTFE).

F. Methods of Application

Many fluid application techniques are available. Selection of a method should be based on optimized lubricant delivery for performance, and on fluid conservation. Overapplication of a fluid will not increase lubrication performance, but may make the workplace so messy that operators will dislike the fluid.

1. Drip Applicators

The application of drawing fluids by drip applicators is as simplistic as the name implies. In many shops, this method may involve a coffee can with a nailhole in the bottom, which allows fluid to drip on a blank prior to the metalforming operation. This technique is one of the hardest to control for accurate delivery. It can also be very sloppy, but has the advantage of being one of the cheapest methods.

2. Roll Coaters

An application using roll coaters has two rollers, one above the other, in which the spacing and pressure between the two rolls are controlled by air pressure. The rolls are partially positioned in a fluid bath. As the metal source is fed through the rollers, the fluid is "rolled" onto the metal prior to the stamping operation. Fluid application can be controlled to milligrams/surface area of metal.

3. Electrodeposition

This fluid application procedure is becoming more prominent. Once reserved for the rolling mills, small units are being tested on the shop floors of metalforming plants. An electrical charge is passed through the fluid, with the opposite charge applied to the metal. As opposite charges attract, the fluid is deposited onto the metal surface. Advantages are the accurate application of a small amount of fluid per surface area, little or no waste, and high-speed application. The major disadvantage is that the high cost associated with an electrodeposition unit puts it out of the reach of the average shop.

4. Airless Spray

Equipment that sprays measured amounts of fluid to precise locations in the drawing press without mixing with air is known as an airless spray unit. Tubing is placed into fluid reservoirs which feed the fluid through a pumping system onto the selected delivery site. Advantages include little or no waste, fluid savings, and a cleaner work environment. Disadvantages include limitations on the viscosity of the lubricant that is applied.

5. Mops and Sponges

Unfortunately, many small stamping shops still use mops and sponges to apply drawing fluids. While these are obviously low-cost application devices, expenses related to this poor fluid delivery system generally overshadow any possible benefits. Disadvantages include poor control of the amount and location of fluid delivery, excessive fluid waste, and sloppy work environment.

G. Removal Methods

In almost all cases, the metalforming fluid will eventually need to be removed from the formed part. In some cases, it is the ease with which this occurs that actually determines which fluid is selected for the operation. Often, problems with plating, painting, or electrocathodic deposition of primers (E-coat) can be traced to poor cleaning or contaminated cleaner reservoirs. There are several basic types of removal methods.

Straight oil-type products have historically been removed through *vapor degreasers*. A vapor degreasing process uses vapors of a solvent such as boiling 1,1,1-trichloroethane. The vapors solubilize and remove the oily residue on the part. The presence of water in the residue will decrease the efficiency of a vapor-degreasing solvent. Today, the use of vapor degreasers is being drastically reduced due to environmental and operator safety concerns. The future availability of these cleaning systems is doubtful.

As an alternative to solvent degreasing, alkaline cleaners are being used. These cleaners can be applied through a dip bath, impingement spray, or a tumbling parts washer, among others. One advantage is that, unlike vapor degreasers, alkaline cleaners are also effective for water-diluted lubricants, even those that contain oil. Alkaline cleaners are often warmed to speed up the processes that cause residue removal.

And finally, many companies install multistage washers designed to clean off stamping fluid residues, acid-etch the surface, and react the metal part to form a *conversion coating* (often a phosphate) for subsequent plating or painting operations. Once installed, these washers are very economical to run and are controlled through specific ion monitoring devices to control washer–chemical concentrations.

H. Premetalforming Operations

In today's stamping plants, there are a variety of experimental processes being evaluated that may be the norm for stamping plants of tomorrow. One of the most interesting is the trend towards "prelubes." This involves the application at the rolling mill of a thin film of lubricant, prior to rolling of the coil. This lubricant may either remain a liquid on the coil or through either evaporation or chemical reaction, convert to what is known as a dry-film lubricant. When this coil arrives at the stamping plant, no further application of lubricant occurs. The operation is completed using only the thin film of mill-applied lubricant. The use of pre-lubricated metal is in its infancy, and has had many growing pains. Since no application of fluid in the press occurs, buildup of metal debris and other residue material is difficult to wash off.

Similar applications are being tried with experimental paints. The paints are applied at an off-site facility prior to forming. When stamped, the paint serves not only as the lubricant, but the part is already prepared for shipping with no cleaning costs. This operation is already in practice for very small parts.

Many times, metal coils arrive with residues of the rolling oil. Any subsequent application of lubricant must be compatible with the mill oils.

In some very rare applications, a metal will be "pretreated" with a conversion material, such as phosphoric acid on cold-rolled steel. The acid will react with the steel to form a phosphated dry-film lubricant. This film serves as an extreme-pressure lubricant and will perform very well on deep-drawing operations.

I. POSTMETALFORMING OPERATIONS

Once the part has been formed, other processes may follow. Cleaning to remove lubricant and mill oil residuals must occur before any finishing operations such as painting or plating. For automobile body parts, E-coating of primer paints is completed before final paint coats are applied. Craters that are visible in final paint coats are attributed to poor surface quality of the metal substrate prior to E-coating. The part must also be free of any contaminants that may affect glue adhesion, welding, and application of modern sound deadeners in automotive applications.

In the electrical industry, postforming operations include the application of resinous insulating materials. If a compatibility problem exists, cracking or removal of the insulation may occur. Most magnet wire is coated with a protective varnish. The drawing lubricant must be compatible with the varnishes to prevent wire surface quality defects.

The can-drawing industry has one of the most involved postforming processes in metalforming. Following the making of a can, the can body will be cleaned, dried, decorated, sprayed with internal and external coatings, and baked. Next, the neck of the can, and a flange for the lid attachment, will be formed. After all of these procedures, the lubricant applied early in the original can-forming process may contribute indirectly to mechanical problems that occur in the final necking or decorating operations.

The making of a part is only the beginning of the metalforming operation. From the original mill-supplied metal to the final manufacture of a formed part, stamping fluids must not only provide lubrication, but ensure final part quality and acceptance for use.

III. METALFORMING PROCESSES

A. Blanking Operations

One of the most common metalforming operations is the blanking process. Nearly all metalforming operations are preceded by a blanking stroke. In this operation, the metal blank is cut from a sheet or strip into a shape desired to affect the final part. The remaining scrap is in the form of a "skeleton" with a hole where the blank had been. The blank can be round as in the formation of metal cans, square as in the making of medicine cabinets, or patterned if needed for an intricate shape or design.

Rough blanking uses a die set with a cut edge. The metal is clamped tightly

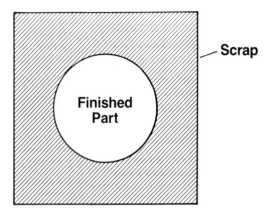

Figure 1 Blanking.

to a specified hold-down force. The punch serves as a moving knife edge that cuts its way through the metal much like a cookie-cutter through dough. The stationary die also serves as a cut edge to complete the cutting operation. In rough blanking the metal is only cut from 35 to 70% through the depth of the metal. Because of the metal flow, the remaining metal edge is torn away. As can be seen in Fig. 2,

Figure 2 Rough blanking.

the top edge of this blanked part is much smoother than the lower edge, which has been broken off.

In fine blanking, the process is very similar to rough blanking. However, fine blanking is done for "finish" quality appearance and size control. As such, the metal is forced 100% through the die cut edges, resulting in a smooth, even appearance over the entire surface edge. (See Fig. 3.)

There are many types of blanking classifications. Cutoff is a class of blanking that describes the use of a punch and die to make a straight or angular cut in the metal. This operation is basically a shearing operation completed in a forming press. Parting is the opposite of blanking in that the scrap is "parted" from the original metal strip, leaving the desired finished piece [9]. Punching (or piercing) is the making of a hole in metal, leaving a round slug of metal as scrap. Notching is similar to punching, only a slit is formed rather than a hole. Shaving uses a cut edge to remove rough edges for precision finishes. Trimming is also a type of blanking. Trimming removes excess metal from a formed part giving the final desired dimension. The *Tool and Manufacturing Engineers Handbook* lists over ten different blanking operations, and over 30 uniquely different metalforming operations [9].

Figure 3 Fine blanking.

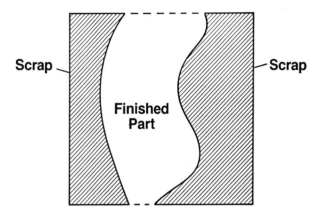

Figure 4   Parting.

B. Drawing

The drawing of metal can be simply described as a punch forcing a metal blank from a blankholder, into and through a forming die. There are degrees of severity of drawing operations, and numerous definitions used to define each. A practical and easy way to define the severity of a drawing operation is to look at the draw ratio. A draw ratio is determined by the ratio of the depth of draw to the original blank diameter [10]:

$$\text{draw ratio (DR)} = \frac{\text{depth of draw}}{\text{blank diameter}}$$

Based on this relationship, draw severity is commonly classified as:

Classification	Draw ratio
Shallow draw	<1.5
Moderate draw	1.5–2.0
Deep drawing	>2.0

In low and moderately severe operations, there is little or no wall thinning. In deep drawing, wall thinning may occur to a greater degree than in shallow drawing. While draw ratios can be used as an indication of difficulty, the severity of a drawing operation is affected not only by the draw depth, but also by the actual amount of wall thinning that occurs within the sidewall.

While the severity of the drawing process can be expressed as a draw ratio, the "drawability" of the metal used can also be expressed as the ratio of the blank diameter (D) to the punch diameter (d) [10]. This relationship is referred to as the limited draw ratio. The metalforming operation is controlled to a large extent on the limits of the metal to be drawn to a desired depth [11].

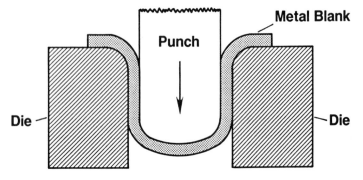

Figure 5 Deep drawing.

$$\text{Limited draw ratio (LDR)} = \frac{\text{blank diameter } (D)}{\text{punch diameter } (d)}$$

For practical use, the LDR is also a measure of the maximum reduction possible in a drawing operation. The LDR is sometimes expressed as the percentage reduction from the blank diameter to the cup diameter as $100 (1 - d/D)$ [12]. The variety of commercial qualities of metal must be considered when a drawing operation is designed. For instance, cold-rolled steel is available in several grades including draw quality aluminum killed (DQAK), which is designed for improved drawability.

Success of a drawing operation is also affected by the ratio of metal thickness to the blank diameter ratio. This ratio can be expressed as t/D, where t = thickness and D = blank diameter to which the shell is being reduced [13].

This ratio is used to approximate maximum allowable reductions for draw and redraw depths. Because the thickness remains constant, and the blank diameter decreases, each successive draw reduction is decreased accordingly. Obviously, other factors including corner radii and metal quality need to be considered.

Blankholder pressure, also known as hold-down force, holds the metal in place under the proper pressure to prevent deformities. With too little hold-down force, the metal may twist in the die, causing wrinkles. Too much hold-down pressure will prevent smooth, even metal flow, causing tear-offs or part rupture.

C. Drawing and Ironing

1. Background

The use of canned foods dates back to Napoleonic days when a French candy maker, Nicolas Appert, conducted food preservation experiments by canning soups and vegetables in champagne bottles [14]. By the early 1800s, the French navy carried bottled vegetables on their ships with reportedly good results. Soon after, a patent for canning in "bottles or other vessels of glass, pottery, tin, or other metals or fit materials" was granted to an Englishman, Peter Durant. In 1811, Durant sold the patent, and the new owners soon were making the first "tin cans" at the phenomenal rate of ten cans per man per day! With the tin mines in England, tin-plated steel was soon being produced in commercial quantities. In spite of the high cost of the canned food, not to mention that the can needed a hammer and chisel to open, canning became a booming industry [14].

In the late 1800s, Luigi Stampacchia received a patent for a process using a double-action press to produce a "double-drawn" can, and the first "ironed" can was produced [14]. A U.S. patent was granted in 1904 to James Rigby for producing cans with the wall thickness reduced by burnishing [14]. Even though other developments took place, and many more patents were issued for the making of cans, it was not until the 1960s that the first commercial beverage cans were produced by the drawing and ironing process first developed in the 1800s. The introduction of the "easy open" end in the early 1960s led to wide consumer acceptance, and now close to 100 billion cans are produced throughout the world each year [15]. In fact, each person in the United States accounts for 547 beverage cans used each year [16]!

Since 1959, when Coors Brewery made the first two-piece aluminum can [17], nearly all beverage cans, as well as many similar types of food containers, are manufactured through the operation known as drawing and ironing. This operation is sometimes referred to as the D&I process. Beverage cans used for soft drinks and beer are called "two-piece" cans because the complete container actually consists of two separate portions. One portion of the two-piece can is the lid, which is produced using a stamping operation. The major section of the two-piece can is the body. The can body is manufactured using the D&I process and will be the focus of this section. The can body and the lid are made independently of each other, and often at completely different locations. The can bodies and lids are then shipped to a bottling plant, which adds the beverage and seals the can.

Steel has the largest marketshare of cans produced. In the mid-1970s, steel cans had dominated the beverage can marketplace. Presently however, aluminum cans have an almost exclusive market share for what is referred to as the "beer and beverage" container. In fact, carbonated beverage cans now represent the largest single use of aluminum in the world [18]. This preference for aluminum beverage cans has been supported not only by years of qualitative taste and flavor success, but also from an ecological standpoint due to the recyclability of aluminum. Today, with increased consumer awareness of environmental issues, and with improved collection of scrap aluminum, over 50% of newly manufactured aluminum cans are from recycled aluminum [19].

In recent years, the use of recycled aluminum has kept the cost of manufactured aluminum cans much lower than the increase observed in the cost of aluminum as a raw material. In fact, during the decade from 1980 to 1990, manufacturing costs for an aluminum can unofficially decreased about 10% in spite of increases in raw material costs and inflation!

Steel producers have devoted extensive research and development efforts in making their coil stock more attractive to can manufacturers as an alternative raw material to aluminum. Because of large subsidies by steel makers, it is actually cheaper to produce a steel can than to make an aluminum beverage can. However, the recycling efforts have generally not been as successful with steel cans as they have with aluminum. In spite of extensive advertising efforts promoting steel as being "recyclable," the low returns per can paid to collectors have so far failed to interest consumers in recycling cans. Recycled steel cans amount to much less than the recycling rate for aluminum cans. In addition, long-term storage of beverages in steel cans may impart what is described as an "iron" taste to products. Technology improvements in interior can coatings have helped to alleviate this concern, however.

For food containers such as those used for fruits and vegetables, steel is the preferred metal. Aluminum cans, however, are making significant inroads into this marketplace and have recently been used for soups, vegetables, fruits, and even wines.

2. Definition

The two-piece can is made in a two-stage process. Whether making an aluminum can or one from tin-plated steel, the process is very similar. The first step is the cupping operation. During the cupping operation, a blank is produced from the coil stock and, in the same punch stroke, is formed into a shallow cup. The cupping operation can be compared to the drawing operations described earlier.

The formed cup is then conveyed to a horizontal metalforming press called a "bodymaker," or "wall-ironing machine." What makes the D&I operation unique is that within the bodymaker, two metalforming procedures are done using one ram stroke of the punch. This is accomplished through the use of a series of forming dies. The first stage of making a can body is the punch forcing the cup through a redraw die. In this die, the cylindrical cup is merely reshaped or redrawn into a longer cup having a smaller diameter.

As the punch is extended, the cup is forced through a series of ironing dies, called a tool pack (see Fig. 7). The tool pack usually consists of three ironing dies, and is used because no single die can make the necessary forming reductions. In the ironing dies, the can sidewall is reduced to the desired wall thickness. The total reduction in thickness going from thickwall to thinwall may be in excess of 70%. As the punch forces the can through the first ironing die, the inside diameter of the can is now the exact diameter of the outside of the punch. The cup is now lodged tightly onto the punch. As the cup goes through the successive ironing dies, the outside can diameter is reduced, and the sidewall thickness is reduced. Controlling the amount of reduction, as well as the volume of metal moving through the tool pack, determines the shape of the can, and to a large extent, the length of the can. Because of the tooling setup in this ironing process, the bottom and top walls of the can are considered as thickwall, while the middle of the can is thinwall.

The nose of the punch is hollow. As the punch continues through the tool pack, it strikes a positive-stop at the end of the ram stroke. This stop is a piece of tooling called the "domer" which is shaped to coincide with the hollow nose. The punch nose forces the can into the domer tooling, shaping the bottom of the can.

As the punch begins to retract from the newly formed can, the can is removed from the punch by segmented pieces of tooling which are spring loaded and known as stripper fingers. The fingers open to allow the can to exit the tool pack, but then close around the outer punch diameter which allows it to be withdrawn from the can as the ram is returned.

Subsequent operations include trimming the can to specified length, washing to remove drawing lubricants, decorating the exterior of the can with the desired brand label, and applying an interior coating to prevent contact of the food or beverage product with the metal substrate.

3. Lubrication Requirements

Lubricant for making aluminum can bodies is applied by flood application through a series of lubricant rings placed between each ironing die in the tool pack. The D&I lubricant is also directed at the cups entering the redraw die. Once the punch has entered the tool pack, only the outside of the can, in contact with the ironing dies, is affected by the bodymaking coolant. The interior of the can, in contact with the punch, must rely solely on the residual cupping lubricant along with any bodymaking coolant retained prior to entering the redraw housing.

An interesting variation in lubrication application allows for a pattern such as crosshatching or longitudinal grooves to be machined onto the punch surface. The lubricant is more easily carried into the work zone. This practice is rarely used with oil-containing or emulsifying products, but is frequently recommended for water-based synthetics.

The ironing process has been compared to similar processes such as wire drawing, extrusion, and tube drawing that involves heavy plastic deformation. S. Rajagopal, from the IIT Research Institute, claims "the reduction in thickness and the resulting generation of new surfaces made wall ironing more severe from the tribological standpoint than deep drawing wherein the surface area remains nearly constant" [20]. He described four basic modes of lubrication encountered in the D&I operations:

a. *Thick film*, in which the surface of the sheet metal is separated from the tooling surface by a film of lubricant which is thick in comparison with the peaks and valleys of the two surfaces.

b. *Thin film*, in which a continuous film of lubricant is interposed between

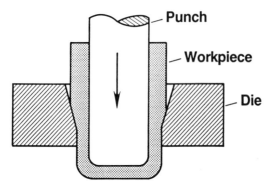

Figure 6 Ironing.

the sheet metal and tooling surfaces, but whose thickness is similar in magnitude to the peaks and valleys of the surfaces.

c. *Boundary*, in which the lubricant film continuity is disrupted by contact between the roughness peaks of the sheet metal and the tooling, with the valleys thereby serving as the lubricant carriers.

d. *Mixed*, which is intermediate between the thin-film and boundary regimes, and in which the contact load is carried partly by asperity contact (roughness peaks) and partly by a thin film of lubricant.

Greater demands are placed on bodymaking coolants than perhaps any other metalworking fluid. Not only must the coolant provide sufficient lubricity for what has already been shown to be one of the most severe forming operations in commercial application today, but secondary fluid requirements may actually be the determining criteria on which the fluid selection is made.

These secondary requirements include adequate detergency to remove dirt, metal fines, and other debris from the punch, dies, and other tool pack surfaces. There must be sufficient cleanliness to carry dirt in the fluid to the filtering system for removal, but not so much so as to prevent lubricant "plate-out" on the punch and can surfaces. Too much detergency will cause removal of residual cupping fluids, resulting in poor interior can quality.

A fluid must also protect the metal surfaces it contacts. For instance, in aluminum can drawing, the coolant must not stain aluminum, should protect slide ways and bodymakers from corrosion, and must not attack cobalt or nickel binders which are commonly used in carbide tooling.

A D&I lubricant may be required to meet specific chemical guidelines established by the end-user. For instance, cans intended for direct contact with foods must be formulated with raw materials that meet FDA criteria for this application. Several breweries require coolants to pass stringent taste and flavor tests before being approved for use in making cans that will ultimately contain their beer.

Bodymaking coolants must be compatible with a variety of cupping fluids, way lubricants, greases, cleaners, and water conditions. They must often handle copious amounts of gear oils or way lubes that leak into the system. While these tramp oils are very effective at helping to remove metallic soaps formed in the tool packs, exorbitant levels may create problems with microbial control, waste treatment, or excessive filter media usage. Any evidence of incompatibility will show up as lower can quality, or decreased production.

Good coolants must also be compatible with some of the antimicrobial additives used to control odors and reduce demands on the filter system. And finally, bodymaking coolants are sometimes even called on to ease waste treatment requirements.

There are as many types of coolants for can drawing as there are varieties of metalworking fluids. Soluble oil emulsions used in the early days of can drawing have given way to synthetic emulsions, which used esters or synthetic base-stocks such as polyaklyleneglycols (PAGs) or polyalphaolefins (PAOs) as replacements for the petroleum mineral oils. The synthetics gained acceptance quickly during the oil shortages of the 1970s. As expected, the synthetics are more expensive, but can save money in the long term through better lubricity, improved cleanliness, and less fluid carry-off per can. Improved work environments are another benefit of the synthetics, as mist levels are lower than for soluble oils.

Water-based, "solution synthetics" have gradually gained acceptance as lubricant technology for can drawing continues to evolve [21]. These fluids have clear to hazy mixes which contrast with the milky appearances of their soluble oil and "synthetic emulsion" predecessors. They are often formulated to reject tramp oils, reduce friction, and lower the need for microbial additives. While cleanliness is very good, there is a fine balance between too much and too little detergency, and overall performance seems to be more difficult to control. In general, greater attention must be paid to selection of filter media, concentration control, tooling surface finish, and tramp or leak oil removal methods. A small amount of tramp oil, in the 1 to 3% range, seems to improve overall performance dramatically with water-based, solution synthetics. This level appears ideal for optimum fines removal, which dramatically reduces friction between the metal contact surfaces within the tool pack. From this reduced friction, tool life is improved.

Lubricity requirements for tin plate are rather unique. Whereas coolants for tin plate must meet the same process requirements as those for aluminum, tin serves as a solid lubricant in addition to the lubricating fluid. Rajagopal showed that under deep-drawing conditions, tin performed as a boundary lubricant, serving as a solid film to inhibit contact between the steel substrate and the punch or dies. This reduces the chance for galling.

Since tin is relatively expensive, any reduction of tin coating can result in considerable savings. Mishra and Rajagopal showed that the load required to draw a cup decreased with increasing tin-coating thickness. In addition, increasing the viscosity of lubricants had the same effect on deep drawing as increasing the tin-coating thickness. They concluded that any attempt to decrease tin-coating thickness must be accompanied by an increase in the viscosity of the liquid lubricant. They determined that total elimination of tin plate could only be accomplished by using lubricants that rely on thick-film regime lubrication.

Byers and Kelly reported success with a synthetic lubricant at not only improving can plant efficiency rates, but in lowering tin weights as well [21]. In their study, the fluid viscosity was relatively low. However, the synthetic fluid did contain lubricants that had very high viscosities and perhaps relied on a thick-film regime. One can-making plant documented in their study was able to reduce tin coatings by 60%, which resulted in a savings of over a million dollars per year in tin-plated steel costs alone.

Figure 7 shows the tooling progression from coil stock to finished can during the drawing and ironing process.

D. Draw–Redraw

Redrawing can describe several operations. The process of lengthening an already drawn part, without sidewall reduction (as occurs in an ironing operation) is known as redrawing. Redrawing can also refer to inside-out reforming in which a cup is placed in the press so that the punch moves through the outside bottom of the cup, through the length of the cup, and out the top, which literally turns the cup inside out.

Redrawing a part to obtain the proper part height or diameter can occur any number of times. Many times, successive stages in transfer presses are simply to redraw the part to a selected depth. However, each draw that occurs limits the successive draws to a percentage of the first draw reduction. The percentage reduction that can be obtained for any successive draw can be calculated by the equation [22]:

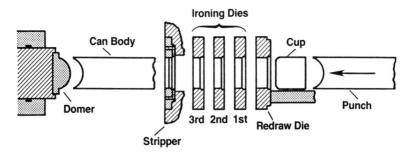

Figure 7 Bodymaker tooling.

$$R = \frac{100\ (D - d)}{d}$$

where

R = percent reduction
D = initial diameter being reduced
d = final diameter to which the blank or shell is reduced

This equation is for a round cup. For square or rectangular parts, the successive draw reductions are controlled by the depth of draw desired, and radii on the punch and die rings at the corners. Since metal flow at the corners is multidirectional for a rectangular part, metal-thickening occurs, which is a major limiting factor in how much reduction can occur in one die [22].

The draw–redraw process has become a very popular method for making containers for food items. A large majority of pet food containers found on grocery shelves are made by the draw–redraw process. Sardine cans, as well as stackable fruit and vegetable cans, are also being made. Due to the noticeable absence of ironing in the process, these cans are generally not extended to the draw depths possible in the drawn and ironed cans.

The lubricant of choice for draw–redraw cans is generally a paraffinic wax. Often, these are food-grade lubricants with FDA approval. This enables the food product to be packed with few after-draw handling processes.

E. Wire Drawing

Donald Sayenga of the Cardon Management Group presented a technical report for the Wire Association International, Inc., at "Interwire 91" [23]. The report was an excellent summary of the history of wire. Mr. Sayenga traced the roots of the word *wire* to the Latin verb, *uiere*, meaning "to plait." He also attributes to the 1771 *Encyclopedia Britannica* the following description: "Wire, a piece of metal drawn through the hole of an iron into a thread of a fineness answerable to the hole it passed through." Two hundred years later, one would be hard-pressed for a better definition. More importantly however, the report showed not only wire's impact on history (where would the phone company be without wire?), but somewhat romanticized the men and women responsible for the development of wire used as a source of inspiration in the making of products such as jewelry, tools and musical instruments, chains, fencing, cables, and even suspension bridges.

The drawing of wire is considered to be one of the most difficult metalforming operations [24]. Wire is made by pulling metal bar stock through a series of reduction dies until the correct shape and size are reached. Most drawn wire is cylindrical, but it can also be drawn to flat or rectangular shapes. Wire can be made

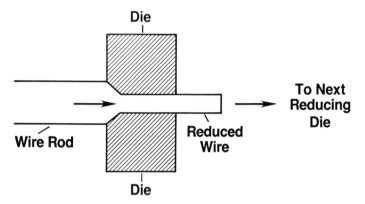

Figure 8 Wire drawing.

from steel, aluminum, copper, or other ductile metals. In addition, copper-clad or tin-coated alloys are used to provide solid-film lubrication. Finished wire can be used in a variety of applications from magnetic wire and electrical applications to high-strength reinforcing cables. In fact, the earliest application of drawn wire recorded is in making chain link armor plate in 44 A.D. [25]. It is believed that this wire was drawn through dies using old-fashioned muscle power. The first lubricant used in wire drawing was developed by Johan Gerdes, who used an accidental combination of urine and urea to draw steel wire [26]. As the story goes, Gerdes, somewhat frustrated, threw several steel rods out a window "where men came to cast their waters" [26]. After "reacting" with the urine, the steel rods were easily drawn.

Since copper wire has the largest volume of use, we will look specifically at copper wire drawing. The rolling mill is where the wire begins. Molten metal alloy is poured into molds and after cooling is cold-rolled into slabs. These slabs are then shaped into continuous thick rods and gathered into large coils ready for the rod-breakdown process.

Wire drawing can be described in three general classes: rod breakdown, intermediate, and fine wire. Rod breakdown is the first wire-drawing process. The rod diameter is reduced through a series of forming dies (see Fig. 8). A take-up reel pulls the wire through the dies, with the wire tension controlled by wraps around a wheellike device called a capstan. During the plastic deformation process, the volume of wire remains constant. As the wire diameter is reduced, the speed of the wire must be increased through each subsequent die to account for the amount of wire produced. As an example, several miles of extra-fine gauge wire can be drawn from a 6 ft section of 100 mm rod. Jan Kajuch of Case Western Reserve University summarized this process [27]. The relationship of the initial volume of wire to the final length of wire after drawing can be shown by some basic principles:

$$V_1 = V_2$$
$$A_1 L_1 = A_2 L_2$$
$$\frac{\pi (D_1)^2}{4} L_1 = \frac{\pi (D_2)^2}{4} L_2$$
$$\frac{(D_1)^2}{(D_2)^2} = \frac{L_2}{L_1} \quad \text{or} \quad \frac{D_1}{D_2} = \sqrt{\frac{L_2}{L_1}}$$

where

π = 3.14159
D = wire diameter (in. or mm)
L = wire length (in. or mm)
A = area
V = volume

This relationship can then be used to determine the basic die parameters for elongation percentage as follows [27]:

$$\% \text{ elongation} = \frac{L_2 - L_1}{L_1} 100$$

$$= \left(\frac{L_2}{L_1} - 1\right) 100$$

$$\% \text{ elongation} = \left[\frac{(D_1)^2}{(D_2)^2} - 1\right] 100$$

Now that we can calculate the percent elongation of the wire, we can determine the amount of reduction that occurs in the dies. We can see from the relationship of elongation to area that [27]:

$$E = \frac{100 A_r}{100 - A_r}$$

where E = die elongation (%) and A_r = reduction of area (%).

The reduction of area (A_r) can be expressed as:

$$A_r = \frac{100 E}{100 + E}$$

and reduction of wire diameter (D_r) in the die is determined by [27]:

$$D_r = 100 \left(1 - \frac{A_r}{E}\right)$$

The preceding equations are elementary for the engineer responsible for making sure the dies are set up properly to draw wire. For the lubrication engineer who must determine if the lubricant is functioning properly, the dies must be set up so as not to exceed the maximum or minimum value for wire elongation, while the machine setup must account for the increased wire speed through each subsequent die. This is all based upon the understanding that the wire volume remains constant.

1. Lubricants

Fortunately for everyone, considerable advances in lubricant technology have been made since Gerdes used urine. As recently as 15 years ago, wire-drawing lubricants were the original combinations of solid lubricants such as metallic stearates, lime (CaO), and calcium hydroxide ($Ca(OH)_2$). These dry powders were applied by drawing the wire through a box containing the lubricant. This box was referred to as a "ripper box." An advantage of the stearate powder was that the lubricant stayed on the wire through the die.

The lubricants used today are complex blends of esters, soaps, and extreme-pressure lubricants. In general terms, the need for lubrication is highest in the rod breakdown operation where the greatest reductions occur. As the wire becomes finer, the need for lubrication decreases, but the requirement for detergency increases. A wide variety of fluid types are used in wire-drawing operations, but are typically oil or polyglycol-based lubricants diluted in water at 10% or higher concentrations in the rod mill. In fine wire drawing, a typical dilution may be as low as 1% of a high-detergency fluid.

2. Metal

The quality of the rod supplied obviously affects the wire-drawing process. Care should be taken by the plant metallurgist to ensure that the metal is the right quality and alloy. Surface defects caused by extraneous alloying agents or other foreign particles may alter the drawability of the wire. Pockets of air in the continuous cast metal rod may result in wire breaks. Surface contaminants such as rolling oil or dirt are also of concern.

3. Dies

The advent of manmade diamond dies revolutionized the dies used in wire drawing. These artificial diamond dies, known in the industry as *compax* dies, hold their tolerance much better than previous natural diamond. The compax dies are much harder, and last many times longer than natural dies, which reduces downtime. Provided the setup for the reduction profile has been correctly made, there are seldom problems with manmade diamond dies.

4. Other Factors

The equipment used in the wire industry ranges from "high-tech" multihead wire machines able to complete much of the rod breakdown and subsequent wire gauges, to small, antiquated but efficient machines that have drawn millions of miles of wire. Regardless of age or ability, the equipment is always well cared for so as to maintain the extremely tight tolerances permitted in wire applications. Wire-drawing rolls and capstans should be periodically checked to make sure that the proper ratios and roll grooves are used. At the same time, a visual inspection for cracks or chipped capstans can be completed. Lubricating nozzles should be placed for optimum application. The take-up system should be inspected for balance and proper tension.

Operators are generally well educated through on-the-job training. It is imperative that an operator know how to "string" a machine quickly and safely. A wire company invests a lot of money to ensure that operators understand their role in the wire operation.

F. Other Metalforming Processes

1. Coining

The world's monetary system would be drastically different without the ability to coin metal. In this operation, metal is thinned or thickened to achieve the desired design [28]. For a coining operation, there is a set of dies, which may or may not be dissimilar, in which the reverse pattern of the final design has been engraved. During the stroke, the metal is forced into the dies, and the defined pattern is imprinted into the metal. Very beautiful and intricate patterns can be created within the dies. This operation is different from embossing, in that there is no change in thickness during embossing.

2. Punching, Piercing, and Notching

The previously mentioned operations are all very similar both in function and in process. When holes are needed, the hole is "punched" out using a punch and die.

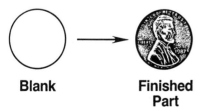

Blank   Finished Part

Figure 9 Coining.

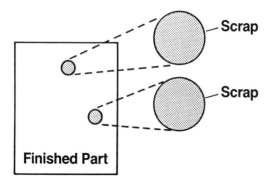

Figure 10 Punching.

Piercing is often used interchangeably with punching, but is really distinguishable in that piercing produces holes by a "tearing" action as opposed to punching's "cutting" action [28]. Notching is actually very similar to punching, but the final shape is a slice in the edge of a finished part. Because part flatness is often important, notching is often done in a progressive die, with multiple strokes [28]. From a lubrication standpoint, most punching, piercing, and notching operations in low-gauge metals are done dry or with a light viscosity mineral oil. In thicker-gauge metals, or in more brittle metals, formulated lubricants may be used to increase tool life.

3. Swaging

Swaging is defined as "a metalforming process in which a rapid series of impact blows is delivered radially to either solid or tubular work. This causes a reduction in cross-sectional area, or a change in geometrical shape" [28]. Perhaps an easier

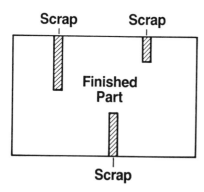

Figure 11 Notching.

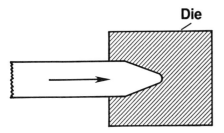

Figure 12 Swaging.

definition compares swaging to the squeezing of tube stock through a die in order to affect its appearance. Cone shapes and tapers on the ends of tooling or tubing are made through swaging. Jet spray devices and welding points are good examples of parts produced through swaging.

IV. CONCLUSION

This chapter has reviewed some of the more common metalforming applications. It would be impossible in this short summary to address all of the operations, so focus was given to a few of the more common techniques. There are excellent books available which give deeper insight into very specific technical aspects of lesser known metalforming procedures.

Included in this discussion have been many examples of the history and tradition of metalforming. These have been presented primarily because of the desire of the author not only to chronicle the significance of the event, but to show that there is more to lubrication topics than merely "making parts."

Also included were practical guidelines for various metalforming operations. These sensible descriptions were not aimed at the engineer responsible for the highly detailed, mechanical setups necessary to form metal into parts. Rather, they were intended as pragmatic descriptions to help better understand metalforming processes in general, and, in turn, obtain a better appreciation for all of the benefits gained in our daily lives from someone shaping metal into formed parts. From opening car doors to opening beverage containers, we see the intrinsic value of lubrication knowledge and practice. Perhaps this chapter will lead to an even greater awareness of the necessity to continue to make advancements in the science, and in the art, of metalforming lubrication.

REFERENCES

1. J. A. Schey, *Tribology in Metalworking*, American Society for Metals, Metals Park, Ohio, pp. 1–3, 134 (1983).
2. J. J. O'Connor, ed., *Standard Handbook of Lubrication Engineering*, McGraw–Hill, New York, p. 2-1. (1968).
3. T. Lyman, ed., *Metals Handbook: Forming*, Vol. 4, 8th ed., American Society for Metals, Metals Park, Ohio, p. 23 (1979).
4. R. C. Weast, ed., *Handbook of Physical Chemistry*, 50th ed., The Chemical Rubber Co., Cleveland, pp. B118–119 (1969).
5. D. R. Hixson, Pressworking lubricants, *Manuf. Eng.* 82(2):56 (1979).
6. T. Lyman, ed., *Metals Handbook: Forming*, Vol. 4, 8th ed., American Society for Metals, Metals Park, Ohio, p. 24 (1979).
7. J. J. O'Connor, ed., *Standard Handbook of Lubrication Engineering*, McGraw–Hill, New York, p. 3-1 (1968).
8. T. Lyman, ed., *Metals Handbook: Forming*, Vol. 4, 8th ed., American Society for Metals, Metals Park, Ohio, p. 25 (1979).
9. C. Wick, J. Benedict, and R. Veilleux, eds., *Tool and Manufacturing Engineers Handbook: Forming*, Vol. 2, 4th ed., pp. 4-1–4-9 (1984).
10. T. Lyman, ed., *Metals Handbook: Forming*, Vol. 4, 8th ed., American Society for Metals, Metals Park, Ohio, p. 163 (1979).
11. C. Wick, J. Benedict, and R. Veilleux, eds., *Tool and Manufacturing Engineers Handbook: Forming*, Vol. 2, 4th ed., p. 4-41 (1984).
12. T. Lyman, ed., *Metals Handbook: Forming*, Vol. 4, 8th ed., American Society for Metals, p. 163 (1979).
13. C. Wick, J. Benedict, and R. Veilleux, *Tool and Manufacturing Engineers Handbook: Forming*, Vol. 2, 4th ed., p. 4-34 (1984).
14. C. Langewis, "Two-piece can manufacturing: Blanking and cup drawing," presented at the Society of Manufacturing Engineers (SME) Conference, Clearwater Beach, FL, pp. 1–12 (1981).
15. News, *CanMaker 4*: 8 (1991).
16. News, *CanMaker 4*: 3 (1991).
17. B. M. Conny, Extracts from Coors: A catalyst for change, *CanMaker 4*: 33–38 (1991).
18. F. Church, Productivity gains head off beverage can shortage, *Mod. Met.* 41(9):68–76 (1985).
19. P. Golding, Aluminum drinks can recycling in Europe: Five successful years of consumer recycling, *CanMaker 4*: 42–45 (1991).

20. S. Rajagopal, A critical review of lubrication in deep drawing and wall ironing, from a conference on metalworking lubrication, American Society of Mechanical Engineers, New York, pp. 135–144 (1980).
21. K. Tucker, "A solution synthetic for 2-piece cans," STLE Annual Meeting, Non-Ferrous Session, Atlanta (1989).
22. J. Byers and R. Kelly, A fluid and tooling system for the production of high-quality two-piece cans, *Lubr. Eng.*, *42*(8):491–496 (1986).
23. C. Wick, J. Benedict, and R. Veilleux, eds., *Tool and Manufacturing Engineers Handbook: Forming*, Vol. 2, 4th ed., p. 4-34 (1984).
24. D. Sayenga, "Wonderful world of wire," presented at "Interwire-91," Wire Association International, Inc., Atlanta, (Cardon Management Group, USA) (1991).
25. G. H. Geiger, Copper drawing agents—Some new ideas, *Wire J. Int.*, *23*: 60.
26. P. Gielisse, "A seminar on copper wire drawing," Cincinnati Milacron, June, 1983.
27. J. A. Schey, *Tribology in Metalworking*, American Society for Metals, Metals Park, Ohio, p. 3 (1983).
28. J. Kajuch, "Basic concepts of wire drawing process," Cincinnati Milacron, March, 1992.
29. C. Wick, J. Benedict, and R. Veilleux, eds., *Tool and Manufacturing Engineers Handbook: Forming*, Vol. 2, 4th ed., pp. 4-4, 4-5, 4-8, 14-1 (1984).

6
The Chemistry of Metalworking Fluids

JEAN C. CHILDERS
Amax, Inc.
Summit, Illinois

I. INTRODUCTION: FLUID TYPES

Throughout the twentieth century, metalworking chemistry has evolved from simple oils to sophisticated water-based technology. The evolution of these products is shown in Fig. 1. Between 1910 and 1920, soluble oils were initially developed to improve the cooling properties and fire resistance of straight oils. By emulsifying the oil into water, smoke and fire were greatly reduced in the factories, thus improving working conditions. With the presence of water in the fluid, tool life was extended by reducing wear since the fluid kept the tools cool. However, water-diluted fluids caused rust on the workpiece, thereby creating the need for rust inhibition.

Synthetic fluids were first marketed in the 1950s because of better cooling and rust protection compared to soluble oils in grinding operations. In the early 1970s, oil shortages encouraged compounders of cutting fluids to formulate synthetic oil-free products that could replace oil-based fluids in all metalworking operations. Synthetic fluids offer benefits over soluble oil technology. These benefits include better cooling and longer tank life because of good hard-water

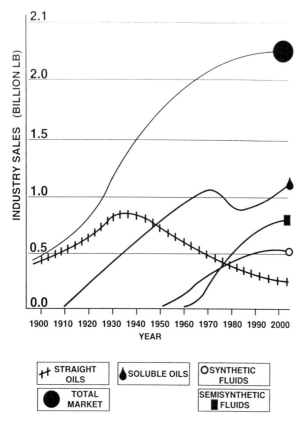

Figure 1 Evolutionary product life cycle.

stability and resistance to microbiological degradation. However, soluble oils, while indeed more susceptible to bacteria growth, provide better lubricity and easier waste treatability than synthetic fluids. These trade-offs encouraged the development of semisynthetic fluids. These water-based fluids contain some oil-based additives emulsified into water to form a tight microemulsion system. These semisynthetic fluids are an attempt to reap the benefits of oil-soluble technology while retaining the good microbial control and long tank life of synthetic fluids.

In the 1980s, synthetic and semisynthetic fluids were growing in a mature market, displacing oil-based technology. However, in the early 1990s, oil prices dropped, placing oil technology at the forefront in pricing. With increasing waste treatment costs, easier to waste-treat soluble oils gained market share over synthetics. Additionally, hazard regulations on ethanolamines commonly used in synthetic fluids for corrosion control further encouraged the use of soluble oils. Therefore, mature straight and soluble oil technology has held its 65% market share.

The chemistry of metalworking fluids is as diverse as a library of cookbooks. Each formulating chemist will develop his own fluid formula to meet the performance criteria of the metalworking operation. But like lasagna, each "recipe" will have common ingredients or raw materials: noodles, cheese, meat, sauce, etc. That is why fluids are at times called "black box chemical blends." No user is fully aware of the exact composition of the fluid used, but the user knows whether it meets certain performance criteria (tastes good). There are many additive blends that will function as metalworking fluids and there is no assurance of the "perfect" fluid for an operation. Misapplication of that perfect fluid could render it unacceptable.

This review of the chemistry of metalworking fluids will identify the building blocks of metalworking fluids, the reasons for utilizing them, and the key parameters for additive selection.

II. FUNCTIONS OF FLUIDS

A metalworking fluid's principal functions are to aid the cutting, grinding, or forming of metal and to provide good finish and workpiece quality while extending the life of the machine tools. The fluids cool and lubricate the metal/tool interface while flushing the fines or chips of metal from the piece. The fluid should also provide adequate temporary indoor rust protection for the workpiece prior to further processing or assembly.

III. ADDITIVE TYPES

The chemical additives used to formulate metalworking fluids serve various functions. These include emulsification, corrosion inhibition, lubrication, microbial control, pH buffering, coupling, defoaming, dispersing and wetting.

Most of the additives used are organic chemicals that are anionic or nonionic in charge. Most are liquids, used for ease of blending by the compounder. Some of the basic chemical types utilized are fatty acids, fatty alkanolamides, esters, sulfonates, soaps, ethoxylated surfactants, chlorinated paraffins, sulfurized fats and oils, glycol esters, ethanolamines, polyalkylene glycols, sulfated oils, fatty oils, and various biocide/fungicide chemical entities.

The functional additives used in metalworking fluids each contribute to the total composition. The effect of the addition of an additive is tested by the chemist to ensure that optimal properties of a fluid are maintained. In general, a fluid should be stable, low foaming, and waste treatable. Many of the properties of additives are mutually exclusive. Typically, if a fluid has excellent biological and hard-water stability, it may be difficult to waste treat. Or if it provides excellent lubricity, it may be difficult to clean. The following reviews the typical properties of additives and the significance to the formulator and user.

A. Stability

The fluid concentrate must be stable without clouding or separating for a minimum of six months storage. The fluid may be tested in cold and hot atmospheres to assess the effect of shipment or storage in winter and summer climates. Some chemists check for gelling, freezing, or "skinning" of the fluid, which may signify handling problems.

B. Oxidative Stability

Some consider the oxidative stability of the additives important. Aerating and heating the coolant can accelerate any destructive oxidation of the chemical additive.

C. Emulsion Stability

In soluble oils, emulsion stability is the most critical property. The emulsifier system must be balanced based on its alkalinity, acidity, and HLB (hydrophilic/lipophilic balance) to ensure a white emulsion with no cream or oil forming at the surface of the fluid.

D. Hard-Water Stability

All fluid types are tested for hard-water stability because of the progressive increase in hard-water salts in the used fluid. As the fluid evaporates, only deionized water is removed, leaving behind water salts like calcium and magnesium. Carry-out of the fluid on the parts also depletes the fluid volume. As more water and fluid concentrate is added, more salts are accumulating in the tank. Calcium and magnesium cations build up in the fluid. Therefore, in soluble oils, the sodium sulfonate emulsifier is changed to calcium sulfonate, an additive that is not an emulsifier. This destabilization of the emulsion causes oil separation and loss of fluid concentration. In synthetic fluids, hard-water stability problems are visible as soap scum formation on the surface of the fluid. Typically, anionic additives may have hard-water stability problems, whereas nonionic-type additives are stable to hard-water salts.

E. Mixability of Fluid Concentrate

The ease of dilution of the fluid concentrate is important from a practical perspective. The oil must "bloom" into the water without gelling to ensure fast and complete mixing. Many times fluid concentrate is not premixed and is added at a point in the tank where there is little agitation. Without good mixability, the fluid concentrate could sink, thereby not contributing to the fluid concentration intended. High soap components can cause mixability problems.

F. Foam

Because of constant agitation, spraying, and recirculation of metalworking fluids, foam can easily form in the tank. Besides being a nuisance, foam interferes with the lubricity and cooling functions of the fluid. Air does not lubricate, so air entrained in the fluid renders a fluid ineffective. Foam also interferes with the worker's view of the workpiece, affecting machining accuracy and measurements. Many emulsifiers and lubricity additives may serve their function very well in a stagnant system but may be marginally useful if they foam excessively.

G. Residue/Cleanability

The fluid should not leave a sticky or hard-to-clean residue on the parts or equipment. Some boron-based corrosion inhibitors can leave a sticky residue. Chlorinated paraffins and pigmented lubricant additives can be difficult to remove in cleaning operations.

H. Corrosion Inhibition

Fluids are tested for their corrosion-inhibiting properties. Since water is the diluent for the majority of fluids, corrosion inhibition is critical. Some additives are film forming (amine carboxylate), some are more like vapor phase inhibitors (monoethanolamine borates), while others actually form a matrix with the metal surface to provide protection (azoles).

I. Lubricity

Additives tested for lubricity can be combined to obtain various types of lubricity, depending on the fluid requirments.

Boundary lubricants like lard oil, overbased sulfonates, esters, soaps, and sulfated oils provide a boundary between the workpiece and tool. This slipperiness is ideal for all systems, especially when machining aluminum. Soft metals need boundary lubricants to allow metal removal with good tolerance by inhibiting the tool from welding onto the aluminum workpiece.

Extreme-pressure additives like sulfur, chlorine, and phosphorus actually form metal complexes with the metal surface at elevated temperatures. Chlorinated additives are the most effective with typically 40–70% chlorine in the product compared to sulfurized additives with 10–15% sulfur, or phosphate esters with 5–15% phosphorus. Each has its problems. Chlorinated additives in general are under scrutiny because of their hazardous nature. Sulfurized materials can stain

Mineral Oils

Paraffinic   Naphthenic

Animal & Vegetable Oils

H_2C-O-R_1
$|$
$HC-O-R_2$
$|$
H_2C-O-R_3

where at least one of R_1, R_2, or R_3 is a fatty acid group of carbon chain length 12 to 22 and the remainder (if any) are hydrogen.

Oleic Acid

$CH_3(CH_2)_7 CH = CH (CH_2)_7 COOH$

Figure 2 Oils and fats.

metals and can quickly cause rancidity. Phosphate esters, the least effective of the three as a lubricant, can cause fungus and mold growth because phosphorus is a good nutrient.

Hydrodynamic lubricity additives provide a variation on boundary lubricity through a high viscosity in the fluid. Typically, this term is used when describing straight oils with viscosity improver added, although some synthetic fluids are formulated with high-viscosity polymer additives that give the fluid a slippery, thick appearance. These elastic polymers may drop in viscosity under shear and heat, but if they are true rheological additives they will regain their viscosity when cooled.

J. Chemical Structures

The chemical nature of most metalworking fluid additives is organic. Figures 2–6 show the chemical structures of some of these additives. Emulsifiers, corrosion inhibitors, and lubricant additives are all of importance in formulating metalworking fluids, as described in Secs. IV–VIII.

IV. STRAIGHT OILS

A straight oil is a petroleum or vegetable oil that is used without dilution with water. It can be alone or oil compounded with various polar and/or chemically active

Petroleum Sulfonate

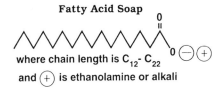

where $R_1 = C_{15}$ to C_{30} alkyl chain
and $R_2 = C_{15}$ to C_{30} alkyl chain or H
molecular weight from 380 to 540

Fatty Acid Soap

where chain length is C_{12}-C_{22}
and (+) is ethanolamine or alkali

Super (1:1) Alkanolamide

$$R-\overset{O}{\underset{\|}{C}}-N(CH_2CH_2OH)_2$$

Kritchevsky (2:1) Alkanolamide

$R-\overset{O}{\underset{\|}{C}}-N(CH_2CH_2OH)_2$ • 25% free diethanolamine

Figure 3 Emulsifiers. (From Ref. 5.)

additives. Light solvents, neutral oils, and heavy bright and refined stocks are among the petroleum oils used.

Paraffinic oils offer better oxidative stability and less smoke during cutting than naphthenic oils. However, most compounded oils contain naphthenic oils because the lubricant additives are more soluble and compatible in naphthenic oils [1].

For environmentally favorable requirements, vegetable oils are the oils of choice. Although considerably more expensive than petroleum oils, they are easily biodegraded for disposal. It follows then that they are more prone to biological deterioration than petroleum oils. Nondrying oils like rapeseed, castor, and coconut oils are best. Rapeseed oil, being the lowest in saturated fatty composition, is the best in lubricity because of its long C_{22} carbon fatty chains. It burns clean and is smoke-free, which is a great advantage over petroleum oils which are frequent fire and smoke hazards.

Straight oils provide hydrodynamic lubrication. When compounded with lubricant additives, they are useful for severe cutting operations, for machining difficult metals, and for ensuring optimal grinding wheel life.

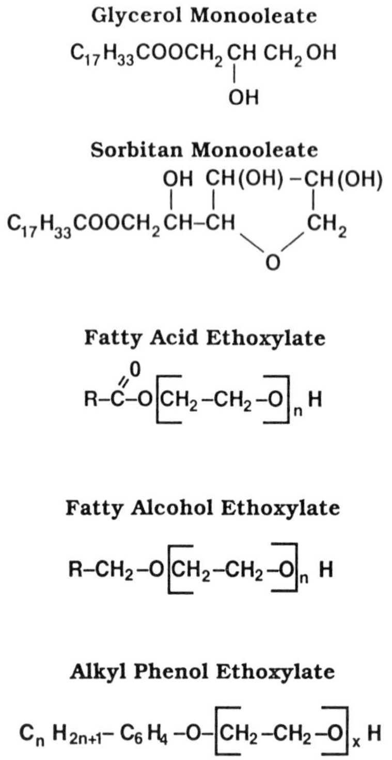

Glycerol Monooleate

$C_{17}H_{33}COOCH_2\underset{\underset{OH}{|}}{CH}CH_2OH$

Sorbitan Monooleate

$C_{17}H_{33}COOCH_2\underset{\underset{OH}{|}}{CH}-\underset{\underset{CH(OH)}{|}}{CH}\diagdown_O\diagup\underset{\underset{CH(OH)}{|}}{CH_2}$

Fatty Acid Ethoxylate

$R-\overset{O}{\overset{\|}{C}}-O[CH_2-CH_2-O]_n H$

Fatty Alcohol Ethoxylate

$R-CH_2-O[CH_2-CH_2-O]_n H$

Alkyl Phenol Ethoxylate

$C_n H_{2n+1}-C_6H_4-O-[CH_2-CH_2-O]_x H$

Figure 3 (continued)

A. Compounded Oils, Mineral Oils, and Polar Additives

One of the basic compounded straight oils is a naphthenic oil with 10–40% boundary lubricants added. These may include animal oils like lard oil or tallow, or vegetable oils like palm oil, rapeseed oil, or coconut oil [2] Oil-soluble esters of these oils are beneficial because they reduce the inherent biodegradation of fatty oils. Examples are methyl lardate and pentaerythritol esters. *Blown oils*, oxygen-polymerized vegetable and animal oils, increase the affinity of the additive for the metal surface, thereby providing added slip between the tool and work-piece.

These polar additives increase the wetting ability and penetrating properties of the oil and provide a slippery boundary lubricant film. The keys to choosing these additives are oxidation resistance, oil solubility, and gumming properties. Petroleum oil fortified with these polar lubricant additives are used in machining nonferrous metals where staining by other additive systems are problematic. The polar additives provide a rust-inhibiting barrier film from the atmosphere thereby

Tolyltriazole Salt

[structure: tolyltriazole sodium salt with CH₃ on benzene ring fused to triazole, N⁻Na⁺]

Calcium Sulfonate

$$\left[C_9H_{19}\text{-naphthalene-}C_9H_{19}\text{-}SO_3^- \right]_2 Ca^{++}$$

Triethanolamine

(HO CH$_2$CH$_2$)$_3$N

Monoethanolamine Borate (7)

[structure: HO-B(OH)-O-CH₂-CH₂-NH₂ cyclic with dative bond from NH₂ to B]

Amine Dicarboxylate

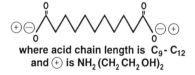

where acid chain length is C$_9$- C$_{12}$
and ⊕ is NH$_2$(CH$_2$CH$_2$OH)$_2$

Figure 4   Corrosion inhibitors.

Chlorinated Compound

$C_x H_y Cl_z$

where X is about 9 to 20 and the weight percent of the chlorine is 20 to 70.

Sulfurized Fatty Oils,
Acids or Esters

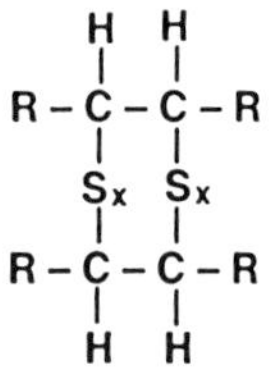

$$\begin{array}{c} HH \\ || \\ R-C-C-R \\ || \\ S_xS_x \\ || \\ R-C-C-R \\ || \\ HH \end{array}$$

Sulfated Castor Oil

∧∧∧∧∧∧∧∧∧∧–C(=O)–O $\ominus$ Na$\oplus$
|
OSO_3 $\ominus$ $\oplus$
Na

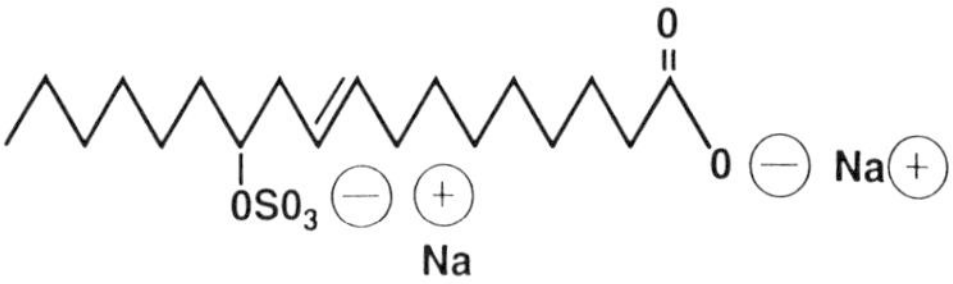

Polyethylene Glycol Esters

$$R-\overset{O}{\overset{\|}{C}}-O-(CH_2-CH_2-O)_n-H$$

Propylene Oxide
Ethylene Oxide Block Polymer

$HO-(C_2H_4O)_x-(C_3H_6-O)_y-(C_2H_4O)_z-H$

Figure 5 Lubricants.

Metalworking Fluids Chemistry

Sulfochlorinated Fatty Oils & Esters

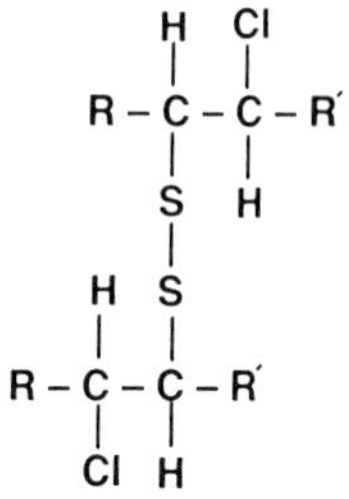

$$R - \underset{\underset{S-H}{|}}{\overset{\overset{H}{|}}{C}} - \underset{\underset{H}{|}}{\overset{\overset{Cl}{|}}{C}} - R'$$

$$R - \underset{\underset{Cl}{|}}{\overset{\overset{H}{|}}{C}} - \underset{\underset{H}{|}}{\overset{\overset{S}{|}}{C}} - R'$$

Phosphate Esters

$$O = \underset{\underset{OR}{|}}{\overset{\overset{OH}{|}}{P}} - OR \quad + \quad HO - \underset{\underset{OH}{|}}{\overset{\overset{OR}{|}}{P}} = O$$

Diester   Monoester

Where R = ethoxylated alcohol or phenol

Dibasic Acid Ester

$R\ O_2\ C(CH_2)_x\ CO_2\ R$

Molybdenum Disulfide

MoS_2

Figure 5 *(continued)*

providing excellent indoor rust protection. These fortified oils are primarily used for light duty cutting operations.

B. Chemically Active Lubricant Additives

For more difficult machining operations, extreme-pressure additives like sulfurized, chlorinated, or phosphated additives are added to the mineral oil. These

Glycol Ethers

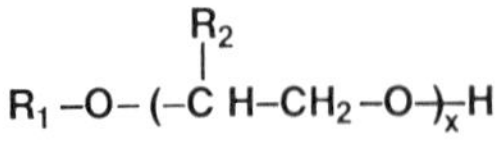

$$R_1-O-(-C H-CH_2-O)_x H$$
$$|$$
$$R_2$$

where R_1 = butyl, etc. and R_2 = CH_3 or H

Propylene Glycol

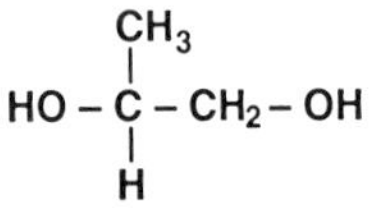

$$HO-\underset{\underset{H}{|}}{\overset{\overset{CH_3}{|}}{C}}-CH_2-OH$$

Figure 6 Couplers.

additives are surface reactive and form metallic films on the tool surface, thereby acting much like a solid lubricant at the metal/tool interface.

These additives are used alone or in combination with one another and paired with polar additives to give a lubricating oil with a wide range of effectiveness at various temperatures and pressures. An oil that contains lard oil, chlorinated paraffins, and sulfurized lard oil can bridge the lubrication needs as follows: At low temperatures and pressures, the lard oil provides good boundary lubrication until temperatures reach 570–750°F. Then the chlorinated paraffin takes over, forming an iron chloride film. Then as temperatures climb to approximately 1300°F, the sulfurized fat takes over, forming the metallic sulfide lubricant film.[3]

The chlorinated additives could be chlorinated waxes, paraffin, olefin, or esters. Chlorinated additives are nonstaining but they can be corrosive, since small levels of HCl can be released. Therefore, inhibitors like epoxidized vegetable oils are often used to inhibit corrosion on the workpiece.

The sulfurized additives are either active or inactive. A sulfurized mineral oil is an active additive in that there is free unbound sulfur that easily reacts as the EP lubricant. However, this free sulfur can stain yellow nonferrous metals. Sulfurized fats like lard oil have a stronger chemical bond and can be less likely to stain metals. Typically, a straight oil that contains sulfurized oils is dark in color and has a pungent odor. There are, however, other sulfurized additives like TNPS (trinonylpolysulfide) that is light yellow in color and ideal for water- and amine-free metalworking fluids. The simplest sulfurized mineral oil formulation would contain approximately 1% sulfur and a fluid for difficult tapping or threading operations would contain approximately 5% sulfur.

There are sulfochlorinated additives where both S and Cl are reacted onto one molecule. These are good for machining low carbon steel and nickel–chrome alloys.

Phosphate esters provide both boundary lubricity from the ester component and phosphorus extreme-pressure lubricity at low temperatures. The effects are less dramatic than with sulfur and chlorine. Phosphate esters must be oil soluble and can be used "as is" in their free acid form, or can be neutralized with an alkaline material. Neutralized phosphate esters are nonstaining and noncorrosive and can provide rust protection properties to the oil blend.

Solid lubricants are used to a limited extent in nonrecirculating systems. Molybdenum disulfide (MoS_2) and graphite are dispersed or suspended into the oil. These additives form metallic sulfide films and flat lubricant structures that provide excellent lubricity for very difficult machining operations.

C. Straight Oil Formulations

Oil formulation	% by wt.
Naphthenic 100 s mineral oil	90
Lard oil	2
Chlorinated paraffin	6
Sulfurized lard oil	2
	100

Straight cutting oils are used in difficult machining and forming operations. They are ideal in recirculating systems with a lot of downtime and where rancidity of the water dilutable fluid is a problem. Straight oils are very stable to degradation, provide good rust protection, and with regular removal of metal chips, are the most trouble-free metalworking fluids from a service aspect. Their limitations are higher cost, smoke and fire hazards, operator health problems, and limited tool life through inadequate cooling.

In drawing and forming operations, oils of high viscosity are valuable. The thicker, more viscous oils provide a tougher hydrodynamic lubricant barrier film. In chip removal operations, however, high-viscosity oils will not clear the chips very well and will act as an insulator, thereby further reducing the cooling properties on the tooling. The viscosity of a finished cutting oil should be low enough to clear the chips and not insulate the heat from the operation but high enough to control oil misting, a common health concern associated with the use of straight metalworking oils.

V. SOLUBLE OILS

With the changeover to carbide tooling and increased machine speeds, water-diluted metalworking fluids were developed. Soluble oils or emulsifiable oils are the largest type of fluid used in metalworking. The product concentrate, an oil fortified with emulsifiers and specialty additives, is diluted at the user's site with water to

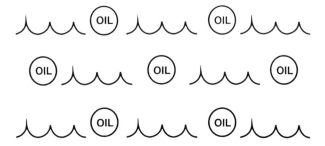

Figure 7 Oil-in-water emulsion.

form oil-in-water emulsions. Here the oil is dispersed in a continuous phase of water.

Dilutions for general machining and grinding are 1–20% in water, with 5% being the most common diluton level. Drawing compounds are diluted with less water—typically 20–50%. At rich 50% dilutions, an invert emulsion is often purposely formed with the oil as the continuous phase. This thickened lubricant has superb lubricating properties and clinging potential on the metal to avoid run-off prior to the draw.

A. Oil

The major component of soluble oils is either a naphthenic or paraffinic oil with viscosities of 100 SUS (Saybolt universal seconds) at 100°F, sometimes termed a 100/100 oil. Higher-viscosity oils can be used with greater difficulty in emulsification, but with possibly better lubricity. Naphthenic oils have been predominantly used because of their historically lower cost and ease of emulsification. Today, naphthenic oils are hydrotreated or solvent-refined to remove potential carcinogens known as polynuclear aromatics.

B. Emulsifiers

The next major class of additives in a soluble oil is the emulsifiers. These chemicals suspend oil droplets in the water to make a milky to translucent solution in water. The size of the emulsion particle determines the appearance. Normal milky emulsions have particle sizes approximately 0.002 to 0.00008 in. in diameter (2.0 to 50 μm), whereas microlike emulsions with a "pearlescent" look have emulsion particle sizes of approximately 0.000004 to 0.00008 in. (0.1 to 2.0 μm)[1]. Some compounders relate the effectiveness of the two types of emulsions to comparing basketballs to small ball bearings. One can visualize more ball bearings entering a tight metal/tooling interface for lubrication than basketballs. Others agree that biostability can be enhanced with a microemulsion. Advantages of a standard milky emulsion are large oil droplet size for forming operations, ease of waste treatability, and lower foam than with microemulsions.

The predominant emulsifier is sodium sulfonate, which is used with fatty acid soaps, esters, and coupling agents to provide a white emulsion with no oil or cream separating out after mixing with water. Nonionic emulsifiers like nonylphenol ethoxylates, PEG esters, and alkanolamides are also used when hard-water stability or microemulsion systems are desired. Many basic soluble oils are complete with this combination of oil and emulsifier system.

C. Value Additives

Many specialty compounders include other additives to add further value to the product. Since the fluid will be diluted with water, the possibility of rust forming is introduced. Normal rust control is usually satisfactory, but this depends on the emulsifier. Some added rust inhibitors used include calcium sulfonate, alkanolamides, and blown waxes. To impart biostability along with rust inhibition, boron containing water-soluble inhibitors are coupled into the formulation.

The pH of the diluted fluid should be 8.8–9.2 to ensure rust protection, metal safety, and rancidity control. This pH should be buffered so the pH is maintained upon recirculation of the fluid. This is more attainable with amines as alkaline sources rather than caustic soda or potash.

To further control rancidity of the fluid from bacteria growth, biocides are often added to the oil. Further tankside additions will be necessary to prolong bacteria control.

The lubricity of a soluble oil comes from the oil emulsion. Because the viscosity of water-dilutable fluids is almost equal to that of water, the film strength or hydrodynamic lubrication potential is negated compared to straight oils. Lubricant additives are commonly added for medium to heavy duty operations. Boundary lubricants like lard oil, esters, amides, soaps, and rapeseed oil are used just as they were in straight oils. Likewise, chlorinated, sulfurized, and phosphorus-based extreme-pressure additives discussed earlier are popular value lubricant additives.

Defoamers are sometimes added if the product foams excessively due to the emulsifier system's properties. Both silicone and nonsilicone defoamers are used, silicone being the most effective at low doses. However, many plants forbid the use of silicone where plating, painting, and finishing surfaces will be affected because of "fish eyes" forming in the painted surface.

The advantages of soluble oils over straight oils include lower cost, since they are diluted with water; heat reduction; and the ability to run at higher machining speeds. Soluble oils are also cleaner, cooler, and more beneficial to the health of the workers because oil mists are no longer inhaled.

The advantages of straight oils over soluble oils include no rancidity, good wettability of the metal surface, good rust protection, and no destabilization problems from emulsions oiling out from hard-water buildup and bacterial attack.

The following is a typical formulation showing the proportions of the additives in a soluble oil product:

Function	Component	% by wt.
Oil	100/100 naphthenic hydrotreated oil	68
Emulsifier	Sulfonate emulsifier oil base	17
EP lubricant	Chlorinated olefin	5
Boundary lubricant	Synthetic ester	5
Rust inhibitor	Alkanolamide	3
Biocide	Biocide	2
		100

VI. SEMISYNTHETIC FLUIDS

A. Oil/Water Base

Semisynthetic fluids are similar to soluble oils in that they are emulsions and similar to synthetic fluids in that they are water-based fluids. The product concentrate usually appears to be a clear solution of additives. However, there is usually 5–20% mineral oil emulsified into the water to form a microemulsion. The emulsion particle size is 0.000004 to 0.0000004 in. (0.1 to 0.01 μm) in diameter [1]. This is small enough to transmit almost all incidental light.

B. Emulsifiers

The emulsifiers used to achieve this microemulsion combine the oil and water to form a clear blend. Alkanolamides are the most commonly used emulsifiers along with sulfonate bases, soaps, or esters as coemulsifiers. A good waste-treatable fluid would contain an amide and sulfonate base or soap package. A hard-water stable product would use a nonionic-type emulsifier along with the amide.

C. Value Additives

Couplers like fatty acids and glycol ethers may be required to regulate the clarity and viscosity of the fluid. Both oil- and water-soluble rust inhibitors are used, keeping in mind that oil-soluble additives must also be emulsified. Alkanolamines like triethanolamine are added to help buffer the pH to a good alkaline level for rust protection.

Lubricant additives can also be either oil or water soluble. Boundary lubricants and extreme-pressure additives like S, Cl, and P can fortify a semisynthetic fluid for more difficult machining operations. Water-soluble chlorinated fatty acid soaps or esters are an example of this type of additive that need not be emulsified into the microemulsion.

Many compounders also add a biocide/fungicide package to protect the product from microbial growth. Because an excess of emulsifiers is required (typically two parts emulsifiers to one part oil), a defoamer may be required. However, selection of defoamers for semisynthetics can be difficult because if the defoamer can be coupled or emulsified into the microemulsion, it will no longer defoam the fluid. If it separates in the drum of product concentrate, it is effective only if totally removed with product concentrate.

Because of the abundance of emulsifiers, the semisynthetic fluids will also emulsify tramp oil. To some users this is a plus, because they have no means of tramp oil removal and their system stays cleaner with this fluid. After time, the once translucent fluid will appear milky, much like a soluble oil. Many feel they are creating an in situ soluble oil. Others believe this acceptance of foreign oil deteriorates the quality of the fluid. Should a formulator want to make a semisynthetic that rejects tramp oil, the formulator should carefully emulsify the oil with alkanolamide. The alkanolamide must be a 2:1 amide with no fatty acid present in order to neutralize the excess mole of diethanolamine, which forms a soap.

D. Semisynthetic Formulation

Function	Component	% by wt.
Emulsifier	Sulfonate base	5
Emulsifier	Alkanolamide	15
Oil	100/100 naphthenic oil	15
Corrosion inhibitor	Amine borate	6
Coupler	Butyl carbitol	1.5
Biocide/fungicide	Triazine/pyridiethione	2
Diluent	Water	55.5
		100

The chemical additives must be mixed together first, then the water should be slowly added to obtain a clear microemulsion. The product should be quality controlled before adding the water. All adjustments should be made at this point to ensure a stable and clear product. Instability will result in a separated product that cannot be reconstituted without removal of the water.

Many users like the "semi" nature of these fluids because of the advantages of both soluble oils and synthetics without many of their individual disadvantages. The advantages of semisynthetic fluids are rapid heat dissipation, cleanliness of the system, resistance to rancidity, and bioresistance. The bioresistance is due to the small emulsion particle size and small amount of oil in the fluid for anaerobic bacteria to feed on. Rust protection and lubricity are better than in a synthetic fluid because the oil and oil-soluble additives provide a barrier film that protects from corrosion and adds lubricity. The disadvantage is foam in grinding operations, acceptance of tramp oil, and less lubricity than soluble oils.

VII. SYNTHETIC FLUIDS

Synthetic metalworking fluids are water-based products containing no mineral oil. The particle size of a synthetic fluid is typically 0.000000125 in. (0.003 μm) in diameter[1].

A. Water Base

The water in the products provides excellent cooling properties, but no lubricity. Water also causes corrosion on metal surfaces. Synthetic fluids are formulated with multiple rust inhibitors and lubricant additives to reproduce the machinability properties of oil-based products.

B. Corrosion Inhibitors

Synthetic fluids usually contain an ethanolamine for general corrosion inhibition and pH buffering capability. Synthetic corrosion inhibitors are amine borates commonly termed borate esters and amine carboxylate derivatives. These low-foaming additives are replacements for amine nitrites, which were discontinued from use due to potential carcinogenicity. Nonferrous inhibitors include benzotriazole, tolyltriazole, and mercaptobenzothiazole. An amine-free inorganic inhibitor is sodium molybdate. Basic amine–fatty acid soaps and alkanolamide also provide excellent rust protection for synthetic systems. They are also good lubricants.

C. Lubricant and Other Value Additives

Other synthetic lubricants include polyalkylene glycols and esters, both of which are low-foaming lubricants with good hard-water stability. Because of their non-ionic water solubility, however, they are difficult to waste treat, which results in high COD (chemical oxygen demand) contents. Boundary and extreme-pressure lubricants must be water soluble. Boundary lubricants include soaps, amides, esters, glycols, and sulfated vegetable oils. Chlorinated and sulfurized fatty acid soaps and esters and neutralized phosphate esters provide extreme-pressure lubricity.

A fungicide is added to protect the synthetic fluid from yeast, fungus, and molds that are prevalent in these fluids. Bacteria are nearly nonexistent due to the high pH and oil-free nature of the synthetic system. Defoamers, wetting agents, and dyes are auxiliary additives found in many synthetic fluids. The wetting agents, or surfactants, reduce the surface tension of the fluid thereby promoting good coverage of the metals for lubrication.

D. Synthetic Formulation

Function	Component	% by wt.
Diluent	Water	70
Rust inhibitor	Amine carboxylate	10
pH buffer and inhibitor	Triethanolamine	5
EP lubricant	Phosphate ester	4
Boundary lubricant	PEG ester	5
Boundary lubricant	Sulfated castor oil	4
Fungicide	Pyridinethione	2
		100

Much new product development is centered around synthetic products in order to produce additive systems that provide optimal lubricity and rust protection in an easily disposed fluid. One such concept is the marriage of semisynthetic technology with synthetic chemistry. By using multiple emulsifiers to couple synthetic water-insoluble lubricants into water, a waste-treatable system is created with petroleum oil absent from the formula.

Synthetic fluids have found widespread use in multiple machining, grinding, and forming operations. They are the products of choice where clean fluids with long tank life and modest lubrication is needed.

VIII. BARRIER FILM LUBRICANTS

A. Drawing, Stamping, and Forming Compounds

In stamping, drawing, cold forming, and extrusions, barrier film-type lubricants are used as the metalworking compound.

The emulsion products used in cutting operations are often formulated differently for drawing operations. The emulsifiers used have a lower HLB value (are more oil soluble), enabling them to emulsify high levels of lubricity additives like chlorinated paraffin. In addition, a thickened emulsion can be formed with amides and esters to give the fluid a higher viscosity enabling it to cling to the metal part during the drawing operation. Blown vegetable oils and lard oils are often used as boundary lubricants because these high-viscosity oils chemically adhere to the metal surface, providing optimal boundary lubrication. Methyl lardate is added to ensure total coverage of the metal prior to the draw. Biocides are not typically used in once-through stamping and drawing applications because bacterial colonies do not grow out of control without recirculation of the fluid.

Honey oils are used in very difficult high-stress draws of heavy gauge metals. These are essentially chlorinated paraffin with surfactant added in order to aid in the subsequent cleaning of parts.

Vanishing oils are an evaporative-type lubricant used to stamp or draw where parts will not be washed. These are typically mineral spirits with a flash point of approximately 140°F with lubricant additives including lard oil, methyl lardate, chlorinated paraffin, or chlorinated solvents. After the draw the mineral spirits evaporate leaving a dry "invisible" residue.

Before chlorinated paraffins became widely used, pigmented pastes were popular drawing lubricants. They are still used in difficult operations or where the use of chlorinated lubricants is not preferred. These are calcium carbonate/fatty acid/oil-based pastes. They may also contain mica or graphite for added lubricity. They are difficult to clean and may contain a surfactant to aid in its removal [4].

B. Wire Draw Lubricants

Solid calcium stearate and other metal stearate soaps are used in wire drawing and cold heading or forming operations. Hydrated lime is mixed with tallow, hydrogenated tallow, fatty acids, or stearic acid to form flake soaps. Borax, elemental sulfur, MoS_2, and talc are added to supplement the lubricity properties.

Dispersions of MoS_2 and graphite in mineral oil are used in cold- and warm-forming operations. After zinc phosphating a metal part, sodium stearate is applied, thereby forming a zinc stearate film on the blanks. MoS_2 will then adhere to the stearate film providing an excellent solid-film lubricant up to 750°F.

C. Prelubes

Prelubes are rust preventatives applied to coil steel that also contain a drawing lubricant package so parts can be formed without cleaning and applying the drawing compound. Polymers and dry-film lubricant packages are used without any extreme-pressure additives that would be released causing staining under the extreme weight of a coil of steel.

With thickened emulsions, solid lubricant dispersions, and pastes, product stability and dispersion properties are important, as are ease of cleaning and high levels of lubricity.

Typical wire draw lubricant formulation	% by wt.
Aluminum stearate	10
Calcium stearate	20
Hydrated lime	66
MoS_2	4
	100

IX. WASTE MINIMIZATION [6]

The waste disposal of metalworking fluids is an issue affecting the choice of metalworking fluid additives. There are three criteria that can be used in assessing the waste minimization parameters. They are waste treatability, hard-water stability, and biostability.

The rising cost of waste disposal and environmental concerns drive the need for waste-treatable additives. Additives that are stable to bacteriological degradation and hard-water salts will promote the longer tank life of a fluid, thereby requiring less frequent disposal.

A. Waste Treatability

In general, anionic additives—those with a negative charge—are the easiest to waste treat because acidification or reaction with cationic coagulants makes removal chemically possible. Nonionic additives—additives with no charge—are difficult to treat because chemical treatment methods are ineffective.

The relative water solubility of the additive also affects its relative waste treatability. The more oil soluble an additive, the more likely it will be removed from the waste stream. For example, a soluble oil that contains oil, an emulsifier base, and a chlorinated paraffin will be easy to treat as long as the emulsifier is anionic. The oil and chlorinated paraffin, having no water solubility, will be removed with the partly water-soluble emulsifier. This phenomena explains why soluble oils are easier to treat than semisynthetic fluids, which are easier to treat than synthetic fluids.

B. Hard-Water Stability

Many additives will react with the calcium and magnesium salts in the water used to dilute a fluid. These calcium complexes are not usually soluble in water, so they

TABLE 1 Waste Treatability of Additives

	Easy	Moderate	Difficult
Emulsifiers	Sulfonates Soaps Sorbitan esters Glyceryl monooleate Alkanolamides Octylphenolethoxylate (HLB 10.4)	Sulfonate base	Nonylphenolethoxylate (HLB 13.4)
Corrosion inhibitors	Calcium sulfonates Amine borates	Triethanolamine	Amine dicarboxylate
Lubricants	Amphoteric	Sulfated oils Phosphate esters	Polyalkylene glycols PEG 600 esters Block polymers Imidazolines

TABLE 2 Hard-Water Stability

	Clear	Stable haze	Precipitate or scum
Emulsifiers	Nonylphenolethoxylate (HLB 13.4)	Octylphenolethoxylate (HLB 10.4) Alkanolamides	Sulfonate Soaps Sulfonate base Soaps
Corrosion inhibitors	Amine dicarboxylates Amines Amine borates		
Lubricants	Block polymers Polyalkylene glycols Phosphate esters PEG 600 esters	Sulfated castor oil Amphoteric salt	

separate from the fluid, thus destabilizing and diluting the effectiveness of the fluid. It can be seen in soluble oils as an "oiling out" or creaming of the emulsion. It shows up as a scum or froth in synthetic fluids. By formulating metalworking fluids with additives that are not destabilized by these salts, tank life can be extended, thereby lessening the frequency of fluids disposal. The additives that are easiest to waste treat are usually the most sensitive to hard-water salts.

C. Biostability

The third criterion for determining which additives contribute to waste minimization is the bioresistance of the additives. This is the ability of an additive to slow the growth of microorganisms in the fluid. The additive essentially does not act as a food source for bacteria or mold or it may interfere with other food sources.

A study was completed that evaluated the biostability of key water-soluble metalworking fluid additives. The test used recirculating aquariums of each additive that were periodically innoculated with bacteria, yeast, and molds from typical fluids. Microbial growth was monitored to determine which additives were biosupportive, biostable (neither supportive nor resistant), or bioresistent, see Table 3.

Bioresistant chemical additives are those that contain boron, are cyclic or saturated, and are branched chained fatty acids or amine-based compounds. These include amine borates, rosin fatty acids, ethoxylated phenols, neodecanoic acid, and monoethanolamine.

Biosupportive or biodegradable chemical additives are typically fatty acids, natural fats and oils, anionics, straight-chained additives, or phosphorus-containing additives. These include soaps, amine carboxylates, sulfonate bases, lard oil, and phosphate esters.

TABLE 3 Biostability of Metalworking Additives

	Bioresistant	Biostable	Biosupportive
Emulsifiers		2:1 Tall oil amide Natural sodium sulfonate Nonylphenolethoxylate (HLB 13.4)	Alkali fatty acid soap Octylphenolethoxylate (HLB 10.4) 2:1 Fatty amide Sulfonate base
Corrosion inhibitors	Amino methyl propanol Amine borate	Triethanolamine	Amine dicarboxylate
Lubricants		600 PEG ester Polyalkylene glycol Block polymers	Sulfated castor oil Amphoteric

Bioresistant additives are difficult to waste treat, and conversely, additives that are biosupportive or biodegradable are relatively easy to waste treat. These mutually exclusive parameters make it difficult to have the best of both worlds. By combining the waste treatability, hard-water stability, and bioresistance of metalworking fluid additives, a matrix is formed (Fig. 8) that directs a formulating chemist to the best choices for a system.

For overall waste minimization the following semisynthetic bioresistant fluid formulation guide applies:

Corrosion inhibitors	Amino methyl propanol Monoethanolamine borate ester
Coemulsifiers	2:1 DEA rosin fatty acid amide Sodium sulfonate
Coupler	Branched diacid
Oil	Napthenic oil
Microbiological aids	Biocide/fungicide
Diluent	Water

X. CONCLUSION

The needs of the consumer, e.g., lubricity, tank life, or water disposability, are paramount in fluids development. For this reason, there are many variations of fluid types within any metalworking fluid compounder's product line. Custom formulations are the nature of the metalworking fluids industry.

Regulatory reporting requirements have opened the doors to metalworking fluid formulations. Once proprietary blends are now identified on safety data sheets and drum labels. New instrumental methods of chemical analysis have unveiled

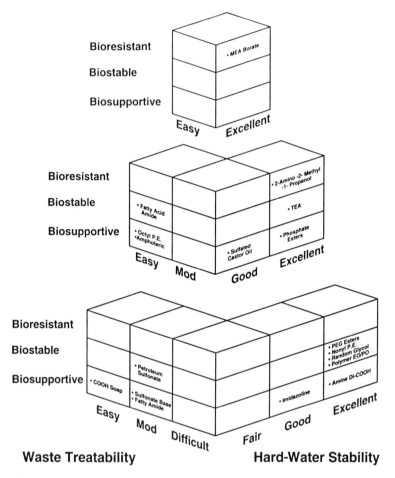

Figure 8 Waste minimization in three dimensions.

what was once closely held, confidential technology. This has placed even more emphasis on the right choice of fluid for an application.

Having developed some formulations designed for a specific task, the next chapter describes laboratory test methods for evaluating the performance and acceptability of the fluid.

REFERENCES

1. J. D. Silliman, *Cutting and Grinding Fluids, Selection and Application*, 2nd ed., Society of Manufacturing Engineers, Dearborn, MI, pp. 35–47 (1992).

2. G. Foltz, Definitions of metalworking fluids, in *Waste Minimization and Water Treatment of Metalworking Fluids*, Independent Lubricant Manufacturers Association, Alexandria, VA, pp. 2–4 (1990).
3. T. J. Drozda, *Cutting Fluids and Industrial Lubricants*, Society of Manufacturing Engineers, Dearborn, MI, pp. 5–7 (1988).
4. W. J. Olds, *Lubricants, Cutting Fluids, and Coolants*, Cahners, Boston, MA, pp. 100–104 (1993).
5. D. R. Karsa, *Industrial Applications of Surfactants*, The Royal Society of Chemistry, London, pp. 61, 64, 65, 183 (1987).
6. J. C. Childers, S.-J. Huang, and Michael Romba, *Lubr. Eng. 46*(6):349–358.
7. K. Niedenzu and J. W. Dawson, *Boron-Nitrogen Compounds*, Academic, (1965).

7
Laboratory Evaluation of Metalworking Fluids

JERRY P. BYERS
Cincinnati Milacron
Cincinnati, Ohio

I. INTRODUCTION

According to a 1989 survey by *American Machinist*, there are 1,870,753 metal-cutting machines and 456,028 metalforming machines in the United States [1]. Some of these will use no metalworking fluid and some will use straight oil, but the majority will use water-based metalworking fluids. Although metalworking fluids significantly affect both the part quality and the productivity of the plant, they account for less than 0.5% of the manufacturing cost of the end product.

Manufacturers will spend hundreds of thousands of dollars on the machines, tens of thousands of dollars on the skilled operator, hundreds or thousands of dollars on the cutting tool or grinding wheel, and only pennies per mix gallon on the metalworking fluid. Yet, if the fluid is not correctly matched to the operation, the result will be scrapped or poor-quality parts and the entire investment will have been wasted. It is precisely because the user receives value over and above the cost of the metalworking fluid that many elaborate laboratory procedures have been developed to aid in the selection of the right fluid for the right application.

A laboratory test method must be meaningful. It must simulate the most important conditions of the metalworking operation. The results must be measurable and must be compared against a standard or a reference fluid. Variables that may affect the test results must be controlled. Unrealistic conditions contrived to accelerate the test may lead to false conclusions and should therefore be avoided.

A given fluid parameter can be measured using several valid methods. Selection will depend on which method best simulates the conditions to which the fluid will be exposed. This chapter provides an overview of the methods available and the meaning of the results. Complete, step-by-step procedures for many of these tests are provided in the references.

II. CHEMICAL AND PHYSICAL PROPERTIES OF THE NEAT FLUID

Tests that are generally conducted on the neat, undiluted fluid-as-sold will be considered in this section. The first two properties, specific gravity and viscosity, give the most fundamental information about any lubricating liquid.

A. Specific Gravity

The specific gravity of a material is the mass of a given volume divided by the mass of an equal volume of some reference material, usually water, at a standard temperature. Specific gravity is also called *relative density*. Density, a closely related value, is the mass of a material divided by its volume. Since the density of water is very close to 1.0 g/ml at normal temperatures, the specific gravity of a fluid is nearly identical to its density. These related properties can easily be determined with an electronic, digital read-out density meter. An equally valid approach is to read the level at which a calibrated, glass hydrometer floats in a cylinder filled with the liquid at a specified temperature. See ASTM method D1298 [2].

B. Viscosity

The single most important property of a lubricant designed to be used neat is viscosity. (This is of much less importance if the lubricant is to be diluted, for example, to 5% in water prior to use.) The viscosity of a fluid is its internal resistance to flow.

Kinematic viscosity is determined by using one of several types of glass viscometer tubes. The time, in seconds, is measured for a fixed volume of fluid to flow, under gravity, through a calibrated capillary tube. This procedure is described in ASTM methods D445 and D446 [2]. During this flow down the capillary, the fluid is under a head pressure that is proportional to its density. Thus, kinematic viscosity is a function of both internal friction and density. The units of viscosity are often expressed as centistokes (cSt), defined as millimeters squared per second.

An alternate unit in common use is Saybolt universal seconds (SUS). ASTM method D2161 [2] provides equations and conversion tables for converting from cSt to SUS.

A Zahn cup is a device sometimes used to determine the kinematic viscosity of opaque fluids that would tend to coat a glass viscometer tube, making the endpoints difficult to detect. The Zahn cup is a small metal cup having a rounded bottom with a hole in the center. The cup is filled with fluid and the time required for the fluid to flow out of the cup in an unbroken stream is measured. The result is expressed in Zahn seconds. Thick-stamping and drawing fluid mixes are often evaluated using this technique. ASTM method D4212 describes the procedure.

The viscosity of a liquid varies with temperature, increasing as temperature decreases. Kinematic viscosities for a liquid at any two temperatures can be used to predict the viscosity at another temperature. This can be done graphically according to ASTM method D341.

The relationship between temperature and viscosity can be expressed by a viscosity index or VI. Using ASTM method D2270, the viscosity index can be calculated from kinematic viscosities at 40°C and 100°C. A high VI indicates a low rate of change in viscosity with temperature, whereas a low VI indicates a high rate of change.

Dynamic viscosity is a function of the internal friction of a fluid and is not related to density. It is reported either as centipoise (cP) or as pascal seconds (Pa·s).

1 cP = 1 mPa·s

One method of determing dynamic viscosity is ASTM method D2983, using a Brookfield viscometer. This instrument measures the resistance of a fluid to the rotation of various shaped spindles at various rotation speeds. The viscosity of many lubricants will vary with the speed of rotation or the *shear rate*. True "newtonian" fluids, however, have a constant viscosity regardless of shear rate.

Kinematic viscosities in centistokes may be converted to dynamic viscosities in centipoise or mPa·s by multiplying by the density (g/cm^3) of the fluid, where both are determined at the same temperature.

C. Flash and Fire Points

The flammability of an oil is extremely important when considering employee safety, manufacturing plant insurability, and transportation requirements. Several different methods exist which will allow the comparison of products, under constant conditions, for their tendency to ignite.

An open cup method such as ASTM method D92 (Cleveland open cup) or D1310 (Tag open cup) can be used to determine both a flash point and a fire point. An open cup of oil is slowly heated at a controlled rate, while a small flame is passed over the cup at prescribed intervals. The flash point is the temperature at which a brief ignition of the vapors is first detected. The fire point is at some slightly higher temperature at which a sustained flame burns for at least five seconds.

Closed cup methods such as ASTM D56 (Tag closed cup tester) and D93 (Pensky–Martens closed cup tester) are run in a similar manner with the cup being opened periodically for introduction of the ignition source. The Pensky–Martens tester incorporates stirring of the sample. Only a flash point is determined with closed cup methods.

D. Neutralization Number

The acid or alkali (base) content of a lubricant can be determined by a simple titration procedure. Acids and bases may be present in the lubricant as supplied, or may develop during use through degradation of product components. Depending upon solubility characteristics, the sample will be diluted in either an organic solvent mixture, or an aqueous solution. A colored indicator is then added in order to detect the neutralization endpoint. A simple acid such as hydrochloric (HCl) or a base such as potassium hydroxide (KOH) is slowly added until the indicator changes color. In the case of an acid number, the results are expressed as the number of milligrams of KOH required to neutralize a gram of sample. In the case of a base number, the titration is done with an acid but the results are expressed as if the base contained in the sample were KOH (again, milligrams of KOH per gram of sample).

E. Lubricant Content

It may be necessary to determine the content of materials that enhance the lubricity of oil, such as fats, chlorine, phosphorous, and sulfur. As a measure of the fat content, ASTM method D94 is used to determine a saponification number. In this procedure any fat present is converted to a soap by heating the sample with a known amount of alkali (KOH). The excess alkali that is not consumed in the conversion process is then measured by titration with an acid. The saponification number is expressed as the number of milligrams of KOH consumed by one gram of the sample.

Chlorine, sulfur, and phosphorous compounds are extreme-pressure lubricants. These can be measured by many wet chemical methods or by instrumental techniques, such as x-ray fluorescence spectroscopy.

III. STABILITY DETERMINATIONS

Oil-in-water emulsions account for a majority of all cutting and grinding fluids on the market, as well as a significant amount of the metalforming fluids. These products are either emulsions as sold in the case of semisynthetics, or they become emulsions prior to use in the case of soluble oils. A dispersion of oil droplets in water is accomplished through the use of surfactants and emulsifiers that rely upon electrostatic or steric repulsive barriers in order to maintain stability. Some synthetic products, containing no oil, are actually microfine emulsions of sparingly soluble synthetic organic surfactants and lubricants. The long-term storage stability of these dispersions is critical, and must be carefully evaluated. Consideration must be given to the product as sold, as well as to the stability of the end-use dilution.

A. NEAT PRODUCT STABILITY

Perhaps the best way to determine whether a product will be stable for one year is to set a sample on a shelf and watch it for one year. Since few formulators can afford that luxury, some means of accelerating the aging process needs to be devised. Heating a sample of the product is a common technique.

> Heating will accelerate most chemical reactions. Of particular concern is the potential hydrolysis of certain emulsifiers under aqueous, alkaline conditions. If this is likely to happen over time, it will happen much faster at elevated temperatures.
> Heating lowers viscosity, increasing the possibility that emulsion droplets will collide and coalesce during Brownian motion.
> Heating to a reasonable temperature will determine whether the "cloud point" of various surfactants used in the formulation is likely to be exceeded during typical storage and transportation conditions.

The temperature selected should not be unreasonably high. Martin Rieger states that "if massive separation of [an emulsion] occurs quickly at temperatures below about 45°–50°C (113°–122°F), the emulsion is clearly unstable. It should be reformulated . . . Similar breakdown at 75°–85°C (167°–185°F) is probably irrelevant . . ." [3]. Dr. T. J. Lin determines emulsion stability by observing the degree of creaming or phase separation after storing samples at room temperature and at 45°C (113°F) [4]. ASTM method D3707 recommends that a 100 ml sample of the emulsion be placed at 185°F (85°C) for 48–96 h, but this high temperature is probably too severe. Some laboratories report that products that are unstable at or above 72°C (160°F) will still have excellent long-term storage stability, but stability at 55°C (130°F) for three to five days is absolutely essential.

Cold-temperature stability should also be considered. Exposure to cold temperatures during winter shipment and storage is unavoidable. Refrigerator temperatures of 5°C (40°F) are not unreasonable. Stability under freeze–thaw conditions is beneficial, but few emulsions will withstand such treatment. ASTM method D3209 specifies three 16 h exposures to 20°F (–7°C) temperatures, which is reasonable. ASTM method D3709 is more severe with nine freeze–thaw cycles over a two week period between 0°F (–18°C) and room temperature.

Antifoaming agents can also affect neat product stability. Antifoams and defoamers function because of their sparing solubility—if they are too soluble, they do not defoam! If a product is formulated with an antifoaming agent, that will be the first material to separate out. Such separation does not necessarily mean that the emulsion itself is unstable, and may have very little effect upon product performance.

B. Dilution Stability

If the metalworking fluid is designed to be further diluted with water prior to use, then this mixture needs to be evaluated for stability. Dilution stability will depend upon both the quality of the metalworking fluid concentrate and the quality of the water used for dilution. Levels of dissolved calcium and magnesium salts are referred to as *hardness*, usually expressed as ppm of calcium carbonate ($CaCO_3$). Total dissolved solids, including sodium chloride and sodium sulfate, will also have an effect. In addition to the initial water quality, consideration must be given to the unavoidable buildup of salts as the fluid is used and water evaporates. It is always best, therefore, to test a product in several different waters, which may be synthetically prepared in the laboratory. Typically, a product will be expected to perform under a variety of water conditions, from soft (75 ppm $CaCO_3$ or less) to very hard waters (400 to 600 ppm $CaCO_3$). Water hardness may also be expressed in other units:

$$
\begin{aligned}
1 \text{ grain hardness} &= 17.1 \text{ ppm } CaCO_3 \\
1 \text{ Clark degree hardness} &= 14.3 \text{ ppm } CaCO_3 \\
1 \text{ German degree hardness} &= 17.9 \text{ ppm } CaCO_3
\end{aligned}
$$

CNOMO [5] method 655202 describes a method for determining the ease with which a metalworking fluid concentrate can be dispersed in water, as well as determining the stability of that dilution. Using a 100 ml graduated cylinder, 5 ml of concentrate is added to 95 ml of water (200 ppm $CaCO_3$). The cylinder is stoppered, inverted 180°, and then returned to upright. The number of inversions are counted until the concentrated material is completely dispersed. Five inversions or less is considered very good, while 40 or more is bad. The cylinder is then allowed to stand for 24 h, and the amount of floating oil or "cream" layer is measured.

DIN [6] method 51367 is used to determine the percent emulsion stability by measuring the relative change in the oil content of the lower portion of a container of emulsion before and after a 24 h static stand. A liter of mix is prepared in a separatory funnel using 20° German hardness (GH) water (358 ppm $CaCO_3$). A special, narrow neck bottle is used to perform an oil break by acidifying a sample of this freshly prepared mix. The remaining mixture is allowed to stand undisturbed for 24 h. The bottom 100 ml is then drained off and a second oil break is conducted. Percent emulsion stability is defined as follows:

$$\% \text{ Stability} = \frac{24 \text{ h oil break results}}{\text{initial oil break results}} \times 100$$

ASTM method D1479 details a similar procedure, but calculates percent oil depletion.

Thus far, only 24 h dilution stability has been considered. M. Smith and J. Lieser have described a much longer stability test using a five gallon aquarium equipped with an over-the-side filter and circulating pump [7]. This test is usually conducted for at least 30 days.

Another means of quantifying emulsion stability is to measure the oil droplet size. This can be done by taking measurements from a Polaroid photomicrograph [4] or by various instrumental methods. The Coulter Counter measures electrical conductivity changes as oil droplets suspended in a salt solution pass through a small hole. Laser light scattering is another common technique. Particle size determinations are a useful measure of emulsion stability if the assumption is made that smaller oil droplets result in more stable emulsions. Dr. T. J. Lin questions that assumption, however [4].

IV. FOAM TESTS

Foaming in a metalworking fluid can lead to higher operating costs due to fluid loss, shorten the life of pumps due to cavitation, and reduce both cooling and lubrication at the chip–tool interface. The metalworking fluid formulator, however, is generally forced to use the same surfactants used by the household products industry which equates foaming with cleaning action. It is important, therefore, to evaluate the foam control of the metalworking fluid being considered.

A. Factors to Consider

A number of factors influence the amount of foam generated in a cutting or grinding operation. These include:

Quality or hardness of the water
Fluid composition as sold
Buildup or depletion of fluid components with age
Type and speed of metalworking operation
Filtration system design
Fluid return trench design
Fluid pressures and flow rate
Fluid temperature
Contaminants such as leak oils, floor cleaners, etc.

With so many factors to be considered, it is obvious that no single foam test will predict the performance of every fluid in every application. Choose a foam test that seems to give the best correlation with past experience, and use the same water for the laboratory test that will be used in the manufacturing operation. With all foam tests, it is best not to run the test on freshly prepared dilutions. Some amount of aging is necessary for the mixture to equilibrate and for reaction of anionic surfactants with water hardness to take place. Aging for at least one hour is recommended, although W. Niezabitowski and E. Nachtman have stated that "the true foaming characteristics of fluids become apparent after one week of standing" [8]. Twenty-four hours of aging is more convenient and is probably sufficient in most cases. Gentle agitation using either a shaker or aeration will accelerate the aging process.

B. Bottle Test

Perhaps the simplest and most commonly conducted test procedure is a bottle test, similar to ASTM method D3601. The bottle should be no more than half full of fluid, and shaking should be at some specified, reproducible rate. The initial foam height is noted immediately after shaking stops, and the time is recorded for the foam to collapse to a predetermined level. The initial foam height and either the collapse time or the residual foam height after a specified waiting period should be used to compare foaming tendencies of various fluids.

C. Blender Test

The blender test is a very severe, although not unrealistic method which simulates the agitation a fluid will receive as it is whirled around by a grinding wheel, cutting tool, or pump impeller. ASTM method D3519 details the procedure. Two hundred milliliters of aged metalworking fluid dilution is placed in the jar of a kitchen blender. The mix is agitated at approximately 8000 rpm for 30 s, and the foam height is measured immediately after the blender is shut off. The time is recorded for the foam to collapse to 10 mm in height. If more than 10 mm of foam remains after 5 min, the residual foam height is then recorded.

D. Aeration Test

ASTM method D892 describes a test method in which air is blown into the fluid to generate foam. The apparatus consists of a 1000 ml graduated cylinder and an air diffuser stone. A fluid volume of roughly 200 ml is aerated at 94 ml of air per minute for 5 minutes. The foam volume is measured immediately after discontinuing aeration and 10 min later. This procedure, designed for lubricating oils, has little relevance to metalworking applications.

E. Circulation Test

The CNOMO test D655212 describes a fluid circulation test using a centrifugal pump and a water-jacketed 2000 ml graduated cylinder with an outlet on the side, near the bottom. A fluid mix prepared in 200 ppm hardness water is added to the cylinder to the 1000 ml level. It is then pumped from the bottom of the cylinder at a rate of 250 l/h and cascaded back upon itself from a height of 390 mm above the 1000 ml mark. The test is run for a maximum of 5 h or until the foam level reaches the 2000 ml mark. The volume of the foam above the 1000 ml mark is recorded immediately after the pump is stopped and 15 min later. This test simulates fluid flow in a machine sump or central system, but is much more severe due to the extremely high turnover rate. Observation of the sides of the graduated cylinder above the fluid level may also be useful as an indication of the cleanliness of the product.

F. Cascade Test

ASTM method D1173 is sometimes referred to as the Ross–Miles foam test, and is widely recognized in the soap and detergent industry. Although the procedure specifies 120°F (49°C) for the fluid temperature, lower temperatures could be used for metalworking fluids. A volume of 200 ml of fluid is allowed to drain at a controlled rate from a glass pipette over a distance of 90 cm into a receiving cylinder containing 50 ml of the same fluid. When all the fluid has drained out of the pipette, the foam height is measured initially, and then again 5 minutes later.

G. Air Entrainment and Misting (or "What Else Can Air Do?")

Air forced into metalworking fluids can be held at the fluid surface as bubbles of foam, or it can create two other phenomena: air entrainment or misting. Air that is held in suspension by the fluid and is slow to rise to the surface is called *air entrainment*. These extremely small, suspended air bubbles can cause a clear mix to become hazy or clouded, and can reduce the machine operator's view of the part being machined or ground. All metalworking fluids entrain some air, but it is more noticeable in very clear synthetics. Products relying on organic corrosion inhibitors tend to entrain more air than those relying on inorganic inhibitors.

Air that is quickly rejected by the metalworking fluid can be propelled above the surface of the fluid causing misting, effervescence, or the "cola effect." This phenomenon is only encountered with very low-foaming synthetics and can be sufficient to cause a foglike cloud to develop near the floor around return trenches or above the central coolant system.

So, there you have it! If air is worked into a metalworking fluid, it will cause either foam, misting, or entrainment (suspension). All three phenomena can be evaluated in the tests described in this section.

V. LUBRICITY

There are a variety of tests for evaluating the lubrication properties of metalworking fluids. Each has its own inherent advantages and limitations. Lubricity tests can be broadly divided into three groups. One group is based upon simple rubbing or rolling action. Another group is based upon metal removal or chipmaking processes. The final group incorporates forming or drawing of a metal sheet. Because of the complexity of field conditions, no single test machine can simulate the lubrication requirements for all in-plant metalworking operations. That is why it is so difficult, or even impossible, to correlate bench test data with actual performance. Therefore, several different lubricity tests should be used to evaluate metalworking fluids. A broad overview of some of these methods is provided below.

A. Rubbing Surfaces

Bench tests that evaluate lubricity in rubbing processes are perhaps the most widely used, and yet of least value with respect to metal cutting and grinding. Evaluation of rubbing action may, however, be of importance in cutting and grinding applications where the workpiece or tool rubs against a support. Examples are blade wear in centerless grinders and tool guides in deep-hole drilling or reaming. Rubbing tests are of greater value for stamping and drawing applications.

1. Pin and V-Block Test

The pin and V-block test is perhaps the most widely recognized of the rubbing tests. Two steel jaws having a V-shaped notch apply pressure to a rotating steel pin immersed in fluid.

Two different tests can be run with this machine. ASTM method D3233 covers a technique for increasing pressure on the jaws until failure, in order to measure the load carrying properties of the fluid. ASTM method D2670 measures the antiwear properties of a fluid as a ratchet mechanism advances in order to maintain a constant load on the pin. The number of teeth advanced by the ratchet during the prescribed testing period is reported as the measure of pin wear.

Table 1 provides data on five metalworking fluids developed using these two ASTM methods. The fluids are arranged with the high oil products at the top of the table, synthetics and water at the bottom. Notice that a very light duty, clear synthetic gave the lowest number of teeth wear and was comparable to the heavy duty soluble oil on failure load. This result is due to the incorporation of a small amount of antiwear additive which allows the light duty product to pass the rubbing test, but will in no way ensure heavy duty cutting or grinding performance. Notice, also, that in the case of the two soluble oils, chlorine and sulfur additives improved the failure load, but did not eliminate the wear.

2. Four-Ball Test

The four-ball tester uses three steel balls held stationary in a cup-shaped cradle while a fourth ball rotates against the others under an applied load. Using this basic

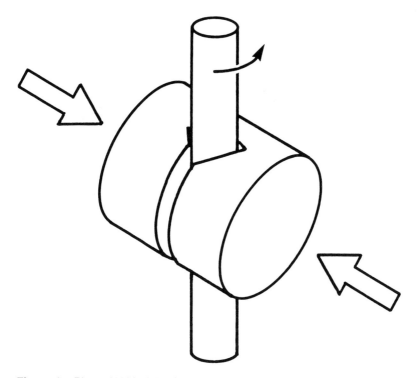

Figure 1 Pin and V-block lubricity test.

TABLE 1 Pin and V-Block Results

Product type	Dilution (%)	Failure load (lb)	(N)	Teeth wear
Heavy duty soluble oil (with chlorine and sulfur)	5	4500+	20,025+[a]	5
Soluble oil	5	2100	9345	28
Moderate duty semisynthetic	3	4500+	20,025+[a]	12
Heavy duty synthetic	5	4400	19,580	100
Light duty synthetic	5	4500+	20,025+[a]	0[a]
Water	100	300	1335	Failure

[a]Indicates the best values, best lubricity.

concept, two different types of tests may be run. One test measures the size of the point contact wear scars on the three stationary balls after a specified time under a constant speed of rotation and load (ASTM D4172). This test is used to determine the relative wear preventive properties of various fluids. The second test measures extreme-pressure capability by using a constant speed of rotation with increasing loads until welding occurs (ASTM D2783). D. Kirkpatrick has used both techniques to compare synthetic, semisynthetic, and soluble oil metalworking fluids [9].

3. Block on Ring

A metal block under an applied load against a rotating steel ring has been used by R. Kelly and J. Byers to compare can-drawing fluids [10], and by A. Molmans and M. Compton to compare cutting and grinding fluids [11]. Several measurements can be made from this test:

1. Frictional force
2. Wear scar measurements on the block
3. Weight loss measurements on the block
4. Failure load at which the lubricant film ruptures

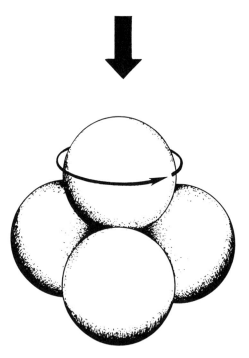

Figure 2 Four-ball lubricity test.

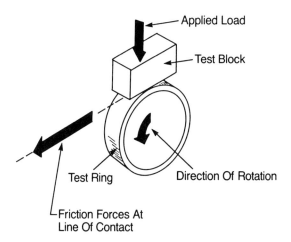

Figure 3 Block on ring lubricity test.

ASTM methods D2714 and D2782 cover these procedures.

4. Soda Pendulum

The friction pendulum or Soda pendulum can be used to measure the coefficient of friction over a wide range of temperatures [12]. The spindle of the pendulum is supported by four balls in a cup containing the test fluid. If no friction were present, the pendulum arm would swing constantly from side to side with no change in the width of swing. Friction, however, makes each swing shorter than the previous one. The coefficient of friction can be calculated from the amplitude of any two subsequent swings. Roehl, Sakkers, and Brand have used this method to compare the lubricity of materials such as isostearic acid and isopropyl myristate [13]. As the graphs in Fig. 5 show, the isostearic acid is the better of the two lubricants.

B. Chip-Generating Tests

The tests described in this section employ machines that actually remove metal and generate nascent metal surfaces which can interact with the lubricants. Some degree of rubbing action also is involved.

1. Lathe Tests

Dr. Charles Yang has described a lathe test using a single point, V-shaped tool that simulates chip-crowding conditions found in heavy duty machining operations [14]. He showed that the vertical cutting force provides a reliable method for predicting tool wear, which is very difficult to measure accurately. Dr. L. DeChiffre has also developed a lathe test in which he measures frictional force, tool wear, and chip–tool contact length [15].

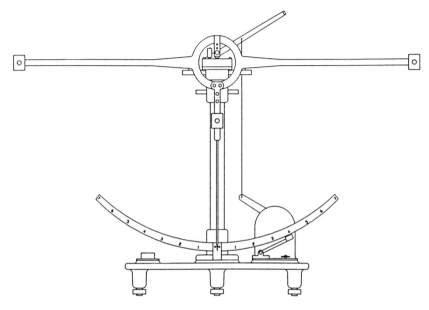

Figure 4 Friction pendulum. (From Ref. 13.)

Using Dr. Yang's method and SAE 1026 steel, cutting force values were determined for the same five metalworking fluids shown in Table 1. Table 2 shows that water performed poorly on the lathe test, followed by the light duty synthetic. A simple soluble oil and a moderate duty, low oil content semisynthetic gave almost identical results. These results show that oil alone is not providing the lubricity. A heavy duty soluble oil with chlorinated and sulfurized additives performed better than the simple soluble oil. Finally, notice that a heavy duty, clear synthetic gave the best results (lowest forces).

2. Grinding Tests

The grinding process also makes chips, but at temperatures and speeds that may be much higher than for a machining operation. A grinding wheel can be considered as a cluster of randomly oriented, negative rake cutting tools [16], which are chemically very different from the tools used in machining. It is, therefore, important to evaluate metalworking fluids for their ability to reduce grinding wheel wear, or increase metal removal rates.

A simple, horizontal spindle surface grinder can be used to evaluate the grinding ratio or G-ratio [11,17]. The G-ratio is obtained by dividing the volume of metal removed by the volume of wheel lost due to wear. High G-ratios indicate

Metalworking Fluids Evaluation

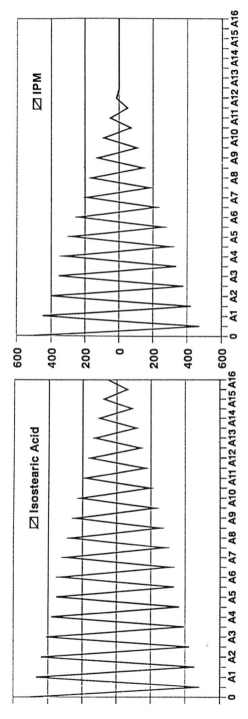

Figure 5 Damping of the friction pendulum for two lubricants at 25°C. Isostearic acid is shown to be a better lubricant than isopropyl myristate (IPM). (From Ref. 13.)

TABLE 2 Lathe Test Results

Product type	Dilution (%)	Cutting forces (lb)	(N)
Heavy duty soluble oil (with chlorine and sulfur)	5	438	1948
Soluble oil	5	464	2065
Moderate duty semisynthetic	3	460	2046
Heavy duty synthetic	5	400[a]	1779[a]
Light duty synthetic	5	480	2135
Water	100	530	2357

[a]Indicates the best values, best lubricity.

low wheel wear and good grinding performance. Surface finish and power consumption may also be measured.

Table 3 shows data from a moderate duty surface-grinding test on SAE 8617 steel using a vitrified bond, aluminum oxide wheel with the same five fluids from Tables 1 and 2. Notice that the heavy duty soluble oil provided the best G-ratio, surpassing both the heavy duty synthetic that looked good on the lathe, and the light duty synthetic that was best on the pin and V-block test. Each condition has a different set of fluid requirements for optimum performance.

3. Drilling Test

Several investigators have used drilling tests to evaluate metalworking fluids. Dr. Herman Leep compared drilling, turning, and milling test methods, and found that testing with high-speed steel drills was "the best method for discriminating between

TABLE 3 Surface-Grinding Results

Product type	Dilution (%)	G-Ratio
Heavy duty soluble oil (with chlorine and sulfur)	5	8.0[a]
Soluble oil	5	5.0
Moderate duty semisynthetic	3	4.0
Heavy duty synthetic	5	5.7
Light duty synthetic	5	2.9
Water	100	2.1

[a]Indicates the best values, best lubricity.

different cutting fluids" [18]. Number of holes drilled, surface roughness, tool wear, torque, and cutting forces have all been used as discriminators by various investigators. W. R. Russell notes that "there are definite performance variables that exist between manufacturing lots (of twist drills), as well as variables that exist in tool performance between tools of the same lot" [19]. Russell's article gives several recommended metallurgical and mechanical considerations in the selection of drills for evaluating coolants.

Chapter 4 provides an extensive summary of drilling research.

4. Tapping Torque Test

Much has been written in recent years about the tapping torque tester [20–22]. Torque values are measured as a tap cuts threads into a predrilled blank which can be made of various metals. The average torque value of five runs is then calculated. Test results are expressed as a percent efficiency, the ratio of the average torque value of a reference fluid to that of the test fluid. The same tap is used on both the reference fluid and the test fluid. L. DeChiffre states that an evaluation of surface finish is also needed [23].

Table 4 lists tapping torque efficiency values for four of the metalworking fluids used in previous comparisons. Two different cutting speeds were used with 1215 steel. At 400 rpm the data shows very little correlation with in-plant experience or with lathe test results. Notice that the heavy duty soluble oil and heavy duty synthetic looked worse than the moderate duty semisynthetic. At 1200 rpm, the light duty synthetic, moderate duty semisynthetic, and the heavy duty soluble oil behave about as expected; but the heavy duty synthetic was a complete failure. This may indicate that the lack of rubbing lubricity seen with this product on the pin and V-block test is an important factor in the tapping test. These data underscore the need for careful selection of the test conditions in order to generate reliable conclusions.

C. Metal Deformation Tests

There seems to be general agreement that no single bench test will give all the information needed to evaluate a metalforming lubricant. C. Wall [24], K. Dohda, and N. Kawai [25], and ASTM standard practice D4173 have all used at least four bench tests to study the various aspects of the metalforming process. Figure 7 illustrates six laboratory test methods commonly used.

Figure 7a is the flat bottom cup or deep draw test. In this procedure a lubricated metal disk or "blank" is forced through a circular die by a bluntnosed punch, forming a cylindrical cup. The maximum drawing force during the test can be used as a measure of lubricity. Another measure is the limiting draw ratio or LDR, defined as the maximum successful blank diameter divided by the diameter

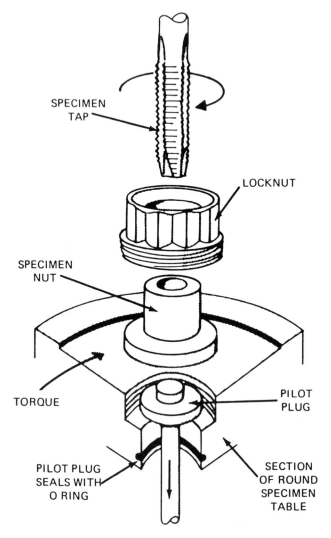

Figure 6 Tapping torque lubricity test. (From Ref. 20.)

of the punch [26]. The cup test combines all aspects of metalforming, including frictional forces and metal deformation forces.

Figure 7b illustrates the dome stretch test. A lubricated metal sheet is stretched over a domed punch with sufficient clamping force to prevent complete cup formation. The maximum drawing force and dome height are measures of lubricity. This test examines stretch forming and the metallurgical aspects of the process.

TABLE 4 Tapping Torque Results

		Percent efficiency	
Product type	Dilution (%)	400 rpm	1200 rpm
Reference fluid (94% naphthenic oil) (6% lard oil)	100	100	100
Heavy duty soluble oil (with chlorine and sulfur)	5	90.6	101.5[a]
Moderate duty semisynthetic	5	103.2[a]	94.6
Heavy duty synthetic	5	100.1	Failure
Light duty synthetic	5	92.3	91.6

[a]Indicates the best values, best lubricity.

Figure 7c is the strip draw test, which uses flat dies and a metal strip to evaluate lubricants under conditions of pure sliding friction. The pulling force is measured at increasing clamping forces. The coefficient of friction is calculated by dividing the average steady state pulling force by twice the clamping force [26].

The ball-on-disk wear test is a modification of the four-ball test described earlier. Figure 7d shows that the three stationary balls have been replaced with a cup holding three disks made of the metal to be evaluated. The cup also contains 5 ml of lubricant. The modified four-ball test can be used for evaluation of drawing lubricants, as well as aqueous rolling fluids [27].

Draw beads are commonly used to control metal flow during stamping, particularly in the automotive industry. They aid in preventing wrinkling and maintaining wall uniformity. The draw bead simulator shown in Fig. 7e evaluates lubricants by pulling a lubricated metal strip through a series of draw beads and grooves (hills and valleys) so that the metal experiences a number of bending and unbending operations. The pulling force is plotted versus the length of travel. All strips from the same lot of metal are tested under the same clamping force [26]. A reference oil is used for a comparison standard.

Figure 7f shows a sheet galling test developed by Bernick, Hilsen, and Wandrei [28]. It is used to evaluate the ability of a lubricant to prevent scuffing and improve die life. The test consists of a flat bottom die and a round top die with a radius of one inch. A normal load is applied hydraulically. By plotting the pulling pressure against time, the static frictional pressure or peak pressure (P_s) and the dynamic frictional pressure (P_d) can be measured. The ratio P_s to P_d can be used to evaluate the ability of a lubricant to prevent galling. A slightly different test for galling is the compression-twist friction test described by Dohda and Kawai [25].

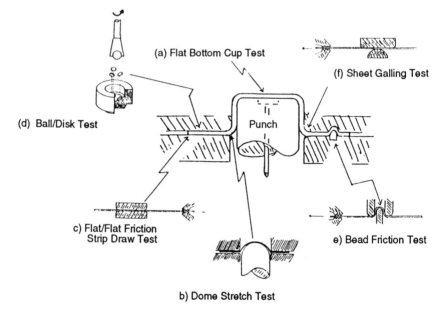

Figure 7 The metalforming process. Separation of deep drawing into areas of interest. (From Ref. 24.)

Each of the basic tests described in this section addresses a different aspect of the total metalforming process. Although the flat bottom cup draw test is, perhaps, the best simulation of a production stamping and drawing operation, no single test can be relied upon as the perfect predictor. It is necessary to select two or three tests that give reproducible results and include the most critical facets of the operation being considered. Only tests using sheet metal stock should be considered as realistic [26].

D. Electrochemical Methods

Metalworking fluids function as lubricants by depositing a thin layer of molecules on metal surfaces that tend to prevent welding of the chip, tool, and workpiece. If the rate or degree of molecular adsorption could be determined, then the effectiveness of a fluid as a lubricant could be predicted. Naerheim and Kendig have used electrochemical impedence measurements as a means of quantifying this chemical adsorption and have shown a relationship between such measurements and metal-cutting forces for three cutting fluids [29]. They anticipate that great time and cost savings could be realized from the use of electrochemical techniques instead of machinability testing.

VI. OIL REJECTION

Leak oil is an unavoidable contaminant to metalworking fluids and may build to significant levels. The actual amount of oil present may never be known if a refractometer or total oil determination is used as the only measure of metalworking fluid concentration. With these methods, all oil present is assumed to have been contributed by the fluid. In some plants, the leak or "tramp" oil level may actually exceed the amount of metalworking fluid concentrate present.

Tramp oil will affect such performance properties as chip settling, foam, misting characteristics, microbial control, wetting action, cleanliness, lubricity, ability to cool, and residue character. Low levels of tramp oil can actually improve some aspects of fluid performance, but high levels will almost always hurt performance.

Metalworking fluids may be formulated either to emulsify leak oil or reject it. If a coolant sump does not have some means of removing oily contaminants, it is best that the fluid be able to emulsify it. Complete oil rejection is probably an unreasonable expectation for products formulated with high oil contents and, hence, a high level of emulsifiers. It is important to note that many hydraulic oils have antiwear agents and detergents which may cause the oil to be somewhat self-emulsifying.

CNOMO test method 655203 offers a procedure for estimating a fluid's trendency to emulsify oil. Ninety milliliters of metalworking fluid dilution plus 10 ml of oil are stirred at 10,000 rpm for 15 s. The mixture is transferred to a graduated cylinder and allowed to stand for 24 h. The volume of floating, unemulsified oil is then read from the markings on the cylinder.

VII. CONCENTRATION CHECKS

Concentration control is extremely important. Every metalworking fluid is designed to perform relatively trouble-free at a definite mix ratio with water. Too weak a mixture can lead to one set of problems (rust, microbial growth, mix instability, lack of cleanliness), while too strong a mixture can lead to another set (foam, skin irritation, high cost, heavy residues). Metalworking fluids are mixtures of ingredients, each performing a definite function. It is wise to check the level of several of these components during use in the plant. Several methods of checking these concentrations are discussed below.

A. Refractometer

Perhaps the most widely recognized concentration test method is the handheld refractometer. A drop of fluid is placed on one side of a glass prism and exposed to a light source. By looking through the eyepiece, one can observe a band of light falling across a number scale. The position of the band of light is determined by the amount of material dissolved or dispersed in the water. The higher the concentration, the higher the band of light will appear on the scale. The value from the refractometer scale is converted to a product concentration value by using a previously prepared graph relating metalworking fluid concentration to refractometer reading. This is a very rapid method, but it has several shortcomings:

1. The refractometer cannot distinguish metalworking fluid components from contaminants. It measures anything that gets into the fluid as if it were the product of interest.
2. As the system ages and contaminants increase, the band of light becomes blurred and its exact position difficult to determine.

Refractometers are very reliable for freshly prepared mixtures, but the accuracy decreases as the fluid is used. Used mixes will tend to give refractometer readings stronger than the actual value. In short, the refractometer is not very accurate, but it is better than no concentration checks at all.

B. Oil Content

If a soluble oil or a semisynthetic product is being used, then oil content is sometimes used as a means of concentration control. A volume of 10 ml of emulsion is added to a Babcock ice cream test bottle graduated on the neck to 20%. It is then

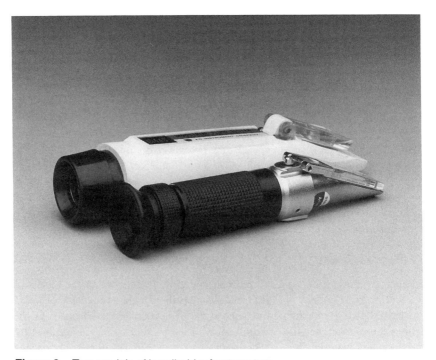

Figure 8 Two models of handheld refractometers.

filled with 30% sulfuric acid and centrifuged for 10 min at 1000 rpm. The volume of floating oil is determined from the graduations. This value is compared to a previously prepared chart or graph showing metalworking fluid dilution versus oil content in order to determine the concentration. The one shortcoming of this method is that leak oil from hydraulics, spindles, and machine ways is indistinguishable from the oil in the metalworking fluid. This causes used mixes to give falsely strong concentration values. Some users try to resolve this concern by allowing the mix to stand quietly or even centrifuge the mix prior to conducting the acid split. This may provide more accurate results, but is not totally effective at eliminating the interference.

C. pH Measurements

Water-based fluids have a chemical property known as pH. Pure water has a pH value of 7.0. Water containing an acid will have a lower pH, while water containing a base has a higher pH. The pH of a solution may be determined by using pH paper and observing a color change, or with an electronic pH meter. Most metalworking fluid mixes are basic or alkaline and have pH values between 8.0 and 9.5. The pH is a very useful number, but it cannot be used as a measure of concentration for metalworking fluids since the pH can remain relatively constant despite wide variations in the product concentration. Generally, aeration of a metalworking fluid during use will cause the pH to drop slightly from its initial value.

D. Alkalinity

The concentration of basic or alkaline components is determined by a free alkalinity titration [7]. A 10 ml sample of metalworking fluid is placed in a beaker or flask with a few drops of methyl orange indicator. Dilute (0.1 N) hydrochloric acid is slowly added with stirring until a color change signals the endpoint. The volume of acid added is related to the concentration of the metalworking fluid. The alkalinity will generally increase as the fluid is used due to accumulation of carbonates from make-up water and the buildup of alkaline materials from the coolant, which are somewhat resistant to depletion. Alkalinity and pH are related, but one cannot be used to determine the other. The concentration of weakly basic components in the metalworking fluid can vary greatly without affecting the pH, but will be readily detected by alkalinity measurements.

E. Emulsifier Content

Many emulsifiers, lubricants, and corrosion inhibitors used in the formulation of metalworking fluids carry a slight negative electrical charge. They are known as anionic surfactants. Examples of these are fatty acids, such as oleic acid, and sodium petroleum sulfonates. Anionic surfactants tend to be depleted over time due to scum formation with water hardness, adsorption onto metal surfaces, and extraction into tramp oil layers. It is, therefore, of critical importance to monitor their presence in the fluid. One method is to determine the amount of cationic or positively charged surfactant required to neutralize all of the negative charges present.

Several such analytical procedures have been described in the literature [30]. Essentially, these procedures call for an exact volume of metalworking fluid containing anionic surfactants to be placed in a clear glass bottle. A colored indicator (bromophenol blue, methylene blue, bromocresol green, etc.) is added to the bottle along with a water-insoluble solvent (carbon tetrachloride or chloroform). A dilute solution of a quaternary ammonium chloride salt (the cation) is then added stepwise. The bottle is capped and shaken between each addition, and then allowed to stand until the solvent layer separates cleanly from the water layer. After all of the anionic charges have been neutralized and a slight excess of cation has been added, a color change is observed in the solvent layer and the titration is stopped. The volume of cationic surfactant solution required to produce this color change is proportional to the amount of anionic surfactant in the sample of metalworking fluid.

These cationic/anionic titration methods are extremely accurate and do not require sophisticated or expensive laboratory equipment. The spent chlorinated solvents, however, are considered hazardous wastes and should be distilled for reuse. Instrumental methods of quantifying the anionic surfactant level include high-pressure liquid chromatography (HPLC) and ion-specific electrodes.

F. Boron Content

Boron compounds are used in some products for corrosion inhibition, and are easily measured using an atomic absorption spectrophotometer (or AA). Borates are very water soluble and are not readily depleted. In fact, Giles Becket reports that in one fluid system the boron level rose to twice the level of other components in a ten week period [30]. It cannot, therefore, be used as the primary concentration method for controlling a metalworking fluid system.

G. Microbicide Level

Microbicides may be present in the metalworking fluid as-formulated or may be added tankside. Since most microbicides tend to be depleted in the process of killing bacteria or mold, it is important to monitor their levels or their effectiveness. Most suppliers can recommend an analytical method for their product. Gregory Russ mentions procedures for triazine bactericides and for phenolic fungicides [31]. E. C. Hill has developed a dip-stick method that measures the capacity for a fluid to control bacterial growth rather than measuring the level of specific compounds [32]. A pad carrying spores of a gram-positive bacteria, dried nutrients, and a growth indicator is mounted on a plastic strip. This is dipped into the sample of metalworking fluid and then incubated at 37°C overnight. If sufficient microbicide is present, there will be no color change on the pad. If the microbial control is weak, the pad will turn a pink or red color indicating bacterial growth.

H. "What Is the Concentration?"

Clearly, there is no one answer to the question: "What is the concentration of my coolant?" A metalworking fluid may contain between ten and 20 different ingredients, each selected to perform a certain function. Some components will increase in concentration over time (alkaline materials, borates, oil, and nonionic detergents) while others will be depleted (anionic surfactants, extreme-pressure lubricants, and microbicides). Figure 9 attempts to depict this reality.

The only circumstance in which we can confidently speak of a single coolant concentration is when a dilution has been prepared in deionized, sterile water and stored in a capped bottle. All is not hopeless, however! An effective technique for monitoring a metalworking fluid is to select its most important functions and check the level of the components responsible for these functions. Frequently, the critical parameters will be alkalinity (responsible for rust control) and anionic surfactants (responsible for emulsion integrity, corrosion control, and lubricity). The "useful" or "effective" coolant concentration will be somewhere between these two values.

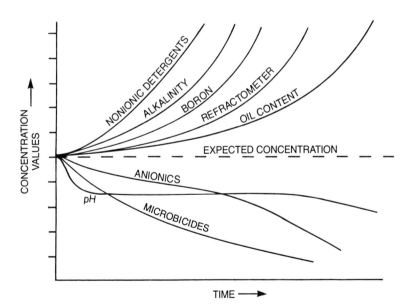

Figure 9   Metalworking fluid composition trends with time.

VIII. STAMPING AND DRAWING FLUID EVALUATIONS

Fluids used in automotive stamping and drawing processes must undergo several unique tests that require a great deal of specialized equipment. Four of the most common tests will be described here, although the exact procedures will vary from one account to another.

A. Phosphate Compatibility

Automotive body parts are given a zinc phosphate treatment before painting. The fluid used to stamp body parts must not interfere with these processes. To check for compatibility, cleaned metal test panels are coated with lubricant using a draw bar. The panels are aged for one week at 50°C and then sent through an eight-stage phosphating process. After drying, the panels are evaluated for uniformity of appearance and size of crystal formation. A small crystal structure is preferred. The phosphate-coating weight, in grams per square meter, is also determined by x-ray fluorescence.

B. Electrocoat

The electrocoat or E-coat process uses phosphated test panels and a gallon of the selected E-coat paint. A cathodic charge is set on the panel, and an anodic charge is set on the paint container. A current of about 1 A and 240 V is applied for two to three minutes in order to obtain a uniform coating thickness of about one mil (0.025 mm). The panel is rinsed and then cured for 20 min at 177°C. At this point, any foreign materials may flash off of the panel causing cratering of the paint. The panels are rated for size and quantity of craters.

Several variations of this test are used. A panel is first coated with lubricant using a draw bar and allowed to dry. The panel may then be placed in the E-coat bath with stirring overnight to determine if any fluid on the panel will be digested by the paint. On the following day, the panel is E-coated and evaluated as before. Alternatively, the E-coat bath may be contaminated with about 0.1% of lubricant and stirred before coating a panel. One final variation on the E-coat test is to prepare a sandwich of two panels with a few drops of fluid between. During the baking phase of the process, the lubricant may spatter out from the joint causing the coating to crater.

C. Adhesive Strength

Various adhesives and sealants are used in the manufacture of automobiles. It is important that any residual stamping lubricant not interfere with these adhesives. One way to address this concern is to conduct an adhesive strength test. Clean, 1 in. × 4 in. coupons (2.5 cm × 10 cm) are coated with lubricant using a draw bar and allowed to dry. Two coupons are then glued together under carefully defined conditions of area and thickness of adhesive. The couple is cured for 20 min at 177°C and allowed to cool. The assembly is then placed in a tensile strength apparatus and pulled apart. Failure load and the amount of extension prior to failure are recorded.

D. Cleanability

Ease of removal from the stamped metal surface is a primary concern for stamping lubricants. Coated test panels are placed in a heated, agitated, alkaline cleaning bath for an appropriate length of time. The panels are then rinsed under a stream of water and observed for uniformity of the water film on the surface of the metal. A uniform film, called a *break-free surface*, is the objective.

Other requirements for stamping and drawing fluids may include ability to weld lubricated steel panels, a staining test run at elevated temperatures (the bake/stain test), and various corrosion tests, described in Chapter 8.

IX. MISCELLANEOUS

A complete discussion of every metalworking fluid bench test would be nearly endless. It is not the intent of this chapter to cover every one, but there are a few more general topics that should be briefly mentioned. Certain other properties are so important and complex that whole chapters have been devoted to them. Corrosion testing will be detailed in Chapter 8, and microbial resistance will be considered in Chapter 9.

A. Waste Treatment

Waste treatment is, usually, the first test required for fluid approval in a plant. Fluids that do not pass this test will not be used. Fluids can certainly be developed to be compatible with a particular plant's waste treatment process, but the term *waste treatable* means different things to different people. Some plants use an acid and alum process in which the pH is lowered with acid and alum is added. The pH is then brought back to neutral with caustic and the floc is separated. Other plants do not use acid, but rely instead upon polyelectrolytes to break the emulsion. A few plants use biological treatment systems with specially acclimated bacteria to digest the chemicals in the water. Some plants use ultrafiltration to separate organics from water, while others evaporate the water. It would be quite difficult to develop a metalworking fluid that could be treated successfully by all of these methods. It is, therefore, important to understand the end user's waste treatment process and the factors that make one product more easily treated than another. Chapters 6 and 13 will be helpful in this regard.

B. Residue Characteristics

Fluids will splash and collect in areas where the water can evaporate, leaving behind product components and any contaminants. All metalworking fluids will leave some type of residue. The deposit can be sticky, particularly in hard water, and will pick up metal fines if it cannot be redissolved by the fluid. Studying the kind of residue produced by the fluid in hard and soft water is desirable. This can be done by allowing some of the diluted fluid to evaporate in a petri dish or beaker. Testing to see how easily the residue will redissolve in the mix may be even more important than knowing how solid or sticky it becomes. An ideal residue would be one that is liquid, nonsticky, and quickly redissolves.

C. Chip Settling

The ability of a metalworking fluid to settle chips is an important property necessary for good performance. The fluid carries the chips away from the work area to a sump or central system tank where they should then settle from the fluid and be removed. If a fluid suspends metal fines, chip recirculation may result, leading to scratched surface finishes. It is also possible for fines to settle too quickly, plating out on machines and filling up return trenches. Chip settling characteristics may be studied by adding a weighed amount of metal fines to the bottle foam test described earlier in this chapter. After shaking, the time required for the bulk of the fines to separate from the fluid should be recorded. Whether the chips settle or rise to the top of the fluid should also be noted.

D. Product Effect on Nonmetals

Plastic and rubber machine components and seals are frequently bathed in metalworking fluid. Major ingredients in the fluid, such as the oil, water, and alkaline materials, are known to affect the integrity of these nonmetal components [33]. ASTM method D471 can be used to evaluate fluid compatibility with such materials. The elastomer is immersed in the fluid for up to 30 days at room temperature or higher. At the end of that time, changes in specimen appearance, weight, volume, hardness, tensile strength, and elongation are recorded. While many waterbased fluids will have similar effects, great differences have been noted in the resistance of various types of elastomers [33].

Another nonmetal to be considered is the machine paint. Improperly prepared surfaces prior to painting and poor-quality paint are the major causes of problems. The compatibility of the metalworking fluid and the machine paint may be checked using steel panels which have been properly prepared and painted. The panels may either be soaked in a dilution of the product being considered, or a few drops of the concentrated product may be placed on the painted surface. After several days of exposure, the surface is examined for signs of discoloration, softening, or bubbling of the paint.

E. Surface Tension

Surface tension is a measure of the inward pull of a liquid that tends to restrain the liquid from flowing or wetting a surface. It is related to such performance properties as cleaning action, lubrication, and foam. Two techniques are used most frequently for this determination:

The du Nouy ring tensiometer, ASTM method D1331; and
The dynamic or "bubble" tensiometer, ASTM method D3825.

The du Nouy tensiometer is a torsion arm balance with a platinum wire ring in a horizontal position hanging from the end of the arm. The liquid to be tested is poured into a shallow cup and placed on an adjustable platform below the ring. The ring is submerged just below the surface of the liquid. By simultaneously adjusting the height of the platform and the torsion on the arm, a measurement is made of the force required to pull the ring away from the surface. Using this procedure, pure water has a surface tension of about 73 dyn per centimeter at 20°C. Addition of surface active agents such as emulsifiers, soaps, and detergents will cause this value to decrease. The surface tension of a water-based metalworking fluid will depend upon the type and concentration of surface active agents present.

Another method of determining surface tension is to measure the pressure required for bubble formation as a gas flows through a capillary tip immersed in a liquid. Such dynamic measurements can be important whenever the surface area of a liquid is changing rapidly, as when a metalworking fluid is pumped out of a relatively quiet reservoir and sprayed into the metal-cutting zone. This technique is also unaffected by the presence of foam on the liquid surface.

X. CONCLUDING REMARKS

This chapter provides a broad overview of the many evaluation methods applied to metalworking fluids. In the interest of space and reader's time, no attempt was made to give complete, step-by-step instructions. Instead, references have been listed for those desiring further detail. None of these procedures should be considered to be the final word on testing methods. The reader should feel free to modify the procedures to meet his or her own needs. Performance in the manufacturing environment is, of course, the ultimate test of a metalworking fluid.

REFERENCES

1. 14th American machinist inventory of metalworking equipment, *Am. Mach. 133*(11): 91–110 (1989).
2. American Society for Testing and Materials (ASTM), 1916 Race Street, Philadelphia, PA 19103.

3. M. Rieger, Stability testing of macroemulsions, *Cos. Toiletries 106*(5): 59–69 (1991).
4. T. J. Lin, Adverse effects of excess surfactants upon emulsification, *Cosm. Toiletries 106*(5): 71–81 (1991).
5. Committee De Normalisation De La Machine Outiels (CNOMO), Service 0927 bat f24, 8-10 Avenue Emile Zola, 92109 Billancourt Cedex, France.
6. Deutsches Institut für Normung (DIN), Burggrafenstrasse 4-10, D-1000 Berlin 30, Germany.
7. M. D. Smith and J. E. Lieser, Laboratory evaluation and control of metalworking fluids, *Lubr. Eng. 29*: 315–319 (1973).
8. W. Niezabitowski and E. Nachtman, Way and gear oil, hydraulic fluids and greases as contaminants in water base metal removal fluids: Corrosion and foam effects, in *Strategies for Automation of Machining: Materials and Processes*, ASM International, pp. 167–170 (1987).
9. D. Kirkpatrick, "Trend to synthetic cutting fluids," *Conference on Lubrication, Friction and Wear in Engineering*, Institution of Engineers, Australia (1980).
10. R. Kelly and J. Byers, Synthetic fluids for high speed can drawing and ironing bodymakers, *Lubr. Eng. 40*(1): 47–52, (1984).
11. A Molmans and M. Compton, "Heavy duty synthetic metalworking fluids are a reality," *Synthetic Lubricants and Operational Fluids*, Fourth International Colloquium at Esslingen, Germany, pp. 40.1–40.5 (1984).
12. F. Thornhill, Other parameters and measurement advantages, in *Monitoring and Maintenance of Aqueous Metalworking Fluids*, K. W. A. Chater and E. C. Hill, eds., Wiley, Chichester (1984).
13. E. L. Roehl, P. J. D. Sakkers, and H. M. Brand, Isostearic acid and isostearic acid derivatives, *Cosm. Toiletries 105*(5): 79–87 (1990).
14. C. Yang, The effects of water hardness on the lubricity of a semi-synthetic cutting fluid, *Lubr. Eng. 35*(3): 133–136 (1979).
15. L. DeChiffre, Laboratory testing of cutting fluid performance, in *Lubrication in Metal Working*, Vol. 2, Third International Colloquium at Esslingen, Germany, pp. 74.1–74.5 (1982).
16. R. K. Springborn, *Cutting and Grinding Fluids: Selection and Application*, American Society of Tool and Manufacturing Engineers, pp. 10–11 (1967).
17. A. K. Mehta, S. P. Dubey, P. D. Srivastav, and F. Waris, "A test technique for the evaluation of grinding fluids," *Proceedings of the Institute of Mechanical Engineers Conference, Tribology—Friction, Lubrication and Wear*, Vol. 1, pp. 517–522 (1987).
18. H. R. Leep, Investigation of synthetic cutting fluids in drilling, turning and milling processes, *Lubr. Eng. 37*(12): 715–721 (1981).
19. W. R. Russell, Cutting tools for cutting fluid evaluation, *Lubr. Eng. 30*(5): 252–254 (1974).
20. W. Faville and R. Voitik, The Falex tapping torque test machine, *Lubr. Eng. 34*(4): 193–197 (1978).
21. T. Webb and E. Holodnik, Statistical evaluation of the Falex tapping torque test, *Lubr. Eng. 36*(9): 513–529 (1980).
22. P. Hernandez and H. Shiraki, Comparison of aqueous extreme pressure cutting fluids on the no. 8 tap torque tester and other cutting methods, *Lubr. Eng. 43*(6): 451–458 (1987).

23. L. DeChiffre, Function of cutting fluids in machining, *Lubr. Eng. 44*(6): 514–518 (1988).
24. C. Wall, The laboratory evaluation of sheet metal forming lubricants, *Lubr. Eng. 40*(3): 139–147 (1984).
25. K. Dohda and N. Kawai, Correlation among tribological indices for metal forming, *Lubr. Eng. 46*(3): 727–734 (1990).
26. ASTM D4173, "Standard practice for evaluating sheet metal forming lubricant."
27. B. L. Riddle, T. E. Kirk, and E. M. Kipp, Reactive additives improve aqueous aluminum foil rolling, *Lubr. Eng. 47*(1): 41–45 (1991).
28. L. D. Bernick, R. R. Hilsen, and C. L. Wandrei, Development of a quantitative sheet galling test, *Wear 48*: 323–346 (1978).
29. Y. Naerheim and M. Kendig, Evaluation of cutting fluid effectiveness in machining using electrochemical techniques, *Wear 114*: 51–57 (1987).
30. G. J. P. Becket, Knowing the true concentration is the key to longer cutting fluid life, in *Lubrication in Metal Working*, Vol. 2, Third International Colloquium at Esslingen, Germany, pp. 105.1–105.12 (1982).
31. G. A. Russ, Coolant control of large central systems, *Lubr. Eng. 36*(1): 21–24 (1980).
32. E. C. Hill, Biocide assays in metalworking fluids as an indication of spoilage potential, in *Industrial Lubricants—Properties, Application, Disposal*, Sixth International Colloquium at Esslingen, Germany, Vol. 2, pp. 21.2-1–21.2-5 (1988).
33. E. Rolfert, The influence of metalworking fluids on common elastomers, *Lubr. Eng. 49*(1): 49–52 (1993).

8
Corrosion: Causes and Cures

GILES J. P. BECKET
Cincinnati Milacron
Cincinnati, Ohio

I. INTRODUCTION

Metalworking fluids can be divided into two basic types: water-free and water-mixed products. Those free from water are mineral oils that contain comparatively small amounts of oil-soluble chemicals to enhance the performance of the product. The second type are water-mixed fluids which may either be solubilized or emulsified in water—the water being approximately 90 to 98% of the total material. Apart from being generally considered a safer material than mineral oil, water does a much more effective job of cooling the tool and the workpiece. Unfortunately, water has a great capacity to corrode the majority of metals. Moreover, it is not just a question of water causing corrosion, bacteria that can come to inhabit water-mixed fluids are capable of causing corrosion through a number of processes. Thus, it is extremely important to have a good understanding of the mechanisms that govern corrosion if it is to be avoided. This chapter also considers corrosion-testing techniques in order to evaluate cutting, grinding, stamping, and drawing fluids. Finally, suggestions are offered to help avoid or correct corrosion problems that can occur.

II. DEFINING CORROSION

Corrosion of a metal is the deterioration of the material because of a reaction with its environment. Looked at another way, corrosion occurs when a metal returns to one of its possible "natural" states e.g., iron oxidizes back to iron ore (oxide) and copper can be corroded by sulfur-containing compounds and returned to its sulfide. Even aluminum corrodes to give a surface layer of oxide that is chemically similar to the bauxite from which it was originally won. Corrosion is a completely natural process, but one that does not suit our modern-day requirement of using metals for structural purposes. In fact, with the exception of only a few metals (notably silver and gold), metals never occur naturally. They are always in the form of compounds, because in this form they are chemically in a lower energy state, which is thermodynamically preferable. In short, if a metal atom can lose one or more electrons from its structure and then go on to combine with other (nonmetallic) elements (e.g., oxygen, sulfur, and chlorine), thereby losing some energy and reaching a more stable (lower energy) state, it will. We view this process as corrosion, which, while generally being a nuisance, is at least made use of in a battery when the freed electrons are channeled to some useful purpose. (A battery may be viewed as "controlled corrosion in a container"). Corrosion is, above all, an electrochemical process, and anything that aids the flow of electrons invariably promotes corrosion. Thus, seawater, which conducts electricity well because of the dissolved minerals in it, is far more corrosive than pure water, which is a relatively poor conductor.

Since corrosion is dependent on the metal(s) involved, the nature of the corroding environment, and physical forces such as temperature, pressure, and friction, it can vary from the mildest surface discoloration to total disintegration of the metal. It is therefore possible to categorize environmental attack not only by whether it is upon ferrous or nonferrous metal, but also by the severity of that attack.

A. Staining

Staining is defined as light corrosion resulting in discoloration or tarnish. This is distinct from more general corrosion in that it is only a surface effect and is unlikely to affect the structural strength of the metal. It is undesirable mainly because it degrades the metal's appearance or because it interferes with electrical contacts in switches and sockets. An interesting aspect of staining is that it does not need a wet environment to occur, which is a common requirement in other types of corrosion. Copper or silver, for example, will discolor even in a dry atmosphere of oxygen, sulfur, or halogen to give the resulting metal oxide, sulfide or halide. The layer formed acts as a solid electrolyte with nonhydrated ions migrating through the lattice. The staining, which is the result of the solid corrosion product, can build to form a coating (or scale) that is thick enough to crack under differential thermal stress, whereupon more intensive corrosion can occur in the fissures. Staining, nevertheless, is even more prevalent in a wet environment, for not only can the agents responsible for dry corrosion still degrade the metal, but many other corrosive process are brought into play.

B. Corrosion

Corrosion is environmental attack on a metallic surface causing changes in metallurgical properties. Whereas staining is a relatively thin layer over the metal surface, corrosion is usually considered to be a more extensive attack.

Aluminum or zinc (amphoteric metals) corrode to a white powdery material, whereas copper gives a typical green product. Low alloy steels tend to show a brown granular oxide layer if the corrosion is brought about by the effects of water and oxygen. Although a corrosion layer does offer some protection against further attack, there is not a significant reduction in the rate of atmospheric corrosion until after about 15 months following the onset. By this time, however, the degree of corrosion, especially with low alloy steels, can be substantial.

Bacterially induced corrosion (anaerobic), which is discussed later, is black and quite different from oxidative corrosion.

C. Rusting

Rusting is the corrosion of ferrous materials, a special case resulting from the importance of ferrous materials. Of all the metals used for structural purposes, iron and steel far outweigh all others. Unfortunately, these ferrous materials are particularly prone to corrosion especially in moist air, which in a polluted environment is also frequently acidic. Iron ore, while being a reasonably concentrated source in nature, requires heating with about four times its own weight of coal or coke to become reduced to iron. We then use the iron in diverse locations until eventually much of it reverts (corrodes or rusts) back to an oxide state similar to that which was originally mined. Unfortunately, the iron oxide is no longer localized, but well distributed throughout the land. Further, we are left with the acidic pollution created during the smelting of the original ore. So rusting is more than just a concern for the loss of outward appearance or structural failure that inevitably results. Its control also serves to reduce atmospheric pollution and retain another diminishing resource.

III. MECHANISM OF CORROSION

A. Ferrous Metals

If we think of chemical reactions at all, it may be that we are used to considering them mainly as some chemical interchange occurring in bulk, for example, a fatty acid being treated with an alkali to form a soap (saponification), or iron ore (hematite) being reduced by heating it with carbon to produce pig iron. Incidentally, the term *reduction* was originally used since there is a distinct reduction in volume observed when going from the ore (oxide) to the metal. When the metal eventually oxidizes it will regain its former natural "rusty" bulk; iron in the metallic form is quite literally unnatural.

In these and all chemical reactions there is a movement of electrons between the reacting species. However, if the reactants are separated in some way, then the reaction can only proceed so long as there is a conducting pathway so the electrons (that form the "currency" of a chemical reaction) can move between them. This then is the situation with corrosion. Electrons and ions pass between a fixed metallic surface and the environment, or between two metallic surfaces through an environment. During this process, the metal is oxidized and some part of the environment undergoes chemical reduction while gaining some energy. The propensity with which any particular metal loses one or more electrons (oxidizes) determines how likely it is to corrode. A naturally occurring passivation layer can markedly retard this corrosion. (This will be discussed later.)

Before an electric current can flow, there has to be a potential difference between two points—more free electrons at one point than the other—and a conducting pathway. In the case of a single metallic surface, this potential difference arises from small changes in local environment. Such a situation occurs when a water drop comes to rest on a ferrous surface and leaves behind a ring of brown oxide. Although pure water is virtually noncorrosive (nonconductive), in practice gasses (O_2, CO_2) and ions (Cl^-, CO_3^{-2}) are likely to be present to increase conductivity and hence corrosivity.

The area in the center of the water droplet is lower in oxygen content than at the rim, thus an ionic differential exists—what is known as a concentration cell. The low oxygen area is termed *anodic* and it is here that the iron loses electrons which flow into the bulk of the metal.

$$Fe \rightarrow Fe^{+2} + 2 \text{ electrons}$$

Nearer the rim of the water droplet (an area termed *cathodic*) the electrons released from the above reaction combine with the water and then reduce some of the more plentiful oxygen atoms to hydroxyl ions (OH^-).

$$1/2 O_2 + H_2O + 2 \text{ electrons} \rightarrow 2OH^-$$

These hydroxyl ions readily combine with the ferrous ions (Fe^{+2}) produced in the anodic oxidation of the iron and iron hydroxide is produced.

$$Fe^{+2} + 2OH^- \rightarrow Fe(OH)_2$$

The iron hydroxide combines with more water and oxygen to form the somewhat more complex oxide we term *rust*, which is generally written as $Fe(OH)_3$ (ferric hydroxide) and can be seen as the brown granular material that rings the pit that had once been the original iron surface.

Corrosion: Causes and Cures

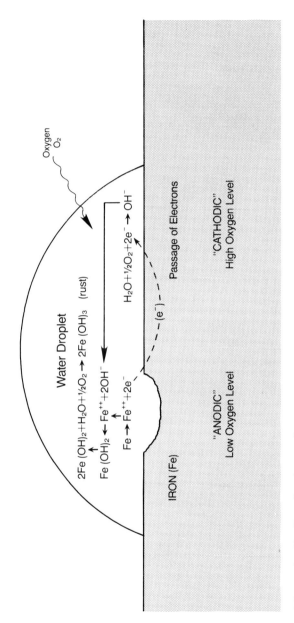

Figure 1 Rusting of an iron surface.

$$2Fe(OH)_2 + H_2O + {}^1\!/_2 O_2 \rightarrow 2Fe(OH)_3 \text{ (rust)}$$

B. Nonferrous Metals

Nonferrous corrosion is the reaction of any metal, other than cast iron or steel, with the environment. Only those metals that are frequently encountered in a metalworking operation will be covered.

1. Aluminum

If a piece of aluminum is cut, it forms an invisible oxide film across the freshly exposed surface almost instantly. This film is only about 0.005 µm thick, but will protect the bulk of the metal from further attack by the atmosphere. However, if the aluminum is in contact with a solution that is continuously able to dissolve away this protective layer, then clearly the attack proceeds and the metal is eventually lost. The advantage aluminum has of being corrosion resistant in air, is somewhat counterbalanced by its being attacked both by aqueous acids and alkalis. In normal use as a structural material aluminum is reasonably unlikely to encounter corrosive solutions. In a metalworking situation, however, the water-based coolant invariably has an alkaline pH, i.e., above pH 7. (Note that the staining of anodized aluminum should not be confused with corrosion. An anodized surface is very prone to the adsorption of any coloring matter, which can give the appearance of staining.)

2. Zinc

Zinc, like aluminum, is also attacked by both acids and alkalis, such metals are *amphoteric*. The best way to prevent corrosion of either of these metals is to keep them dry or to hold them at a pH of between about 8 to 9.

3. Copper

Copper is only slowly corroded by acids and alkalis, though fatty acids present in many cutting fluids can react to form pale green soaps. However, fatty acids are unlikely to cause corrosion or discoloration of copper during the short time they are in contact while machining. Copper is discolored though, by the formation of copper sulfides due to a reaction between certain sulfur-containing compounds that can be present in cutting fluids. The problem is not insuperable however, since there are several extremely effective copper corrosion inhibitors that can be incorporated into a cutting fluid. These generally act by forming a molecular layer of an insoluble organic compound over the entire copper surface.

IV. TYPES OF CORROSION: BOTH FERROUS AND NONFERROUS

A. Uniform Attack

As discussed earlier, a drop of water on a metal surface results in areas that are either anodic (where electrons are given up) or cathodic (where electrons are used to reduce oxygen to hydroxyl ions, OH^-). If the water, or other aqueous fluid is not just a drop, but a complete coating over the metal, then these anodic and cathodic areas are continually changing, resulting in a fairly uniform degree of corrosion. We may consider this to be the most common form of corrosion encountered in everyday life, the slow attrition of metallic (especially ferrous) materials. Since it is a long-term process, it is only of significance in a metalworking environment as far as the machine tool is concerned. The component is in contact with the metalworking fluid for far too short a period. Where it does impinge on the workpiece is when it is stored wet with coolant, and this is invariably a mistake since metalworking fluids (coolants) should never be thought of as long-term rust preventatives.

B. Effects of Electrolytes

Pure water ionizes only slightly to form hydrogen and hydroxyl ions (H^+, OH^-, though it is more correct to say that the H^+ goes on to recombine with a water molecule to form the hydroxonium ion H_3O^+). This degree of ionization is low and as a result pure water corrodes most metals at a very low rate. The addition of either an acid, or an alkali, or a salt, greatly increases the ionic content of the water and promotes corrosion. The obvious example is seawater, which causes much faster corrosion than fresh water.

The electrolyte can play one of several roles. It can increase the electrical conductivity of the corroding fluid, thus bringing about faster dissolution of the metal. It may also react directly with the metal surface to form a soluble compound that is washed into the fluid. However, should the electrolyte react with the metal surface to form an insoluble film that not only prevents dissolution of the metal, but also further attack by the electrolyte itself, then we say the metal has become passivated and the rate of corrosion can be considerably decreased.

C. Differential Aeration

An example of differential aeration is the water drop on an iron plate, which we used earlier in describing the mechanism of corrosion. Here areas low in oxygen are anodic to areas high in oxygen and it is in the anodic areas that the metal begins to corrode away. It is not just the simple example of the water drop where differential aeration occurs. Consider the slide way of a machine tool that had pools of oil lying over an area damp with water-based metalworking fluid. The area well under the oil film would be oxygen deficient compared to the more exposed aqueous material. As a result, it is quite possible for a differential oxygen cell to be set up and for there to be corrosion *under* the oil since that would be the anodic area.

D. Bimetallic

If pieces of iron and copper are placed in water, it is possible to measure an electric current flowing. In fact, a simple battery has been constructed, and in a short time it will be noticed that the iron begins to become discolored and soon pitting sets in. This is corrosion involving two metals joined together and wetted by one liquid— commonly known as bimetallic corrosion.

A typical example of this type of corrosion is where the steel rivet in a copper sheet is rapidly corroded. However, should the metals be reversed and the rivet is copper and the sheet steel, then, so long as a large section of the sheet is wetted, the corrosion occurring to the steel will be widely spread and thus hardly noticed. If it is only the area around the copper rivet that is wet then naturally the steel in that localized part of the sheet will experience the full degree of the corrosive action, and severe pitting (and a loose rivet) will result.

Since one metal will generally promote corrosion in another connected to it, it is important to know which couples are best avoided. In fact, metals can be listed to form a series, the electrochemical series. If two metals are brought in contact and wetted, then the metal that is *higher up the list* (more electropositive) will corrode in preference to the other. The metal is more electropositive, after all, since it is more inclined to lose electrons (which are negative). Referring back to the

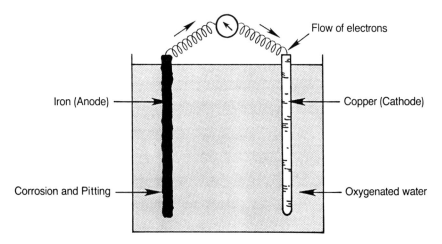

Figure 2 Bimetallic corrosion.

diagram of the water drop will make it clear that the pitting of the metal occurs where the metal loses its electrons.

Thus, in Table 1 when any two of the metals are in contact, and wetted, that which is above the other on the list will tend to corrode. Since pure metals are not generally used in practice, it makes more sense to include common alloys in the list as well.

E. Erosion Corrosion

Whereas erosion is just the mechanical wearing away of a surface by a fast-moving stream of fluid, erosion corrosion has the added dimension that the abraded material is not usually the metal, but the protective film. Many metals and alloys develop a protective film, frequently an oxide layer, which when abraded away reforms from the parent metal below. Gradually therefore, the metal is lost if the film is continuously being worn away. Materials such as stainless steel, which depend heavily on a protective layer to maintain good corrosion resistance, are particularly vulnerable to this type of corrosion. The attack causes the formation of characteristic smooth groves and holes in the surface of the metal. Laboratory corrosion testing of a metal under *static* conditions will not show this form of corrosion if the material is ultimately for use in an environment where, although it will be subjected to the same chemicals, it will be under dynamic conditions. Erosion corrosion is particularly important in pipe work and in pumping equipment.

TABLE 1   Electrochemical Series for Selected Metals

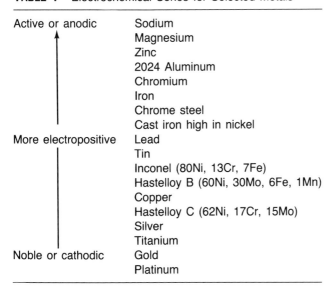

Active or anodic	Sodium
↑	Magnesium
	Zinc
	2024 Aluminum
	Chromium
	Iron
	Chrome steel
	Cast iron high in nickel
More electropositive	Lead
	Tin
	Inconel (80Ni, 13Cr, 7Fe)
	Hastelloy B (60Ni, 30Mo, 6Fe, 1Mn)
	Copper
	Hastelloy C (62Ni, 17Cr, 15Mo)
	Silver
	Titanium
Noble or cathodic	Gold
	Platinum

F. Pitting

Pitting, as the name suggests, is the localized attack of a metal surface that leads to small holes. The pit occurs when the corrosive site (the anodic area) remains small and in one spot. As discussed earlier, whenever there is a local difference on the metal surface, due to differences in relative oxygen concentrations, or indeed an impurity in the metal, or more or less any morphological difference, then a potential gradient will result between these two areas. This difference in electrical potential means that one area is anodic compared to the other (cathodic) area, hence a galvanic cell is set up and thus corrosion can begin. A typical example is when there is a slight scratch in the protective (oxide) film of a metal. The minute area of metal exposed becomes anodic to the (comparatively) enormous area of cathodic oxide surface, which is more noble, and as a result corrosion occurs in the scratch, forming a pit. It would appear that corrosion products resulting from the reaction prevent the metal in the scratch from reforming the protective film and thus stopping the corrosion. Chlorides (and the other halides) are particularly prone to causing pitting corrosion, especially in stainless steels.

G. Fretting Corrosion

Fretting corrosion is similar to erosion corrosion, except in this case the mechanism by which the oxide, or other protective layer, is worn away is by direct contact of another solid material, not the flow of a fluid. There has to be relative movement between the two surfaces, typically vibration, in order for fretting corrosion to occur.

H. Intergranular Corrosion

Although metals appear to the naked eye to be uniform, in fact they consist of grains of metal with boundaries between them. These are the intergranular boundaries. Should corrosion occur preferentially along the boundaries then the metal is weakened and can fail in service. Even though the boundaries are considered to be anodic (prone to corrode) compared to the grains proper, this fortunately does not result in significant corrosion. If it did, then certainly metals could never have been used for structural purposes and technology would have remained in the era of wooden wheels and ships! However, this insignificant corrosion between the grains can result in a real problem if the nature of the boundary is changed to make the interface considerably different from the parent grains. This can occur in austenitic steels (a solid solution of one or more elements in face-centered cubic iron) when they are heated to between 1000–1400°F. Under this condition the carbon in the steel tends to migrate to the boundaries, react preferentially with the chromium in the alloy, and precipitate out of solid solution. In effect then, the steel in the boundary area is considerably lower in chromium than it is in the interior of the grain. This is a situation where the boundary is much less noble (more anodic) than the grain and causes rapid corrosion. There are various ways to prevent this from occurring, for example by ensuring that the carbon content of the steel is particularly low (<0.02%) or by adding small amounts of exotic metals (columbium or tantalum) which react more strongly with the carbon than the chromium.

High temperatures present during welding can unwittingly cause this type of heat treatment and thus intergranular corrosion, which could lead one to believe that a weld had failed, whereas in fact this more involved process had occurred.

I. Stress Corrosion

There are two stages in stress corrosion: the initiation of a microfissure (crack), and then the propagation of the crack. The initiation can result from many causes, for example pitting corrosion or intergranular corrosion, whereupon a small, sharp, anodic area is surrounded by a large cathodic area. Under stressed conditions this could be enough to cause the crack to propagate since it has been shown that there is a point of maximum stress just ahead of the point of the crack into which the crack moves. Of course this stressed area continues to move ahead with the crack following until either the metal parts, a soft area, or a hole is reached wherein the energy of crack propagation can be relieved. However, with stress corrosion there is good evidence that the propagation of the crack is accompanied by a corrosive action actually occurring within the crack.

Situations where stress corrosion is common are in parts that have been welded and not stress relieved, in heat exchangers where there can be a buildup of corrosive deposits, or in metals that show a strong chemical susceptibility to a particular material that was used when cutting the metal. Examples of the latter are brass that was cut with a coolant high in amines, or aluminum or titanium that was cut with a coolant high in chloride ions. Because of the intermetallic nature of the corrosion, it is not usually possible to wash off the offending material with any degree of certainty.

J. Bacterially Induced Corrosion

So far we have covered "chemical corrosion," or more precisely, "electrochemical corrosion," and although we have chosen to differentiate between, say, bimetallic and intergranular corrosion, when it comes down to it the process is essentially the same: an electric current flowing between two dissimilar regions eating away the surface that gives up its electrons most easily. Bacterial corrosion, however, is substantially different; to start with, it is biochemical in nature. A biochemical reaction, just as any other chemical reaction, depends on the movement of electrons, except in this case the initiating step is biological in nature.

In metalworking fluids mainly composed of water, there are two groups of bacteria that can flourish. Firstly, those that require an oxygenated environment, termed *aerobic* bacteria (from the Greek *aeros* meaning air), and secondly those that proliferate in the absence of oxygen, termed *anaerobic* bacteria.

1. Aerobic Corrosion

A water-mix cutting fluid provides a reasonably favorable environment for culturing bacteria. The warmth, water, dissolved or emulsified organic materials (oils, corrosion inhibitors—always a good source of nitrogen), and areas of high and low oxygenation, all encourage rapid bacterial (and mold) growth unless some material that retards microbial action is incorporated into the mix. Such antimicrobial agents are termed *biocides*.

Aerobic bacteria can influence corrosion in a number of ways, some indirectly, others directly. The more obvious way bacteria can affect the corrosion control of a metalworking fluid is simply to metabolize (destroy) the chemicals that were originally included in the mix to confer corrosion protection on machines and workpieces during usage. These anticorrosive agents are essential if rusting is to be prevented since the metalworking fluid may easily be 95% water. Moreover, anticorrosive agents are frequently rich in nitrogen, which is a prime source of energy for the majority of bacteria. Apart from just "eating out" the components of the cutting or grinding fluid, the microbes produce acidic waste products which themselves can be corrosive to metals. A third way the bacteria can harm the mix is by degrading a useful constituent by breaking it into two or more smaller molecules, some or all of which can be corrosive.

2. Anaerobic Corrosion

The other type of bacteria which can cause corrosion was originally investigated early this century by two Dutchmen, Von Wolzogen Huhr and Van der Vlugt, while studying the corrosion of buried pipes in the polder regions of Holland. They noticed that black staining (iron sulfide) occurred not only as an adherent corrosion product on the pipes themselves, but also in the soil in the vicinity of the corroded pipes. From this they ultimately deduced the presence of sulfate-reducing (or anaerobic) bacteria.

Many microbes are able to reduce small amounts of sulfate for the synthesis of sulfur-containing substances, however, comparatively few are able to utilize sulfate reduction as their major energy-producing activity. By far the majority of living organisms derive most of their energy—to go on living—by oxidizing sugars, particularly glucose, and this can only be done if oxygen is taken in at the same time and reduced to water.

Not so with sulfate-reducing (anaerobic) bacteria. They do not use sugars, but a simpler type of chemical known as a lactate, which they oxidize to acetate, thus deriving the energy to live. However, instead of using oxygen (like most other organisms) to carry out their main energy-producing process, they reduce sulfates to sulfides. Unfortunately for metalworking fluids, sulfides (especially iron sulfide) are black in color and smell sulfurous. Although there are comparatively few types of bacteria that reduce sulfate on a "large scale," they are, unfortunately, very widely distributed, for example, in fine metallic swarf (especially cast iron) in a machine tool or system. With oxygen present there will be no problem from these sulfate-reducing bacteria—no discoloration or foul smells and no corrosion, but the bacteria will not die. Even a few parts per million of dissolved oxygen will prevent anaerobes from feeding and multiplying—which is about all bacteria can do with the energy they obtain—and thus they will not be noticed. However, other bacteria that require oxygen to live (aerobes) can rapidly deoxygenate static areas and inadvertently provide the necessary conditions needed by the anaerobes.

Where does the sulfate necessary for anaerobic life come from in a cutting fluid? Firstly, from the emulsifiers. Many widely used emulsifying agents are based on sulfated and sulfonated long-chain molecules which are able to provide an excellent energy source. However, even without this source, many metals (and most mineral oils) have sufficient sulfur, or sulfur-containing impurities, to supply the bacteria.

Furthermore, water invariably contains a certain amount of calcium and magnesium sulfate dissolved out of the gypsum and other rocks that it percolates through before being held in a reservoir. The anaerobic bacteria, of course, are ubiquitous and are present in water, on rags, in dandruff. These bacteria wait in a dormant state for the oxygen level to decrease to allow them to flourish.

So it appears that with or without oxygen, bacteria can cause corrosion. The problem is generally seen as rust if it is due to aerobic bacteria that have created conditions likely to cause corrosion, or as black staining if the culprit is the anaerobic type of bacteria which has actually taken some sulfur from the metal.

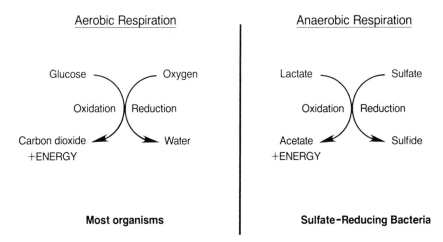

Figure 3 Bacterial respiration.

V. CORROSION PREVENTION METHODS

A. Inhibitors

1. Passivators

If you look back to Table 1, listed in decreasing order of electropositivity—where the metals higher up are more likely to corrode than those below—then you will note that chromium is in fact above iron. Thus, you would expect that when chromium is added to iron, it would render the resultant alloy *less* corrosion resistant; however, this is clearly not the case, since chrome steel is prized for its corrosion resistance. Indeed, the iron is now "stainless steel." Plain and simple iron, or carbon steel, will react with damp air to form a surface oxide layer that is porous, thus more water and oxygen can penetrate through to the virgin metal underneath and continue the corrosive process. Chromium, on the other hand, also forms an oxide film, but it is not porous, it is impervious to further penetration and so the initial incredibly thin (transparent) oxide layer acts as a self-sealing barrier to continued oxidative corrosion. The alloy is unreactive to moist air; it has become passivated. Aluminum (even higher up the table) exhibits the same property of self-passivation, and even high silicon cast iron can form a film of protective silica (SiO_2). Sodium, however, does not form a passive layer. Indeed, a piece of sodium metal left in moist air is so reactive that it will probably burst into flames. Clearly, the nature of the film or corrosion product that forms on a metal is much more important than its relative position on the electrochemical series. Although the most likely film to form on a metal naturally is the oxide, the same principle of a thin protective film formed *from the substrate metal itself* applies to other passivating agents. For example, iron becomes passivated when immersed in solutions of nitrites (NO_2^-), chromates (CrO_4^{2-}), molybdates (MoO_4^{2-}), tungstates (WO_4^{2-}) or pertechnetates (TcO_4^-). In fact, iron can even be temporarily passivated by dipping into concentrated nitric acid solution. However, none of these methods are applicable to metalworking fluids (with the possible exception of molybdates) because of toxicity considerations, even though nitrites were used extensively in the past. Thus, there are few, if any, options open to formulators of metalworking fluids if they are seeking to use oxidizing agents as a means of passivating a workpiece and thus preventing corrosion.

2. Organic Film Formers

This method of corrosion control is much more akin to painting the metal surface. Fatty acids and similarly configured molecules have a long water-repelling hydrocarbon "tail" and a "head" that has a strong affinity for the metal surface.

    The long, thin molecules line up roughly parallel to each other and perpendicularly to the metal surface, forming a fatty monolayer that is essentially impervious to water and oxygen. Although these fatty molecules are excellent

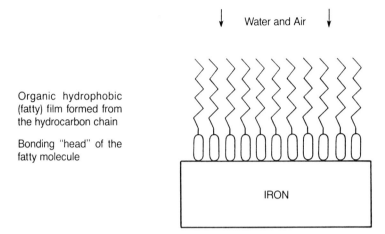

Figure 4 Organic corrosion inhibitor film formation.

corrosion inhibitors for use in water-based metalworking fluids, they are not really tenuous enough to provide long-term corrosion protection during, say, storage. They were never designed with that in mind, and since it follows that any substance that can be deposited from an aqueous solution can almost certainly be washed away just as easily by water, long-term corrosion protection of machined parts usually requires an oil-based anticorrosive coating to be applied. However, the machine tool will be protected by the anticorrosion agents in a metalworking fluid, so long as it is kept at a sufficient concentration and in good condition.

B. Inhibitors in Microbially Induced Corrosion

If microbes are causing corrosion, then the remedy is to destroy the microbes. This can be done in several ways, depending on the severity of the problem.

If the metalworking fluid has had a long history of bacterial odors and the associated corrosion problems, then there is little that can be done except to dump the fluid, clean the machine, and refill. To prevent the problem from reoccurring, several steps should be addressed, not least of which is ensuring that the fluid is maintained at the correct concentration. A mix that is too lean will soon fall prey to microbial contamination. Pollution of the mix with waste materials is another common cause of bacterial problems. If the rise in bacterial population is a new occurrence, then just the dosing of the system with the correct biocide may be sufficient. However, a little "detective work" on why the problem occurred in the first place is important to prevent its reoccurrence.

With anaerobic contamination the problems of foul odors and black staining can frequently be relieved by oxygenating the mix, simply by keeping it *continu-*

ously circulating and checking to make sure there are no stagnant areas, or heavy deposits of sludge in the base of the tank or system. These latter places are ideal breeding grounds for anaerobic bacteria, especially if the sludge is from cast iron or other low-grade ferrous material that is rich in sulfur. Removal of the sludge is essential. In short, more bacterial problems and the associated corrosion can be avoided if the mix concentration is held at the recommended level, the fluid is continuously circulated, stagnant areas are designed out of the system, and finally, the filtration and drag out equipment is working efficiently.

VI. CORROSION TESTING METHODS FOR METALWORKING FLUIDS

Before filling an expensive machine or central system with an unfamiliar metalworking fluid, it is only right that engineers should ask for some data from the supplier regarding the ability of the fluid to prevent corrosion when you consider that the final mix might consist of up to 98% water filled into what is essentially a cast iron structure. Many users do their own testing and may even have evolved their own test methods. However, testing a water-based metalworking fluid to evaluate its anticorrosive properties is so fraught with problems that the customer may end up missing a good "in-use" product or selecting something that turns out to be unsuitable, unless care is taken. Consider the following test results:

A semisynthetic metalworking fluid was tested to determine its break point using five different (but widely used) ferrous corrosion tests. The values shown below are corrosion break points for the fluid, or the minimum concentration in water that will prevent corrosion, according to that particular test. The test material in each case was cast iron.

Test method	A	B	C	D	E
Break point	2.5%	4.5%	>5.0%	1.25%	5.0%

Since the recommended use concentration of the fluid was 3%, and field trials showed that it did not cause corrosion at this level, it is probable that people using only test methods B, C, or E would reject the product as having poor corrosion control. Thus, slavish adherence to a corrosion test method will neither guarantee that the best fluid is finally selected nor that it will be used at the most economic dilution. However, it is equally obvious that no one can risk an unknown fluid in an expensive machine without at least some testing. It is as well therefore to have some knowledge of the various test methods so at least an informed interpretation of their results can be made. The three metal types that are usually involved in corrosion testing are:

Ferrous: cast iron and steels
Aluminum alloys
Copper alloys

Testing against ferrous alloys is by far the most common.

A. Ferrous Metals

Ferrous corrosion tests generally originate from one of three sources:
Customer (end user)
Manufacturer (or supplier)
National or international testing organization (who tend to establish "standardized tests")

Generally the customers' tests will be a variation on one or more of the standardized tests. Although metalworking fluid manufacturers will use these standard tests as well, they frequently have a number of self-devised methods to highlight factors they consider important, especially during the development of new products.

Any test method chosen *must* be cross checked against a machine trial, preferably in a typical, yet noncritical, machine, using available water.

1. Chip Test

a. Steel Chips on a Cast Iron Plate   This method is the basis of the Herbert test (UK), The Institute of Petroleum IP 125 (UK), and the DIN 51360 part 1 (Germany).

The Herbert test was the forerunner of them all, but the specifications are so imprecise that its value is somewhat questionable. Basically with all these tests, a cast iron plate is cleaned and polished up with a fine abrasive paper. Four small piles of clean steel chips are positioned on the plate and are then wetted with the test mix. The four piles are treated with four different dilutions of the same mix, or conversely, with four different products. The plate and its chips are placed in a closed container for 24 h, after which the chips are removed and the degree of corrosion on the plate examined.

The DIN 51360 part 1 gives highly specific instructions, even to the point of specifying the level of salts in the water used to make the mixes. Therefore, it alone can be considered as an absolute test allowing comparison between different testers and localities.

b. Cast Iron Chips on a Steel Plate   This test method is analogous to that described above, but now the plate is steel and the chips are cast iron. The test procedure is essentially the same and is used as a standard in France (CNOMO) and widely accepted in Italy.

c. Cast Iron Chips on Filter Paper   This test has gained wide acceptance since it can be quick (as little as 2 h), is "reasonably reproducible," is simple (there is no metal plate to clean and polish), and the final paper (test result) can be fixed into a notebook for future reference. Typically, about 2 g of clean (dry cut) cast iron chips are spread onto a filter paper in a Petri dish. The diluted mix is pipetted on to the chips and the dish covered. After a set period of time the chips are removed and the paper examined for staining, if there is any it is usually graded depending on its severity. Test methods that use this general technique are the IP 287 (UK), DIN 51360 part 2 (Germany), and ASTM D4627 (US). A variation of this test is to soak the chips in the metalworking fluid first, and then drain the mix and pile the chips onto the paper. The damp chips then cause staining of the paper if the corrosion control of the fluid is insufficient.

2. Flat Surface Tests

a. Open Cast Iron Cylinders Small cast iron cylinders, about 1 in. in diameter and 2 in. long are ground flat and then lapped to a polished finish. These are stood in a 100% humidity cabinet and a film of the metalworking mix is pipetted on to the virgin surface of the metal. The cabinet is closed and left usually overnight. The next day the cast iron surface is examined for signs of corrosion. Generally, a series dilution of the product will be used to determine the "break point" for that particular metalworking fluid.

b. Stacked Steel Cylinders This test is analogous to that described above, except that steel cylinders are used and after the mix has been placed on the top surface of the steel, a second steel cylinder is mounted on top of the first. The majority of the mix is squeezed out, but a thin film remains between the disks. The 100% humidity cabinet is closed and left overnight. The next day the top cylinder is removed and the resulting corrosion of the steel (if any) examined and rated. Various finished steel parts, such as bearing races, may also be used.

3. Panel in Closed Cabinet

Though not normally applied to metalworking fluids used for cutting and grinding, there are special tests required for fluids used for stamping and drawing products. These tests take the form of dipping the component part in the tests fluid, allowing it to drain and then hanging or clamping the part in an environmentally controlled cabinet. The reason for these tests is that parts stamped or formed in some way from sheet metal are often placed in bins after being cut from the roll of metal and may wait considerable time before being used. Thus, it is particularly important that the fluid used in the metal fabrication process leave a coherent film of corrosion inhibitor over the metal surface. Typical parts would be automotive panels cut and formed from mild steel. Any subsequent corrosion to these panels would involve either scrapping them or expensive cleaning processes.

a. Humidity Cabinet The simplest panel test involves a closed cabinet where the humidity is maintained at or near moisture saturation levels and at an elevated temperature. Often the heating circuits in the apparatus will be programed to cycle through heating and cooling periods so that the moisture in the enclosed air has a chance to condense out onto the component from time to time. This falling below the dew point clearly mimics conditions that can occur with components stored in factory areas where the temperature fluctuates throughout the day. Incidentally, the air in factories that use metalworking fluids in significant amounts is generally saturated with water vapor.

b. Acid and Salt Atmospheres Since the damp air in factory environments is frequently contaminated by acid materials (exhaust from furnaces and heat treatment facilities), components can be subjected to dilute acid droplets condensing out on them. To ensure that the metalworking fluids used in stamping and drawing leave behind sufficient rust protection under these harsh conditions, humidity testing in an acid atmosphere is carried out. It is very similar to ordinary humidity testing, but with the added challenge of acid vapors in the chamber. Salts can also be introduced into the test atmosphere as a further variation of this test.

4. Stress Corrosion Tests

In-service failure of aerospace components generally has catastrophic consequences. Therefore, metalworking fluids used in these fields are checked to ensure that they will not induce any form of point corrosion that could lead to propagation, or initiation, of a crack through the component. A typical material that has been linked to metal cracking is chlorine, when present in the fluids used to work titanium. Since cracking can occur months or even years after exposure to the causative agent, more subtle testing is required than in simple ferrous corrosion testing.

Generally the test for stress corrosion consists of bending a coupon, cut from the subject metal, into a U-form and holding it in a clamp. The metal, which is under considerable stress, is then soaked for a short while in the metalworking fluid under investigation. The metal, still held in the clamp, is then subjected to high temperatures in a special oven.

After removal from the oven, the component is etched and polished, and examined for metallurgical defects. Using this technique, it is possible to screen out those metalworking fluids that could cause stress cracking in particular metals.

B. Nonferrous Metals

Corrosion testing of nonferrous metals is usually a simple matter of taking a coupon of the metal in question, cleaning and polishing it, and then partially immersing it in the test fluid. After a time (typically 24 h) the component is removed and examined for discoloration, beneath, on, and above the fluid line.

Another test procedure, that is in essence similar, is to half fill a small clear glass bottle with nonferrous turnings, then measure in enough test fluid to half submerge the pile. Stopper the bottle and leave it undisturbed. Check the turnings, especially at the fluid/air interface, every day for corrosion or discoloration. Typically, such a test would be left to run for 5 to 7 days.

Even if no corrosion is observed, it is good practice to filter the test fluid and to measure the concentration of dissolved metallic ions in it. A metalworking fluid suitable for nonferrous metal should show little or no corrosion of the material and should not dissolve more than a few part per million of it over a period of about a week at room temperature.

C. Multimetal Sandwich

Many of the components used in aircraft components are fabricated from several metals and it is important to know if during the machining of these multimetal components there could be corrosive interaction brought about by the metalworking fluid. A test for this relies on taking pieces of the various metals and clamping them together and bringing them in contact with the metalworking fluid. Some tests simply require that the pieces are clamped and then soaked (probably at an elevated temperature) in the fluid. Other tests are more severe and involve clamping the metals and then drilling and possibly reaming through them with the test fluid.

D. Galvanic Test Methods

Since corrosion is always accompanied by the passage of electrical current, considerable efforts have been made to try to utilize this phenomenon as a means of testing for corrosion. The method differs from those so far discussed because in conventional stain or rust tests the degree of corrosion observed on the test piece is taken as a direct measure of what is likely to occur in real usage. Thus, if a particular cutting fluid causes significant rusting of cast iron test pieces or even chips on a filter paper, then one is naturally hesitant about filling an expensive machine tool with the product, since the machine tool is made largely from just an alloy of iron. However, galvanic, and similar electrochemical tests try to *predict* possible corrosion problems by making a voltaic cell out of two (usually dissimilar) pieces of metal and measuring the current that flows between them when immersed in the test fluid. Once again we come back to the idea of a corrosion cell being a "battery" except here we are trying to relate the magnitude of the current generated to future possible corrosion problems. There are numerous variations of this theme, such as first joining the two coupons together and then soaking them in the fluid, before separating them, and then measuring currents produced by the now corroded components. However, no matter how much use one makes of these corrosion tests, they still have to be related to "field results" to be really useful.

E. Other Electrochemical Techniques

More involved electrochemical techniques than that discussed under galvanic test methods are potentiostatic polarization and AC impedance spectroscopy.

Potentiostatic polarization is where a known DC potential is applied between a metal electrode and a standard (for example a calomel) electrode. The electrolyte, for our purposes, would be the metalworking fluid under test. The magnitude of the current that flows is dependent on the applied potential and the corrosion-inhibiting characteristic of the metalworking fluid. By applying a potential that increases from typically -2 V (a high cathodic potential), through zero to an anodic potential of typically $+2$ V, and measuring the absolute current flow, it is possible to distinguish at least four distinct electrochemical regions. These are (going from cathodic to anodic potentials) reduction of water, reduction of oxygen, passive region, and oxidation of water. The passive region is the area that shows little or no change in current flow for an increase in applied potential. A metalworking fluid that offers good corrosion inhibition will typically have a significantly lower current flow in this passive region than a fluid with poor anticorrosive properties.

AC impedance is a similar technique, but uses an alternating current (AC) whose frequency gradually changes. The real and imaginary components of the impedance of the system are determined and modeled as if they were part of an RC (resistance/capacitance) circuit. This method is preferred in systems where the metal is particularly well inhibited, for example if the metal is painted.

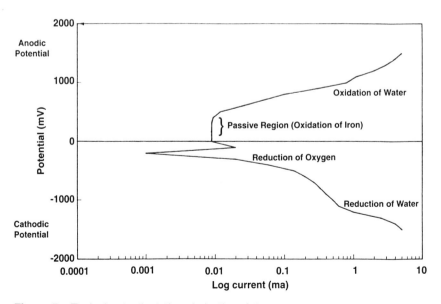

Figure 5 Typical potentiostatic polarization plot.

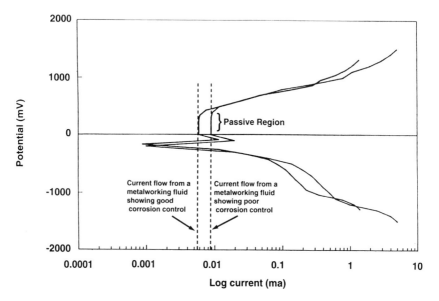

Figure 6 Potentiostatic comparison of two fluids.

VII. PRACTICAL STEPS TO PREVENT CORROSION WHEN USING METALWORKING FLUIDS

A. Choice of Metalworking Fluid

Clearly this is the starting point for solving (or avoiding) corrosion problems with metalworking fluids. When considering ferrous corrosion control (rust), there comes a time when further dilution of the product will no longer provide sufficent inhibitor to prevent corrosion. That, after all, is how we generally test these products; how much can we dilute them until the test pieces rust? However, there is another aspect these "break point tests" do not show. Most ferrous corrosion inhibitors are based on nitrogen, and this is an energy source fervently pursued by bacteria that may grow in the diluted mix. Thus, even if a particular product shows good "break point" test results, if it is likely to become microbially infected quickly, the bacteria will rapidly deplete the inhibitors and leave behind corrosion by-products. A product with good break points *and* good microbial control is essential.

B. Water Quality

As stated earlier, dissolved ions can greatly increase the corrosivity of an aqueous solution by either interacting directly as a corrosive agent or by simply increasing the electrical conductivity of the fluid. Thus, if water containing high chloride or sulfate levels is used as mix water for a metalworking fluid, a degradation of the corrosion control of that product will surely occur. What actually constitutes high ionic levels depends on the nature of the fluid and the metals it comes into contact with; however, chloride levels higher than 100 ppm and sulfate levels higher than 200 ppm should, if possible, be avoided. As water is lost through evaporation from the machine tool sump or central system, these ions will naturally concentrate in the metalworking fluid mix and could then cause corrosion. Unfortunately, there is no really practical way of removing these ions from the mix and so in areas where dissolved ions in the water are high two solutions are possible: either use dionized water for makeup or, if this is not possible, increase the mix concentration to combat their corrosive effects. Tankside additives may also be considered (as discussed in Sec. VII.E).

C. Dilution Control

The majority of problems experienced with metalworking fluids can be traced back to problems of not holding the mix at the recommended concentration level. A mix that is excellent at 5% may give considerable problems at 3%, and surely will at lower strengths. Not only can you over-dilute the corrosion inhibitors, but microbial damage (as discussed earlier) can set in. Holding the concentration too high will still provide good ferrous corrosion control, but it is uneconomic, it may lead to skin irritation problems, and the associated rise in alkalinity could stain nonferrous materials.

D. Treatment of Metalworking Fluids When in Use

If rust is experienced when using a water-mix cutting and grinding fluid, do not immediately blame the fluid. First investigate the situation. Where is it occurring? When is it occurring? Are the production pieces or just the machine showing rust flecks? There are numerous questions that have to be asked. If the corrosion is occurring on areas of the machines well away from where cutting fluid is being used, then it would appear to be water vapor in the air that is the problem, increasing mix concentration would not be an answer. Workpieces can suffer a similar fate if they are taken from the machine and stacked on cardboard or on any other absorbent material. The metalworking fluid (with its associated inhibitors) is mopped up by the absorbent material which then acts like a wick giving back high water vapor levels into the surrounding air. The water vapor is devoid of inhibitors and causes the components to corrode.

If the problem is one of insufficient corrosion control because of over-dilution of the mix, then clearly the remedy is to add more concentrate. Adding more concentrate if the mix is microbially contaminated is only curing half the problem. The addition of biocide to the mix should be considered first.

E. Tankside Additives

Additives to prevent and cure problems largely take the form of just the corrosion inhibitor package from a metalworking fluid, or a biocide to keep the microbial count low. While these are useful (especially the biocide), it may often be more useful to add more concentrate than just the corrosion inhibitor. Remember, many metalworking fluid problems are concentration related and adding "bits and pieces" makes determining the "real" concentration of the mix very difficult. Not all corrosion inhibitors will increase the apparent concentration of a cutting fluid, but some will. Biocides generally do not affect concentration measurements.

9
Metalworking Fluid Microbiology

L. A. ROSSMOORE
Biosan Laboratories, Inc.
Ferndale, Michigan

H. W. ROSSMOORE
Wayne State University
Detroit, Michigan

I. INTRODUCTION

Ever since water was introduced as an integral component of metalworking fluids (MWFs), microbial growth in these fluids has become a major concern among formulators and end-users alike. The problems associated with unchecked microbial growth are well documented and include corrosion of tool and workpiece, loss of lubricity, rancid odors, slime formation, and the plugging of delivery lines. Recently, certain public health concerns have been raised regarding the effect of large amounts of aerosolized bacteria on the exposed population of workers.

This chapter addresses some of the effects bacteria and other microbes can have on MWFs. It also looks at ways to monitor and control the growth of microorganisms. Laboratory testing plays a key role in biocidal treatment selection and that topic will be examined as well. Some health effects of microbes in MWFs as well as current research in various related fields will also be discussed.

Because most readers are assumed to be nonmicrobiologists, this chapter will begin with a basic primer on microbiology. As a science, microbiology deals with living things too small to be seen with the unaided eye, hence the term microorganisms. There are many types of microorganisms, or microbes, but for the purposes of this chapter the main ones of concern are bacteria, yeasts, and molds.

II. DEFINITION OF MICROBIAL TYPES

A. Bacteria

Bacteria are single-celled organisms which, unlike cells of so-called higher animals, contain cell walls. Bacteria are categorized as either gram positive or gram negative on the basis of the cell wall. There are three basic shapes of bacteria. Spiral bacteria, some of which are called spirochetes, include *Treponema pallidum*, the causative organism of syphilis. There are also round bacteria, which are called cocci. Medically significant organisms such as *Staphylococcus* and *Streptococcus* are included in this group. The third type is rod-shaped and these are called bacilli. There are scores of examples in this category and included in the group is *Pseudomonas*, an important organism in the degradation of MWFs. Bacteria form colonies on solid culture media. A colony is the visible growth that results when a single cell replicates in a dish of agar media.

By definition, of course, bacteria are quite small. In fact, for an average-sized organism, you would have to lay 25,000 side by side to equal one inch in length. To put it another way, one cubic inch of solid bacteria would contain 9 trillion cells.

B. Yeast

Yeast, like bacteria, are single-celled organisms. They are somewhat larger than bacteria and have less variation in shape; they are either round or oval. Yeast are also able to form colonies on solid culture media. Often these colonies are indistinguishable from bacteria and require further analysis to confirm their identity.

C. Molds

Molds, along with yeasts, comprise the group of organisms called fungi. Unlike bacteria and yeast, molds can be composed of more than one cell. Molds may not always be thought of as a microorganism because we may see molds (e.g., mushrooms) without the aid of any magnifying instrument. However, when we see growths such as those on a moldy piece of bread, what we really are seeing is a colony containing millions of mold cells. Molds also form colonies on solid culture media and these are easily distinguishable from bacteria and yeasts by their fuzzy and filamentous appearance.

III. MICROBIAL GROWTH RATES

A. Generation Time

The growth rates of all microorganisms are defined in terms of an organism's generation time. A generation time is the time required for a population to double in number. Growth must be described this way because increase in cell number is always geometric in that each cell gives rise to two cells, each of which gives rise to two more cells, and so on.

Microbial generation times can vary from several minutes to several hours. Of the bacteria isolated from MWFs, 20 min would be an optimal doubling time. At this rate, a theoretical culture starting with one cell would grow to eight cells in one hour. In eight hours, there would be about 1 million cells. However, by 48 h, the culture will have grown to 44 times the mass of the earth!

B. Growth Kinetics

In practice, of course, bacterial populations never reach this size. The main reason is because microbial growth occurs in four well-defined phases (Fig. 1), and only for a relatively short period of time does logarithmic growth occur. In a closed system such as a laboratory culture, cell division begins with a lag phase in which the organism acclimates to the environment and little or no increase in population is seen. This is followed by the log (logarithmic or exponential) phase in which the cells divide at their fastest rate. Next is the stationary phase, when the total number of living cells in the population no longer increases. Finally, the death phase occurs and the cells die exponentially.

Unlike laboratory situations, where microbial growth can be tightly controlled in a so-called *closed system*, growth in a metalworking operation or other similar environment can better be described as occurring in an *open system*. In an open system there are almost constant changes, such as influx of nutrient, addition of toxicants (e.g., biocides), fluctuation in temperature, and changes in pH. In open systems the length of any of the growth phases can be altered depending on environmental conditions.

IV. EFFECTS OF MICROBES ON MWF PERFORMANCE

A. Corrosion

There is a law of physics which states that for every action there is an equal and oposite reaction. This axiom can be extended to cover MWF systems in that as

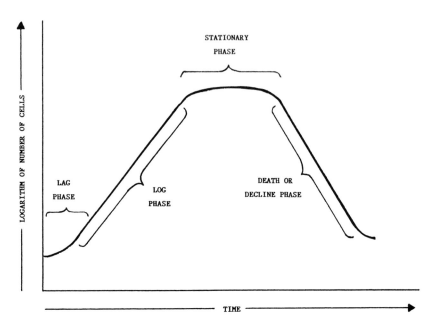

Figure 1 Phases of growth in a bacterial culture.

microbes are affected by their environment, they in turn have an effect *on* their environment. One area where microorganisms can have influence is in a phenomenon called microbially influenced corrosion, or MIC. Certain kinds of bacteria are capable of either establishing an electrolytic cell or stimulating anodic or cathodic reactions on metal or other surfaces. These electrochemical reactions in metalworking systems can cause corrosion of both tool and workpiece. Bacterial contamination of MWFs can also indirectly cause corrosion. This is done two ways, either by consuming corrosion inhibitors and/or by producing corrosive by-products such as organic acids [1].

B. Changes in Fluid Chemistry

The loss of anticorrosion components is not the only chemical change that microbes can exert on a fluid. Several well-controlled studies have shown that other fluid components are partially lost and sometimes completely depleted in the presence of viable microbial populations [2–4]. This change in fluid chemistry can be directly correlated with loss of fluid function, particularly in the case of soluble oils where the hydrocarbon is degraded. A concurrent loss of lubricity can occur with this type of microbial activity.

Sophisticated instrumentation such as high-pressure liquid chromatography

(HPLC) and infrared spectroscopy (IR) can detect even the most subtle changes in fluid chemistry brought about by microbial metabolic activity (Fig. 2). There are other less subtle changes in the fluid which are also caused by microbes, changes which are very well known to most workers in the field. Microorganisms are responsible for a variety of unpleasant odors, some of which are described as musty,

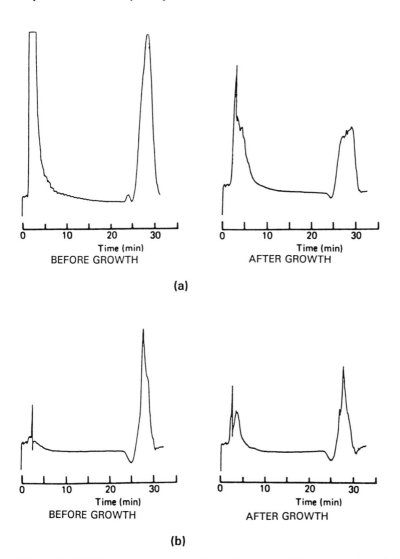

Figure 2 HPLC scan before and after bacterial growth in metalworking fluids: (a) 5% soluble oil, (b) 5% synthetic fluid. For details see Ref. 4.

or as having the smell of an outhouse or locker room, or the "Monday morning" smell of rotten eggs. It is not uncommon for production to be halted by workers who refuse to continue until such odors are brought under control.

C. Physical Blockage

Some microbes, particularly fungi, can form slimy deposits in the working fluid. If these growths are not properly treated they can become so massive as to clog delivery lines and filtration equipment. Again, production must occasionally stop in order to physically remove this material.

V. METHODS TO ASSESS MICROBIAL GROWTH

There are many ways to assess the level of microbial contamination in MWFs. There are qualitative methods such as the presence of odors and slime formation, as discussed earlier. There are other gross changes which would be presumed to have microbial origins such as color changes, emulsion splits, corrosion evidence, and general losses in fluid function such as lubricity. Ideally, microbial levels should be monitored before they reach a level where they are capable of causing such qualitiative changes.

A. Conventional Methods: Plate Count and Dipslide

Methods for monitoring microbial levels in MWFs can either be conventional or unconventional. The conventional method to which all other methods are compared is the standard plate count [5]. This method employs the use of serial tenfold dilutions of the fluid into sterile petri dishes. After all the appropriate dilutions are made, the plates are filled with a type of microbial growth medium called agar. Agar contains basic nutrients and is liquid above 45°C. Once it cools to room temperature, it hardens to a gel. Upon incubation for a day or two, the bacteria form visible units called colonies. Because the colonies can be seen with the naked eye, they can be counted and the number obtained from the count is multiplied by the dilution factor in order to get the organism count per milliliter.

The plate count method, although widely recognized in applied microbiology as being quite reliable and effective, can also be time consuming and costly. A very similar method was developed in the early 1970s and was modified from a test used to measure bacterial levels in urine specimens. This is the so-called dipslide method which, rather than using petri dishes, coats plastic paddles with agar medium. There are no dilutions to make or culture media to prepare. The dipslide is immersed into a sample for a few seconds, then withdrawn and placed into an airtight holder and incubated. Most dipslide media contain triphenyltetrazolium chloride, or TTC, which is colorless but the bacterial colonies turn it red as they grow (a reduction), thus making them easier to see (Fig. 3). Also, because dipslides have two sides,

Metalworking Fluid Microbiology 253

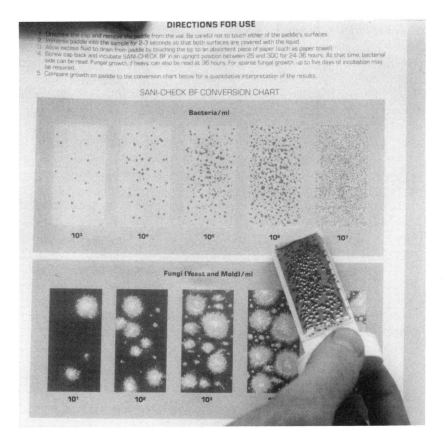

Figure 3 Reacted dipslides and comparison chart.

many times one side of the paddle is coated with a medium selective for the growth of bacteria while the other side can contain a fungi-specific growth medium. This means that with a single dip in the fluid sample, the user can test for both the presence of bacteria and fungi.

B. Unconventional Methods

There are, as mentioned, some less conventional methods used to measure microbial populations in MWFs [6]. One such method involves the measurement of the enzyme catalase, a substance present in almost all microbial cells. Catalase converts hydrogen peroxide into water and oxygen gas. In the method used for microbial enumeration in MWFs, a sample of fluid is mixed in a test tube with a small amount of hydrogen peroxide, after which the tube is closed with a diaphragmed stopper. A probe is placed in the stopper which measures gas pressure. In theory, the gas pressure is the result of the catalase enzyme breaking down the peroxide and releasing oxygen, and the amount of gas given off should be directly proportional to the microbial load of the sample. In practice, however, this is not always the case. There are several interfering factors, not the least of which is that this method cannot distinguish between living and dead microorganisms. A dead microbe may still contain signifiant levels of catalase. Therefore, it is possible that a biocide addition, although quite effective in reducing microbial levels, may not be indicated as such by the results of catalase measurement. This test would still likely show high levels of the enzyme, leading to the erroneous conclusion that a significant number of viable organisms are still present [7].

Another relatively unconventional method involves the measurement of dissolved oxygen in the used fluid. This technique is based on the fact that living bacteria and fungi consume oxygen as they grow and metabolize. In principle, this means that a very low dissolved oxygen level in a fluid would indicate a very high microbial population and vice versa. As with catalase measurements, this method does have its interferences and drawbacks. One major disadvantage is that it cannot distinguish between bacteria and fungi. Still, measurement of dissolved oxygen is widely used and is a fairly good "quick and dirty" technique for measuring microbial contamination. It is a particularly effective tool when used in conjunction with some more conventional method like dipslides (Fig. 4). Both dissolved oxygen and catalase measurement may be affected by the physical and chemical condition

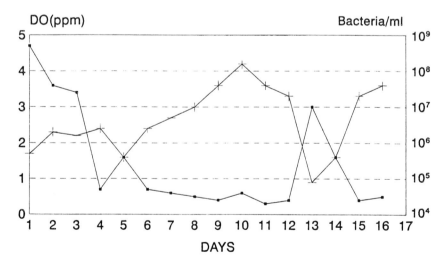

Figure 4 Comparison of dissolved oxygen vs. plate count for a semisynthetic fluid: ■, DO (ppm); +, bacteria/ml.

of the fluid. The presence of certain biocides or metal fines can interfere with these tests.

Regardless of what method is used to enumerate microbes, the main goal in performing this task should be to establish some type of treatment regimen. If the fluid user is not prepared to take some type of corrective action after microbial growth grows out of control, there is really no need to take any measurements.

VI. CHEMICAL BIOCIDES

One very popular way of controlling microbial growth in MWFs is through the use of chemical biocides. Before getting into more specific information about biocides, it would be appropriate to define the term. *Biocide* is a term that has gained popular acceptance over the past 30 years, although a literal definition is redundant (i.e., *bio* means life and *cide* to kill, what else *can* be killed?). Biocides are commonly defined as chemical agents used in industrial systems to control or prevent deterioration. They may or may not *kill* all microbes present. Rather, the key word here is *control*.

Universally, these chemicals are under the purview of some governmental agency. In the United States this is the Environmental Protection Agency (EPA), given authority under the Pesticide Act of 1972. This act transferred responsibility to the EPA for administering the Federal Insecticide, Fungicide, and Rodenticide Act (FIFRA). Pesticides, including biocides, must be registered with and by the EPA. Registration implies satisfying certain specific requirements of the EPA. These include toxicological testing and environmental impact. Efficacy testing (i.e., does it work effectively?) is only required on submission if health claims are made for the biocide (such as preventing the spread of a communicable disease).

In 1987 the EPA issued a chronic and subchronic toxicological data recall for all the products previously registered as active antimicrobial agents. The amount of data required is dependent upon the application. This means that not only is the nature of the chemical structure important but also the level of exposure, which determines risk category. Thus, a "biocide" can be registered for cooling tower applications, a low-exposure situation requiring minimal toxicology, and not be registered for MWFs use, which is classified as a high-exposure application requiring maximal toxicological testing. This emphasizes a very important point, biocides are registered for *each* application. The other point to remember is that if microbial growth is controlled by molecules that are not considered biocides (i.e., not registered) and no antimicrobial claims are made, then they may not legally be biocides.

This is a situation in which each practitioner must decide whether de facto function is equivalent to de jure recognition. What practitioners should also bear in mind is that registration and its attendant required toxicological testing certainly protects the worker from excessive exposure as well as protecting the practitioner from the legal consequences of ignorance.

A. Types of Approved Biocides

Registration does not guarantee effectiveness. The marketplace makes that decision. The 1987 EPA data recall mentioned earlier has not been totally satisfied, although the important requirements have been met for all of the biocides in use at the time of this writing. Some compounds, because of their broad application in human contact areas such as cosmetics and toiletries, have already satisfied the maximum exposure toxicological test requirement. Table 1 lists by chemical category most of the products EPA has registered for use in MWFs [8].

By far the most numerous of these products are the so-called formaldehyde condensates. These materials have been used worldwide for about 30 years. Although some of them are proprietary (Tris Nitro, Bioban P-1487, Bioban CS-1246, Bioban CS-1135, Vancide TH), the most popular [hexahydro-1,3,5-tris-(2-hydroxyethyl)-s-triazine] is practically a commodity. In the United States, it is made and/or sold by several specialty chemical companies including Olin, Buckman, ANGUS, Stepan, and U.S. Professional Laboratories. For the most part, this product has widespread compatibility but it does have drawbacks, as do all the formaldehyde condensates. These will be discussed later.

There is perhaps at this time no more successful single biocide product than methyl chloroisothiazolone, originally synthesized and patented by the Rohm and Haas Company and currently sold for metalworking under their trade name, Kathon 886 MW. When this biocide is compatible, it is the most effective product of its type, both in antimicrobial efficacy and cost effectiveness. Notice that compatibility is a very important consideration. Specific incompatibilities for all biocides will be discussed in the next section.

Any biocide that stays on the market has probably found some type of niche, however small. For example, of the three phenols listed, two (OPP and PCMX) are registered for use in the United States. Until recently, these biocides have been pretty much ignored because of a very conservative interpretation of EPA point discharge limitations in the early 1970s. They are now back in the good graces of industry and are used in compatible applications. The Buckman polymeric quaternary compound, Busan 77, is another biocide that has sought and found its own useful but small niche in the marketplace. Although this product has only limited effectiveness when used by itself, it is very effective combined with other products when the desired effect is to remove and kill biofilm. This is an area of extreme concern in systems that have become excessively fouled.

Biocide testing in the laboratory is very important before treatment in the field can be adequately instituted. This subject will be discussed in some detail later in the chapter.

TABLE 1 Biocide Variety: Based on EPA Registration, Efficacy, and Marketing Activity[a]

A. Formaldehyde condensates
   1. Aminals
      a. Grotan, Onyxide 200, Triadine 3, Bioban GK, Busan 1060
      b. Vancide TH
      c. Dowicil 75
      d. Bioban CS-1135, Bioban N-95
   2. N-Methylols: Grotan HD (not available in the U.S.)
   3. C-Methylols: Tris Nitro, Bioban BNPD
   4. Nitromorpholine: Bioban P-1487
B. Isothiazolones
   1. Benz: Proxel CRL
   2. Chloromethyl: Kathon 886MW, Kathon MWC
   3. N-Octyl: Kathon 893
C. Thiocyanobenzothiazole: Busan 1030
D. Bromonitriles
   1. Dibromonitrilopropionamide (DBNPA)
   2. Dibromodicyanobutane: Tektamer 38
E. Glutaraldehyde: Uconex 345
F. Pyridinethione: Omadine
G. Dioxanes: Givgard DXN
H. Phenols
   1. Dowicide A
   2. PCMX
   3. Dichlorophen
   4. PCMC
I. Polymeric quats: Busan 77
J. Mixture
   1. Triazine-pyridinethione: Triadine 10, Triadine
   2. MBT-dithiocarbamate: Vancide 51

Addresses of Listed Biocide Manufacturers

ANGUS Chemical Co. (Biocides A1a, A1d, A3, A4), 1500 E. Lake Cook Rd., Buffalo Grove, IL 60089
Buckman Laboratories, Inc. (Biocides A1a, C, I), P.O. Box 8305, 1256 N. McLean, Memphis, TN 38108
Calgon Corp. (Biocide D2), P.O. Box 1346, Pittsburgh, PA 15230
The Dow Chemical Co. (Biocides A1c, D1, H1), Designed Products Dept., Midland, MI 48640
Givaudan Corp. (Biocides G, H3), 100 Delawana Ave., Clifton, NJ 07014
R & F Products (Biocides A1a, A2), 245 Edwards St., Aurora, Ontario L4G 3M7, Canada
Miles, Inc. (Biocide H4), Organic Products Div., Mobay Rd., Pittsburgh, PA 15205
NIPA Laboratories, Inc. (Biocide H2), 3411 Silverside Rd., 104 Hagley Bldg., Wilmington, DE 19810

TABLE 1 (continued)

Olin Corp. (Biocides A1a, F, J1), Olin Chemicals, 120 Long Ridge Rd., Stamford, CT 06904
Rohm and Haas Co. (Biocides B2, B3), Industrial Chemicals Div., Independence Mall West, Philadelphia, PA 19105
Stepan Co. (Biocide A1a), Northfield, IL 60093
Union Carbide Corp. (Biocide E), Specialty Chemicals Div., 39 Old Ridge Rd., Danbury, CT 06817-0001
R. T. Vanderbilt Co., Inc. (Biocides A1b, J2), Industrial Minerals and Chemicals Div., 30 Winfield St., Norwalk, CT 06855
Zeneca (Biocide B1), Specialty Chemicals Div., Wilmington, DE 19899

[a]This list contains trivial and trade names. Consult manufacturers for CAS numbers, MSDS, and EPA registration information.

Biocides approved for use in MWFs can either be categorized as primarily antifungal, primarily antibacterial, or broad spectrum (meaning more or less equally effective against both). Oftentimes the categories become blurred and the effectiveness becomes dose dependent. For instance, formaldehyde donors such as the triazines can have very good activity against fungi at high doses, but when levels are lowered, that effectiveness disappears. In fact, studies have shown that low doses of triazines actually stimulate fungal growth. On the other hand, fungicides such as octyl isothiazolone and sodium mercaptopyridine can be bactericidal at very high doses.

B. Factors Affecting Biocide Activity

One of the most important factors affecting biocide potency is the issue of how and when a biocide is added. There are three ways a biocide can be added for use in a MWF. First, it can be put into the fluid concentrate by the formulator at their facility. Second, the biocide can be added to the fluid at the site of use in the premix tank. Or last, it can be introduced sumpside right into the use-diluted fluid.

Dosing a biocide into the MWF concentrate is a marketer's dream and a lubrication engineer's nightmare. An otherwise compatible biocide at its EPA-registered use concentration may fail for a number of reasons. For one, even though a biocide may be soluble at its use concentration (e.g., 0.15% or 0.015%), it may not be soluble in the neat fluid at 20 times that concentration, the level which would be required in the concentrate in order to achieve the desired strength in the use dilution. Also, chemical reactivity at the higher levels in MWF concentrates can destroy the biocide molecule. It should be made clear that concentrate pH and extreme storage temperatures as well as MWF composition are all factors in incompatibility. For example, methyl chloroisothiazolone loses activity above pH

9 and/or 60°C as well as in the presence of secondary amines (Fig. 5) [9]. These facts severely limit the concentrate dosing of this product. A method of overcoming some of these incompatibilities will be discussed in the section on mixtures.

Instability and subsequent loss of efficacy upon adding biocide to concentrated fluid is not limited to isothiazolone biocides. Triazines as well as other types all tend to lose some activity upon storage in MWF concentrates. It is imperative that stability testing be conducted to determine the degree of biocide loss in these situations.

Another important consideration in biocide dosing is the biocide concentration coefficient η; $K = C^{\eta} T$, where C = concentration and T = time. Essentially this formula tells us what to expect when a biocide is diluted. Note that η is an exponential function. For formaldehyde, $\eta = 1$, which means if C is halved, activity

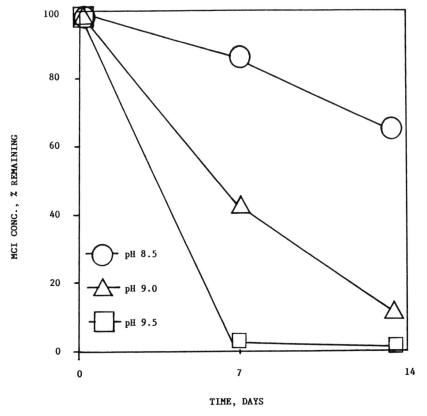

Figure 5 Methyl chloroisothiazolone stability in two metalworking fluids. (Modified from Ref. 9.)

is also halved; for phenols, $\eta = 6$, which means that if C is halved, activity is reduced by 2^6 or to $1/64$ of the original activity. Extrapolate this scenario to concentrate dosing in which the biocide has been added at 20 times its use dilution based on a 1:20 dilution of said concentrate. If a user dilutes to 1:40 accidentally or on purpose, depending on η, the results can be disastrous.

Concentrate dosing has already been discussed. However, even in the diluted form, some formulation components contribute to biocide activity and longevity. Increase or decrease of pH affects a number of MWF biocides. As mentioned, methyl chloroisothiazolone is more stable at acid and neutral pH, and although there is molecular breakdown above pH 9, the rate of microbial death (i.e., chemical halflife is longer than biological halflife).

Two commonly used formaldehyde condensates, Tris Nitro and Hexahydrotriazine, are respectively increased and decreased in activity at pHs of 9–10. Corrosion inhibitors, major ingredients in MWFs, can interact with biocides in a number of ways in which both corrosion protection and biocidal activity can be affected. The number of permutations is extensive and can be positive, negative, or neutral. Several examples are worth mentioning. Sodium molybdate, a corrosion inhibitor, increases the activity of isothiazolones versus sulfate-reducing bacteria. Hexyldiethanolamine, another corrosion inhibitor, increases the biocidal activity of hexahydrotriazines [10]. Conversely, some biocides can influence the performance of corrosion inhibitors (Table 2). Both chlorine dioxide and isothiazolone can antagonize several corrosion inhibitors and negatively impact their performance (Table 3).

Because interactions between fluid components can vary depending on the additives and the biocide chemistry, it is suggested that all new MWF formulations be tested for both biocidal efficacy and corrosion protection. Along the lines of biocide–corrosion inhibitor interaction is the subject of metal–biocide interaction. Depending on the metal with which a biocide comes in contact, there can either be synergism or antagonism. Once again it is suggested that testing be done in order to determine the reaction of these two materials.

C. Biocide Mixtures

There are many benefits which can be derived from mixing two or more biocides together. Such mixtures can occasionally produce synergistic activity. Also, mixtures can be broad spectrum (i.e., effective against both bacteria and fungi), where the individual components have only single activity. Effective mixtures, by lowering the overall amount of biocide required, can reduce MWF toxicity to the worker. Finally, mixtures can either be used to overcome microbial resistance to a particular biocide or, more importantly, keep resistance from developing.

Proof of synergism (i.e., where $1 + 1 = 3$ or more) is not always easy. There are a number of published cases in which relatively nonbiocidal compounds are

TABLE 2 Effect of Corrosion Inhibitors on Biocide Activity

Biocide:	Polymeric quat[a]		Chlorine dioxide[b]		Glutaraldehyde[c]		Nitromorpholine[d]		Hexahydrotriazine[e]		Hexahydrotriazine + pyridinethione[f]		Methyl chloro-isothiazoline[g]	
Corrosion inhibitor	B	F	B	F	B	F	B	F	B	F	B	F	B	F
Amine borate (2%)	S	S	A	N	A	N	N	N	S	S	S	S	N	N
Amine carboxylate (2%)	A	A	S	S	N	S	N	N	N	N	N	N	N	N
Cyclic alkanolamine (0.1%)	N	A	N	S	N	N	N	S	S	S	S	S	N	N
Amino propanol (2%)	N	N	S	N	N	N	N	N	N	N	N	N	N	N
Sodium molybdate (2%)	N	N	N	N	N	N	N	N	N	N	N	N	S/A	N
Aryl sulfonamide carboxylic acid (2%)	N	N	N	N	N	N	S	S	N	N	N	N	N	N
Sodium tolyl triazole (2%)	S	S	N	N	N	N	N	N	N	N	N	N	S	N

B, bacteria; F, fungi; S, synergistic effect; A, antagonistic effect; N, no effect.
[a]ppm commercial biocide = 100, 500, 1000
[b]ppm commercial biocide = 1000, 2000, 3000
[c]ppm commercial biocide = 500, 1000, 2000
[d]ppm commercial biocide = 250, 500, 1000
[e]ppm commercial biocide = 375, 750, 1500
[f]ppm commercial biocide = 250, 500, 1000
[g]ppm commercial biocide = 10, 25, 50

TABLE 3 Effect of Biocides on Corrosion Protection: ASTM Cast Iron Chip Test (ASTM Designation D4627)

Corrosion inhibitor	No biocide	Biocide					
		PQ	CLO	GL	NM	TR	TR+P	MCI
Water	2+	3+	4+	3+	2+	2+	2+	3+
Triethanolamine	1+	1+	3+	1+	1+	1+	1+	1+
Soluble oil	1+	3+	2+	3+	0	0	0	3+
Amine borate (2%)	2+	2+	3+	3+	1+	1+	1+	2+
Amine carboxylate (2%)	1+	1+	2+	2+	1+	0	1+	2+
Cyclic alkanolamine (0.1%)	2+	2+	4+	3+	1+	2+	2+	3+
Amino propanol (2%)	2+	2+	4+	2+	1+	1+	2+	3+
Sodium molybdate (2%)	1+	1+	2+	2+	1+	1+	1+	1+
Aryl sulfonamide carboxylic acid (2%)	3+	3+	3+	3+	2+	3+	3+	3+
Sodium tolyl triazole (2%)	2+	1+	3+	1+	1+	2+	2+	2+

Key: 0, no corrosion on filter paper; 1+, <25% corrosion on filter paper; 2+, 25–50% corrosion on filter paper; 3+, 50–75% corrosion on filter paper; 4+, >75% corrosion on filter paper.
PQ, polymeric quat; CLO, chlorine dioxide; GL, glutaraldehyde; NM, nitromorpholine; TR, hexahydrotriazine; TR+P, hexahydrotriazine + pyridinethione; MCI, methyl chloroisothiazolone.

effective in increasing activity of known biocides [10–12]. For example, EDTA, Cu^{2+}, and hexyldiethanolamine all increase the activity of hexahydrotriazine, while the latter compound and other formaldehyde releasers as well as Cu^{+2} apparently increase the activity of isothiazolones [13].

A most popular mixture of sodium mercaptopyridine and hexahydrotriazine (marketed by Olin Corporation under the name of Triadine 10) has enjoyed commercial success for about 15 years. The mixture is effective against both bacteria and fungi. Not only does it give the product broad spectrum properties, it is also somewhat synergistic in that the triazine's antibacterial properties are actually potentiated by the introduction of the pyridine [14].

The use of mixtures to reduce the development of resistant populations is a biologically reasonable approach since it is rare for resistant mutants to evolve to two biocides simultaneously. However, in the presence of an underdosed mixture, mutants resistant to two biocides can develop sequentially. The preferred treatment would be to dose systems alternatively with the two products [15].

D. Mode of Action

A detailed presentation on mode of antimicrobial action is beyond the scope of this chapter. Suffice it to say that where known, the information enables practitioners to make more informed decisions on biocide selection. For example, more than ten chemically different products all based on formaldehyde in their structures owe their activity totally or in part to formaldehyde [16]. If one fails it would be fruitless to try another of the so-called formaldehyde condensates simply because their mode of action is the same. Knowing that isothiazolones react with cell nucleophiles (amino acids, nucleic acids) helps us to understand why these biocides lose effectiveness in the presence of certain chemicals.

E. Guide for Biocide Selection

At this juncture it should be clear that all selections must be based on some test protocols. The following list summarizes the sequence for selection.

1. Product must be EPA (or equivalent) registered. This ensures that environmental and toxicological concerns are covered.
2. Product must be compatible with the MWF. That is, it must not interfere with any of the functional attributes of the fluid (e.g., corrosion protection).
3. Product must be readily waste treatable.
4. Product should be cost effective.
5. Product must be broad spectrum in controlling all microbes responsible for deterioration problems.

VII. OTHER METHODS FOR MICROBIAL CONTROL

"Other" here is meant to describe biocidal control by nontraditional, nonchemical methods. Most of these procedures represent either physical or mechanical methods for killing or separating microbes from the MWF, or in some cases merely neutralizing the results of microbial activity.

A. Physical Methods

1. Heat Pasteurization

Pasteurization, especially as it is used in the dairy industry, implies temperatures below the boiling point of water employed to reduce but not eliminate microbial populations. There is no question that such temperatures (Table 4) will drastically reduce the numbers of living bacteria. About 24 h after heating, the survivors are able to reproduce to the point where the original level has been reestablished. With successive heatings, heat-resistant survivors will reproduce more rapidly [17].

When MWFs are pasteurized, heat is applied to the fluid in a bypass setup. Bearing in mind that since one MWF function is cooling, there is a practical limit to both the size and rate of bypass heating. Note that after heating and subsequent

TABLE 4 Effect of Biocides on Pasteurized Metalworking Fluid Bacterial Survivors

Biocide/ppm[a]	Log$_{10}$ CFU/ml aerobic bacteria											
	Time zero			24 h after pasteurization			48 h after pasteurization			72 h after pasteurization		
	25°C	60°C	70°C	25°C	60°C	70°C	25°C	60°C	70°C	25°C	60°C	70°C
Control	7	<3	<3	6.7	6.1	6.8	6	6	6.1	7.7	6.1	7.1
Bioban P-1487[b]/250				7.1	<3	<3	7.1	<0	<0	6.3	<0	<0
/500				4.8	<3	<3	4.8	<0	<0	6.8	<0	<0
/750				4.3	<3	<3	4.3	<0	<0	7.1	<0	<0
Kathon 886MW[c]/10				4.7	<3	<3	5	<0	<0	6.4	<0	<0
/25				<3	<3	<3	<0	<0	<0	1.3	<0	<0
/50				<3	<3	<3	<0	<0	<0	<0	<0	<0
Uconex 345[d]/250				7.4	<3	<3	6.7	4.1	>3	6	5.6	5
/500				7.2	<3	<3	5.4	3.0	1.4	6.1	5.4	6
/750				7.0	<3	<3	7.9	<0	<0	7.4	1.4	<0
Traidine 10[e]/250				7.2	<3	<3	5.8	2.4	1.6	4.7	5.5	1.6
/500				6.8	<3	<3	5.8	2.7	<0	4.0	<0	<0
/750				6.1	<3	<3	5.8	3.0	<0	5.9	5.6	<0

[a]ppm of commercial product.
[b]4,4'(2-ethyl-2-nitrotrimethylene) dimorpholine (20%) + 4-(2-nitrobutyl) morpholine (70%).
[c]5-chloro-2-methyl-3(2H)-isothiazolone (11%) + 2-methyl-3(2H)-isothiazolone (3.5%).
[d]glutaraldehyde (45%).
[e]1,3,5-tris(2-hydroxyethyl) hexahydro-s-triazine (63.6%) + 2-pyridinethiol-1-oxide (6.4%).

cooling, there is no residual microbial control as there would be with a more conventional chemical biocide treatment. However, studies have shown that the use of heat in combination with chemical biocides can create a synergistic effect, thus lowering the amount of biocide normally needed.

2. Microwave

The use of microwave in water-based fluids is simply a different way of heat treating for microbial control [18].

3. Sonic Oscillation

High-energy sound (i.e., 25,000 cycles/s) will kill microbes in MWFs, but again the mode here is primarily from heat generation.

4. Radiation

There is some history of the use of ultraviolet radiation in the range of 260 nm in an actual application. However, UV light has poor penetrability even under the best conditions, such as in clean water, and in a MWF it could hardly penetrate at all. Unless a system could be set up where a very thin film of fluid could pass by and recirculate over a UV source, practical effectiveness should not be expected. Although impractical for obvious reasons, ionizing radiations (e.g., pulse-applied ^{137}Ce gamma rays) could control bacterial growth in a contaminated soluble oil emulsion. A laboratory study was performed in which this type of radiation was applied to MWFs. The results of the experiment were published in 1968 [19]. However, no field trials with this method have been done, again for obvious reasons of safety. Imagine exposing workers to potentially high doses of ionizing radiation!

B. Mechanical Methods

Mechanical methods are those procedures which physically remove the microbes from the fluid.

1. Filtration

Filters employed in MWFs do not have the inherent pore size to filter out microorganisms directly. These mechanisms are instead designed to take out larger particles such as fines, swarf, chips, and metal oxides. More often than not, these particulates will have microbes adhering to their surfaces, so when they are filtered out, microbes are removed as well. Depending on the biological condition of the system, the particulates so filtered may remove 50–90% of the microbes which adhere to their surfaces.

2. Centrifugation

The same statement can be made for centrifugation as for filtration. The forces are not sufficient to separate microbes from MWFs, but those attached to particulates will be removed.

In summary, mechanical methods do remove some microbes but not a significant amount. More importantly, in studies on pasteurization efficacy it was found that systems which have been cleaned by centrifugation responded more favorably to both heat and biocide treatment [17].

VIII. LABORATORY TESTING

A. The Need for Testing

In order for biocides to be used effectively it is imperative that they be tested in the laboratory prior to introduction into the field. Naturally all new biocides are put through rigorous efficacy trials by the chemical companies that develop them. However, simply because someone has made a decision to register a product for use in MWFs does not necessarily mean that it will perform well under all conditions and in all situations, hence the need to test fluid–biocide combinations for specific applications.

Microbiological efficacy testing must be done by fluid formulators in the fluids they wish to evaluate. End-users not only test the biocide in the fluid they plan to use or are currently using, they must also attempt to develop a microbial challenge based on microorganisms endogenous to their plant.

B. ASTM D3946

There are two biocide efficacy tests for MWF which are sanctioned by the American Society for Testing and Materials (ASTM). One has the ASTM designation D3946 and is called, "Test Method for Evaluating the Bioresistance of Water-Soluble Metal-Working Fluids" [5,14]. This test was originally designed to measure the relative bioresistance of MWFs without added biocide. It can still be used for this purpose, particularly by fluid formulators who attempt to develop fluids which are relatively bioresistant. However, for the most part ASTM D3946 is used as a short-term biocide efficacy test. The procedure involves setting up 1 l samples of use-diluted MWF. The samples are inoculated with bacteria and fungi at time zero, and pH and microbial measurements are made at that time. Samples are then aerated with compressed air for five days, at which time pH and microbial measurements are again made. Aeration is shut off for two days to encourage anaerobic growth and for a third time microbes and pH are measured. Finally, aeration is resumed for five days and a last microbial and pH measurement are made. This is a good short-term test as the aeration and inoculation offer a significant challenge to the fluid–biocide package and gives the experimenter a significant amount of data in a relatively short period of time.

C. ASTM E686

The other ASTM test involved with MWF biocides is designated E686 and entitled, "Method for the Evaluation of Antimicrobial Agents in Aqueous Metal-Working Fluids" [14,20]. This test is very similar to ASTM D3946. One liter samples of use-diluted fluid are inoculated with bacteria and fungi at time zero. Microbial counts are made and pH is also measured at this time. Biocides are added as well. The aeration scheme is also similar to the D3946 test in that it follows a five-day on/two-day off regimen. The schedule is purposely set up to simulate field conditions wherein a plant is running during the five weekdays and shut down on weekends. Such a scenario would tend to favor the growth of anaerobic bacteria, the kinds of organisms responsible for "Monday morning" odor. By ceasing aeration in the laboratory test for two days, the biocide is challenged in its ability to deal with this specific problem.

ASTM E686 is scheduled to last for six weeks and it follows the specific on/off aeration regimen for that entire length of time. Unlike the two week bioresistance test (D3946), E686 calls for reinoculation of the fluid each week. Thus the test offers a much greater challenge to the biocide, both in time and in severity of microbial challenge. What may typically happen in the lab is that a fairly large number of samples will be run using the short-term (D3946) test protocol. Successful samples from that series will then be given the additional challenge of exposure to the long-term test.

D. Other Microbiological Tests

Almost without exception, microbiological testing of MWFs in the laboratory deals with concerns of microbial control. If a biocide is working, laboratory analyses can determine just how well and what biocide level is required to maintain effective control. Field samples can be checked for bacterial and fungal count by traditional methods such as plate counts and biocide level measured, either by wet chemistry procedures or more automated tests such as HPLC.

The most important information these tests can reveal is the nature of a biocide failure. Assuming the right dosage has been applied, a biocide can fail for two major reasons. First, a biocide can either be incompatible with the fluid or with some other element in the metalworking system (e.g., tramp oil, sludge, metals). Analysis for residual biocide level will determine if this is the case. If, for example, 100 ppm of product has been applied and an analysis several hours after dosing shows only 20 ppm left, obviously something is destroying the biocide. On the other hand, if the microbial count in the fluid is elevated and an analysis shows that what should be an effective level of biocide is detectable, there is a chance the microbes have "built up" resistance to the biocide. This is the second major reason a biocide can fail. In order to prove this, fluid–biocide incompatibility must be ruled out. The next step would be an isolation of the various organisms in the fluid followed by an exposure of the pure cultures to varying biocide levels. This type of experiment can specifically determine if true resistance has developed. If this has occurred it is important to switch biocides immediately and begin a program of alternating biocide type every other month so that resistance is less likely to develop again.

Another microbiological procedure which is often performed on MWFs is the isolation and identification of the microorganisms present in the sample. This is usually done to trace the cause of a particular problem such as slime buildup, corrosion, change in fluid color, or the development of a unique odor. Also, there may be occasion to suspect that microorganisms in the fluid will have an adverse effect on the health of exposed workers. Identification of bacteria and fungi in the fluid is done to see if any harmful organisms are present. However, as will be discussed in the next section, the survival of human pathogens in MWFs is extremely rare.

IX. HEALTH EFFECTS OF MICROBES IN MWFs

A. MWF Bacteria Rarely Cause Disease [21]

Regardless of which biocide is chosen or what type of fluid is used, it is simply unrealistic to expect to run sterile MWF systems. The existence of bacteria in water-based coolants and lubricants is a fact of life. What needs to be controlled and what can be controlled with proper maintenance are the levels of bacteria. Typically, well-maintained systems should be running bacterial counts of less than one million per milliliter.

With the acknowledgment that some bacteria *must* exist comes the inevitable question of just how safe is it to work in the presence of literally billions of microorganisms. The answer to that question is that, except under rare conditions, bacteria capable of surviving in MWFs cannot cause harm to the worker. True pathogenic bacteria (that is, those able to cause human infection and disease) need a very nutrient-rich environment for growth and metabolism, an environment which cannot be found in MWFs. Since it is very unlikely that disease-causing bacteria can survive in MWFs, it is equally unlikely that bacteria in fluids can cause dermatitis among those people routinely exposed to them.

B. Occasionally There Are Exceptions

In certain instances, microbes in MWFs can prove harmful to workers. One case would be when a worker is severely immunocompromised, that is, the person's natural immune system is somehow damaged and therefore fails to provide natural resistance to otherwise harmless organisms. This could be caused, for example, by a disease such as AIDS or by certain immunosuppressant drugs such as those taken by some cancer patients. In these cases even the "harmless" bacteria in the fluids could cause infection and even death.

Another example of illness presumed to be from MWF microbes occurred in 1980 at a Ford Motor Company manufacturing facility in Windsor, Ontario, Canada [22]. In that case over 200 workers were stricken with a malady called Pontiac fever, which is a flulike illness caused by the same bacteria that causes the more serious Legionnaire's disease. The Centers for Disease Control (CDC) investigated this outbreak and concluded that the illness was caused by certain bacteria which were isolated from a very poorly maintained MWF sump. By poorly maintained they meant the system was stagnant for a long period of time, had a very low pH (below 7.0), and had significantly elevated populations of aerobic and anaerobic bacteria. The CDC recommended that in order to avoid any future outbreaks, bacterial populations in general should be controlled and sumps should not be allowed to stagnate for extended periods of time. Laboratory studies have shown that low pH, unaerated fluids, and high populations of all types of bacteria make ideal breeding grounds for *Legionella*, the organisms responsible for Pontiac fever. In fact, the same study showed that *Legionella* actually cannot survive in virgin, use-diluted MWFs. The only way this organism can stay alive in MWFs is for the fluid to be rancid, the more rancid, the better (Table 5).

C. Endotoxins

Although it is rare for bacteria in fluids to directly cause illness, there is a suspected indirect relationship between certain maladies and microbial load. Most bacteria

TABLE 5 Percent Die-Off of *Legionella* Species in Selected Metalworking Fluids After 18 h Exposure at 25°C

Metalworking fluid		*Legionella pneumophila*		*Legionella feeleii*	
		Stagnant	Aerated	Stagnant	Aerated
Soluble oil:	New	99.9	99.99	99.9	99.9
	Used	97.0	99.6	99.0	99.99
Synthetic:	New	50.0	50.0	99.9	99.9
	Used	0	0	99.0	99.9
Semisynthetic:	New	99.99	99.99	99.99	99.99
	Used	99.99	99.99	99.99	99.99
Controls:	Phosphate buffer	0	0	0	0
	Hard water	0	0	0	0

found in metalworking systems are gram negative. These bacteria possess in their outer membrane a substance called endotoxin. It is well known that endotoxins are capable of causing illness, particularly a short-term flulike illness with fever and chills. Although endotoxins have most certainly been identified in fairly high levels in MWFs, the relationship they have to human disease is still not proven [23]. It is interesting to note, however, that certain commonly used MWF biocides whose activity is related to formaldehyde have an ability to neutralize the effects of endotoxin. Since endotoxins are present in both living and dead bacteria it is possible for some biocides to reduce bacterial count without affecting endotoxin activity. If endotoxins are proven in the future to cause harm as they become aerosolized in MWFs, then one criterion for biocide selection may turn out to be whether a biocide can neutralize endotoxins [24].

X. FUTURE TRENDS AND CONSIDERATIONS

The logical question at the end of chapters such as this is, Where do we go from here? What does the future hold for the microbiology of MWFs? One very current issue is the subject of endotoxins, which has been discussed.

Another topic on the forefront of research is the potential for formulating fluids that by their nature resist microbial growth and therefore require no biocide. Some studies have already shown that the inclusion of certain alkanolamines in the fluid formulation can lessen the ability of microbes to grow in the fluid [25]. This effect is especially pronounced at pHs above 9.5.

Other formulators have had success with the use of amine borates as a fluid component. Fluids containing this material do show relative bioresistance compared to similarly formulated fluids without it. The key to all the attempts at biostable or bioresistant fluids is that efforts thus far have yielded only relative bioresistance or relative biostability. The problem is that microbes are very industrious and eventually even the most hostile environments can be adapted to by certain organisms. The problem is further exacerbated by overdiluting the fluids. While a formula may be fairly biostable at 20:1, dilution to 30:1 or beyond might render the fluid completely defenseless to microbial activity.

Many new and innovative ways to enumerate bacteria are constantly being developed. Most of these methods originate from clinical bacteriology as methods for counting bacteria in urine. Some involve testing for specific deoxyribonucleic acid (DNA) (as is being done in forensics), adenosine triphosphate (ATP), or other components of the microbial cell [6]. Thus far, all of these methods are cumbersome, expensive, or lack the sensitivity required for accurate testing. The ideal test would obviously be the one that is cheap, very fast, and always accurate. If such a test were available, it would undoubtedly be very, very popular.

REFERENCES

1. H. W. Rossmoore and L. A. Rossmoore, *Practical Manual on Microbiologically Influenced Corrosion*, National Assn. of Corrosion Engineers (NACE), Houston (in press).
2. G. H. M. Holtzman, H. W. Rossmoore, E. Holodnik, and M. Weintraub, *Dev. Ind. Microbiol.* 23: 207–216 (1982).
3. R. Almen, G. Mantelli, P. McTeer, and S. Nakayama, *Lubr. Eng.* 38(2): 99–103 (1981).
4. H. W. Rossmoore and L. A. Rossmoore, *Int. Biodet.* 27(2): 145–156 (1991).
5. American Society for Testing and Materials (ASTM) Designation D 3946-92, in *Annual Book of ASTM Standards*, Vol. 5.03, ASTM, Philadelphia (1992).
6. ASTM Designation E 1326-90, in *Annual Book of ASTM Standards*, Vol. 11.04, ASTM, Philadelphia (1990).
7. J. E. Gannon and E. O. Bennett, *Tribol. Int.* 14: 3–6 (1981).
8. H. W. Rossmoore, in *Comprehensive Biotechnology 3*, M. Moo-Young, C. L. Cooney, and A. E. Humphrey, eds., Pergamon, New York, pp. 249–269 (1986).
9. G. L. Willingham, and G. L. and R. L. Derbyshire, *Lubr. Eng.* 47(9): 729–732 (1990).
10. D. Oppong and E. O. Bennett, *Tribol. Int.* 22: 343–345 (1989).
11. M. Sondossi, V. F. Riha, and H. W. Rossmoore, *Int. Biodet.* 26: 51–61 (1990).
12. V. F. Riha, M. Sondossi, and H. W. Rossmoore, *Int. Biodet.* 26: 303–313 (1990).
13. H. W. Rossmoore, *Int. Biodet.* 26: 225–235 (1990).
14. H. W. Rossmoore, J. F. Sieckhaus, L. A. Rossmoore, and D. DeFonzo, *Lubr. Eng. 35*: 559–563 (1979).
15. M. Sondossi, V. F. Riha, H. W. Rossmoore, and M. Sylvestre, *Int. Biodet. Biodegrad.* (in press).
16. M. Sondossi, H. W. Rossmoore, and J. W. Wireman, *J. Ind. Microbiol.* 1: 86–96 (1986).
17. H. W. Rossmoore, L. A. Rossmoore, and A. L. Kaiser, in *Biodeterioration 7*, D. R. Houghton, R. N. Smith, and H. O. W. Eggins, eds., Elsevier, New York, pp. 517–522 (1988).
18. E. C. Hill and R. Elsmore, *Biodeterioration 5*, T. Oxley and S. Barry, eds., Wiley, New York, pp. 462–471 (1983).
19. H. W. Rossmoore and J. G. Brazin, in *Biodeterioration of Materials, Microbiological and Allied Aspects*, A. H. Walters and J. J. Elphick, eds., Elsevier, New York, pp. 386–402 (1968).
20. ASTM Designation E 686-91, in *Annual Book of ASTM Standards*, Vol. 1104, ASTM, Philadelphia (1991).
21. H. W. Rossmoore, *Lubr. Eng. 50*: 253–260 (1993).
22. L. A. Herwaldt et al., *Ann. Intern. Med. 100*: 333–338 (1984).
23. S. M. Kennedy et al., *Amer. J. Ind. Med. 15*: 627–641 (1989).
24. H. Douglas, H. W. Rossmoore, F. J. Passman, and L. A. Rossmoore, *Dev. Ind. Microbiol. 31*: 221–224 (1990).
25. M. Sandin, S. Allenmark, and L. Edebo, *Antimicrob. Agents Chemother. 34*: 491–493 (1991).

10
Filtration Systems for Metalworking Fluids

ROBERT H. BRANDT
Brandt & Associates, Inc.
Pemberville, Ohio

I. INTRODUCTION

Metalworking fluid chemistry and its uses have gone through many changes over the years. With the greater performance requirements for both direct metal removal attributes and indirect functional attributes, upsets in metalworking fluid integrity will impact performance. If a once-through use was practiced, contamination from the operation itself would not cause difficulty. However, even once-through use could be compromised by the water used to mix the concentrate and the containment or delivery methods chosen. But for the most part, the metalworking fluid would contain all the active ingredients and performance packages required.

In the real world, the practice of once-through fluid use is not acceptable except in some specific processes. Therefore, the metalworking fluid is reused. The reuse consists of collecting the used fluid in some sort of container and then recirculating the fluid back to the tool–workpiece interface. Units as simple as a tank and pump may constitute a recirculation system.

When reuse is instituted, a variety of interactions from an assorted number of contaminants begin to occur. The metalworking fluid is subjected to the metal chips and fines of the process, airborne contamination from cascading fluid over a part and the machine, machine leakages, residues left on the part from previous operations, water, operators, etc. [1]. This list can be quite long.

In order to provide metalworking fluid in an acceptable operating condition after it has been subjected to these and other degradation contaminants, the impact from these contaminants needs to be minimized. This can be accomplished chemically by adding new concentrate or additives to the working solution. However, some contaminants may not be appreciably affected by additives or concentrate additions. These include metal chips, fines, and free oil.

Whenever possible, contaminants need to be removed from the metalworking fluid and the system. The removal process is generally some means of separation or filtration. Many aspects of the process need to be discussed before a final approach or system can be selected. It is not just a matter of saying we want to remove the metal chips. Various criteria for design need to be addressed and a number of questions need to be answered. These criteria include: material being worked, type of machine processing, chip shapes produced, amount of material removed, production rates, machine horsepower, metalworking fluid type, amount of fluid required, and a floor plan layout. As we begin to develop a system, typical components will be addressed and some answers will be provided. Typical system components include: return troughs, chip conveyors, filters, supply line pumps, makeup systems, and electrical and pneumatic controls.

Application of ultrafiltration, nanofiltration, or reverse osmosis is generally not applied to in-plant metalworking systems. These filtration regimes are used in selected processes such as incoming water preparation and metalworking fluid waste treatment. These membrane systems, if used as the metalworking fluid filtration on-site, would have a deleterious effect on the fluid by selectively removing certain ingredients.

II. PARTICULATE

Before moving to a discussion of various particulate, let us first address a key issue—How free of particulate should the fluid be kept? How clean should the fluid be? Various answers are given and various approaches taken to these questions. A piece of equipment may be purchased with an implied guarantee of particulate cleanliness. However, each piece of equipment will provide cleanliness to a certain equilibrium level. We can always expect some residual, equilibrium level of metal particulate in the metalworking fluid. There is no absolute filtration in the metalworking fluid industry such that all metallic particulate will be removed. Given this fact, the best that can be attained is to minimize the metallic contamination equilibrium level being maintained by one or a combination of separation or filtration devices. Considering only the usual filtration and separation processes then, let us look at the equilibrium level of the metal fines in the metalworking fluid.

The equilibrium level occurs because the filtration device will not remove one-hundred percent of the metal removed in the metalworking process each time the fluid passes through the filter [2]. This residual quantity stays in the system and builds until the filter reaches the equilibrium level. This level is different for each process and filter system. The level is related to the fluid used, the maintenance of the fluid and machines, the filter, and a host of other outside influences. To determine the equilibrium level which is reasonable and/or necessary, tests can be run on existing filtration systems and then correlated [3]. These tests will provide a direction toward the particulate levels which should be maintained. Table 1 provides some information on two aspects of the metal particulate to consider at the equilibrium level. The first is the amount of metal fines which would be tolerable, i.e., the quantity of the dirt in the system. It is not sufficient, however, to describe only a quantity of dirt. This quantity of dirt, expressed in mg/l or parts per million (ppm), gives only an indication of the amount by weight of dirt at equilibrium. For example, a quantity represented by 10 mg/l (ppm) would seem small. Extrapolated to its meaning in a large volume system, however, it could be quite significant. In a 10,000 gal (38,000 l) central filtration system, the weight of metalworking fluid may approach 83,000 lb (38,000 kg). This would mean approximately 0.83 lb (0.38 kg) of metal fines circulating in the system. If a system

TABLE 1 Equilibrium Averages for Suspended Solids from Various Equipment

	Quantity[a] (ppm)	Quality[b] (μm)
Cast Iron		
Machining	20	15
Grinding	30	30
Steel		
Machining	25	20
Grinding	12	16
Aluminum		
Machining	10	15
Grinding	10	15
Glass		
Grinding	100	<5

[a]Metallic particulate only.
[b]Most probable size: measurement by microscope and electrical sensing zone instrument.

contained 100,000 gal (380,000 l), the quantity would be 8.3 lb (3.8 kg). Increasing the parts per million would mean higher levels of recirculating metal fines.

The other necessary number to deal with at the equilibrium point is the quality of the recirculating metal fines. The quality refers to the size of the fines being circulated. How large are the particles reaching the tool–workpiece interface? It has been suggested that particulate in the 3 to 8 μm range have more effect than had been previously suspected [4]. The size of the particle, however, is open to some discussion because there are a number of ways to describe and also determine the particle size. Typically, particle size is considered in terms of spheres. Therefore, the particle size number may be referred to as the spherical diameter. This, however, is generally different than in the real world. Chips and particulate come in a variety of shapes including flat platelet, cylindrical, parts of a broken helix, etc. Rarely is the particle a sphere. But a guide or common reference is necessary and therefore we talk in terms of only a single micron size—a linear dimension—to describe the particulate. Figure 1 shows some linear dimension comparisons. Agreement on an acceptable level of fluid cleanliness is one of the first requirements of system design. There should be two numbers presented: a quantity of recirculated dirt at equilibrium and the average particle size.

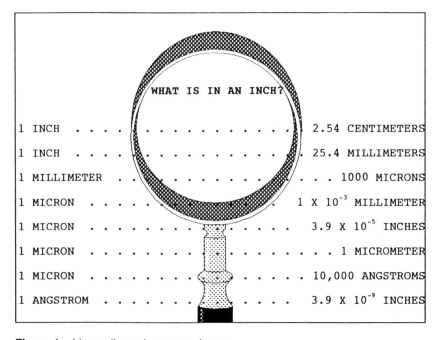

Figure 1 Linear dimension comparisons.

Filtration Systems 277

Figure 2 Varieties of steel-machining chips.

The particulate being recirculated at equilibrium is only the resulting end of the particulate produced by the metal removal operation. The particulate produced consists of a large number of different shapes, sizes (up to feet in length), and volume. We will talk about volume in a subsequent section. The shapes and sizes vary according to the metal being worked and the tooling on the machine. If the metal is steel and it is being machined, the chips can be long and stringy or curled and small (see Fig. 2). If the steel is being processed by grinding, the chips produced are generally referred to as "fish hooks" because they lock together in a steel wool pad arrangement (see Fig. 3). Aluminum forms a variety of shapes. The shape of

Figure 3 Varieties of steel-grinding chips.

Figure 4 Varieties of aluminum machining chips.

aluminum machining chips are in general the same as aluminum grinding chips (see Fig. 4). Cast iron produces a much different chip than either steel or aluminum, more of a flat type and the chips form a more dense mass (see Fig. 5). Under usual machine operations, the metals produce a variety of large and small chips. In certain operations, however, such as honing, finish grinding, lapping, and others, the average size of the chip is small and presents a different filtration requirement.

Figure 5 Varieties of cast iron machining chips.

With different alloys, different machines and processes, and different tooling, it is necessary to explore the chip configuration closely in order to apply the best filtration system available to reach the desired effect.

III. TRANSPORT SYSTEMS

While the metalworking fluid is being delivered to the machine at the tool–workpiece interface, work is being done in the form of metal removal. The fluid flushes chips produced in the operation and carries them as part of the fluid flow. This fluid mixture flows off the machine and into a variety of devices which will transport the fluid and chips back to a point of separation or filtration. These methods include H-chain, chain and flight, push bar (harpoon), metalworking fluid, and in some cases overhead troughs and sumps.

A. H-Chain

The H-chain or rubbish chain is not commonly used because of potential repair difficulty when the chain breaks under machines (see Fig. 6). In use, the H-chain is applied to cast iron-type materials which have been machined. This system does provide one advantage, it can deliver the chips to a tote-box prior to the separation or filtration equipment. A problem with this method is the need for additional fluid-holding capacity in the reservoir of the filter in order to accommodate the "draw down" resulting from the retention of fluid in the conveyor trench.

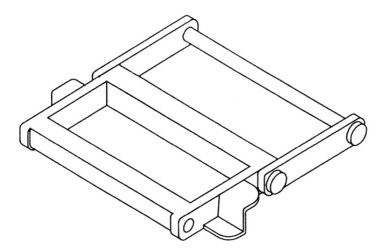

Figure 6 Typical H-chain section. (Courtesy Clarmatic Industries, Inc.)

B. Push Bar (Harpoon)

The push bar conveyor or oscillating system is typically applied to steel-machining systems (see Fig. 7). Steel chips can be delivered to a tote-box before the fluid is separated or filtered. The typical installation requires a large amount of mechanical apparatus under the floor, usually in troughs. This system requires maintenance of the system's hydraulic components as well as the push bar itself. In typical setups, the fines and metalworking fluid are allowed to exit the system through panels of perforated plate into a filtration system. Although these conveyors can be used for other metals, steel machining is the most common application. The same draw-down considerations are needed with this system as with (the H-chain) because of the retention of metalworking fluid in the troughs.

C. Chain and Flight

Another type of conveyor system applied in special cases is referred to as the chain and flight system (see Fig. 8). This system is composed of two continuous loops of chain between which has been bolted or welded iron bar stock or angle iron. This chain and flight system is put into a trough and the movement of the conveyor removes heavy settled solids up a ramp to a discharge point. The trough usually has an overflow opening so the liquid can flow out of the trough, along with fine metal particles, and into a filter system. The advantage of this system is that, it removes

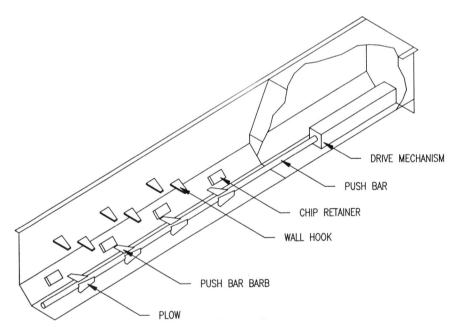

Figure 7   Typical push bar (harpoon) section. (Courtesy Clarmatic Industries, Inc.)

Filtration Systems

Figure 8 Chain and flight conveyor.

bulk solids before the filtration process system. The metals usually machined or ground when this type of system is employed are cast iron or nodular iron. Draw down is also a concern with this system as with the two mentioned earlier.

D. Flume System

The most commonly used method for moving chips and fines generated during the machining or grinding process is the metalworking fluid itself. This is referred to as a velocity flume system (see Fig. 9). The fluid and the metal particles fall from the machine into a trough in the floor under or alongside the machine. This trough

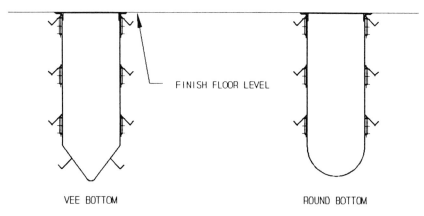

Figure 9 Troughing advantages and disadvantages. Primary applications include: (a) cast iron machining, grinding, honing; (b) aluminum machining, grinding; (c) steel machining, grinding (most universal return system). Advantages include: (a) flexibility in layout, (b) readily adapts to machine wet decks and foundations, (c) adapts to changes in material machined, (d) cost. Disadvantages include: (a) requires additional filtration capacity to supply flushing capacities required, (b) velocity flushing may extenuate foaming tendency of any given metalworking fluid, (c) misdirected, plugged, or incorrect flow may produce "dead spots" and plugging may occur.

contains nozzles which deliver metalworking fluid under pressure to the flume or trough system. The momentum of the fluid discharged from the nozzles is transferred to the cascading machine fluid and metal particles. This means the fluid and particles are moved down the trough and into the filtration system. Typically the velocity of the fluid in the flume system varies between 6 and 12 ft. (1.8 m and 3.6 m)/s. Table 2 shows the different velocities needed to maintain chip and fluid movement down a trough with a slope of one-quarter in. (6.4 mm)/ft (0.3 m).

The flume system is provided to transport material to a central process point.

TABLE 2 Metalworking Fluid Velocities in Fluming Systems[a]

	Machining	Grinding
Cast Iron	8 (2.4)	8 (2.4)
Aluminum	6 (1.8)	6–7 (2.0)
Steel	12 (3.6)	10 (3.0)

[a]In units of ft/s and (m/s).

It should not retain particles or fluid when the system is turned off. There are different opinions as to the slope of the trough and number of nozzles used which directly impacts the gallons of fluid needed for the flushing process. A slope of one-half inch (12.8 mm)/ft (0.3 m) would require less gallons and fewer nozzles. If this steep slope is used, the usual practice is to place a nozzle at the end of each trough run with a minimum number or no additional nozzles in the trough. The difficulty with this system may be that for long trough runs, the invert or depth of the trough at its discharge point into the filter is twice as deep as the usual one-quarter inch (6.4 mm)/ft (0.3 m) slope. This will mean a deeper pit and more steel for the flume. An advantage of the one-half inch (12.8 mm)/ft (0.3 m) slope system is the reduced requirement of flushing gallons from the filtration system.

Typically, a transfer line requiring 1500 gal (56,775 l)/min of metalworking fluid on the machine tool may require an additional 1500 gal (56,775 l)/min for flushing the fluid down the trough to the filter system. Usually the filter supplies both the machine and flushing requirements. This can substantially increase the size of the filter. However, a compromise could be used. It is possible to conceive a system with separate pumping and piping systems for the machine and flushing system. This would allow complete flexibility of the gallons used for flushing and yet ensure clean filtered metalworking fluid at the tool–workpiece interface where the best filtered liquid is needed. This type of system would use the initially received fluid after some settling or preseparation for the flushing. The flushing system usually consists of stream-directing nozzles such as fire hose nozzles, which have three-eighths to five-eighths inch openings. It is not necessary to finely filter fluid which will be delivered to these size nozzles. The typical size used is one-half inch. The other liquid would then be further processed through a positive filter to remove fines and be delivered to the machine tool.

The pressure of the fluid at the nozzles will vary based upon the pumping volume of the pumps and the amount of fluid allowed to flow. In systems where the pump supplies both the machine and flushing system, the velocity can vary appreciably. This variation can contribute to higher than designed pressures at the nozzles, resulting in higher velocities of exiting fluid. When this happens, the nozzle discharge liquid becomes a venturi-type device drawing air into the stream and causing or enhancing foam conditions. A dual system would alleviate this fluctuation in pressure. At too low a pressure, the velocity may be decreased enough to cause inadequate flushing, leaving chips in the trough system. Although the fluid flowing over these chips may appear aerated and turbulent, the deposited chips may become stagnant and contribute to metalworking fluid microbiological control problems.

IV. BULK CHIP SEPARATION SYSTEMS

Some metalworking operations produce a volume of chips which would interfere with the normal cycling or indexing of a filter system. These operations are steel machining, aluminum grinding, and machining. When these operations are performed, consideration should be given to removing the bulk of the chips by a distinct separation means. Some of the bulk separation can be accomplished by using the push bar or conveyor-type transport systems. However, because velocity flume flushing is the more commonly used, another separation device needs to be placed at the end of the trough. These devices are usually referred to as primary separators. These units are tanks which are equipped with a perforated plate hinged belt conveyor or a chain and flight conveyor. These conveyors travel in an inlet trough or over a stainless steel wedge wire panel, respectively. The perforated plate hinged belt conveyor system allows the passage of fine particles and fluid into the filter for further processing (see Fig. 10). Most of the large particles of stringly steel machining chips are retained on the conveyor belt and deposited into a tote-bin. The same process occurs for aluminum grinding and machining chips moving into and through wedge wire panels. The bulk of the aluminum chips are removed as the chain and flight conveyor moves over the wedge wire screen. The openings in the wedge wire can vary but are usually one-eighth in. (3.2 mm) (see Fig. 11). This provides for large particle and large volume removal of the aluminum. These types of primary separation systems should be applied on most operations producing large chips or large volumes of chips.

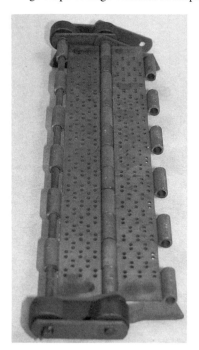

Figure 10 Hinged belt conveyor.

Filtration Systems

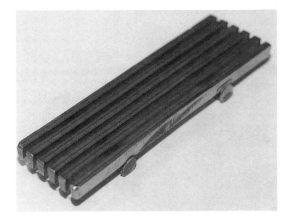

Figure 11 Wedge wire for primary separation.

V. RECIRCULATION SYSTEMS

After the transport of the chips and fluid has been accomplished along with primary separation, if needed, the fluid needs to be further processed. This processing takes place by a process of clarification. Usually the word "filtration" is loosely applied to metalworking fluid and particle separation processes, even if the process relies on a physical characteristic and does not involve a filter. Recirculating systems are just that; they receive dirty fluid and, after processing, send it continuously back to the machines for further use. The time for this to occur may be a minimum of 3 min or as long as 1 h. Whatever the time, it is a circular flow of metalworking fluid and a continuous removal of particles.

These recirculation systems can be divided into two groups and are discussed here in two main categories: Separation systems and filtration systems (see Table 3). These categories contain a variety of equipment developed to accomplish the same thing, cleaner fluid. The driving force found in these categories is either gravity, vacuum, or pressure.

A. Separation Systems

The physical characteristics used in separation processes are specific gravity differential, foam bubble inclusion, or ability to be magnetized. These various characteristics are put to use in a variety of equipment.

1. Settling Tanks

The basic separation device for the individual machine is the settling tank (see Fig. 12). This unit is generally set alongside a machine and receives liquid from

TABLE 3 Clarification Chart

Separation	Simple setting
	Flotation
	Centrifugal
	Cyclone
	Centrifuge
	Magnetic
Filtration	Disposable media
	Bags
	Cartridges
	Rolled media
	Precoats
	Fiber
	Diatomaceous earth
	Permanent media
	Metal mesh
	Fabric belts
	Wedge wire screens

the process. The fluid capacity of the tank provides for a retention time. The retention time is an indication of settling that will take place in the tank. The type of metal and size of chip or particulate formed in the metal removal process as well as the fluid and retention time will determine what equilibrium point will be reached. Typically, the retention time for the average tank is 5 min. Less retention time will usually mean a large quantity of fine recirculation. In large central systems where settling is the only method of particle removal, the retention time should not be less than 10 min. In these systems, the time can range between twelve-and-a-half minutes to 15 min. These settling devices are applied to cast iron machining with some application to other noncake-forming particles, such as glass grinding and silicon sawing. The particles produced by grinding cast iron may also be separated by settling, but the retention time is customarily double that of machining. This is due to the fine nature of the particles produced and their "lightness" compared to the fluid used in the process. The application of settling systems to cast iron machining and grinding does not produce metalworking fluid "clean" enough for the requirements of most critical metal removing processes.

The settling system can provide primary separation of the larger chips and particles. However, separation processes can be added to the settling system in the form of centrifugal and magnetic devices.

Filtration Systems

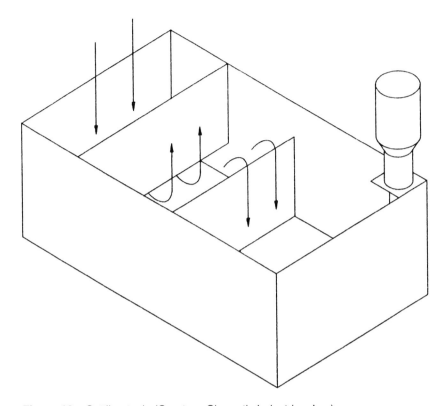

Figure 12 Settling tank. (Courtesy Clarmatic Industries, Inc.)

2. Foam Separators

In some separation systems, particle removal has been reported as enhanced by the occurrence of foam. The generation of foam or small air bubbles tends to entrap the fine particles causing them to float with the foam. The fines brought to the surface of the tank can be removed along with the foam by a conveyor or bar mechanism. This particle-filled foam is generally moved into a tank which settles the fines when the foam breaks. This method has been used but is limited in application to metalworking fluids which can sustain a foam condition. For fines removal, this method has proven difficult to control in a consistent manner.

3. Centrifugal Separators

The addition of centrifugal devices could be in the form of a hydrocyclone and/or centrifuge. Hydrocyclones are devices which, when supplied tangentially with metalworking fluid, move the fine particulate to the outside wall of the hydrocyclone device (see Fig. 13). The particulate concentrates on the outside wall of the device and moves down the cone wall. A portion of the liquid and concentrated particulate flows out of the bottom of the cone unit. The rest of the liquid moves back toward the top of the hydrocyclone through a vortex-finding device. The liquid processed in this manner reaches an equilibrium based on the differential pressure across the hydrocyclone and the specific gravity differences between the fluid and the particles. Under the same conditions, a smaller diameter hydrocyclone can remove finer particulate than a larger diameter hydrocyclone. Larger diameter hydrocyclones are used because they tend to produce more gallons per minute of cleaned fluid. Smaller diameter hydrocyclones are usually manifolded together to produce a common inlet and outlet for a cluster of units. The best performance will be obtained by maintaining proper differential pressure and applying the units to particulate where specific gravities are highly divergent from the liquid. An example of a system which may not reach a satisfactory equilibrium level would be cast iron grinding fines suspended in a heavy opaque soluble emulsion product. Better

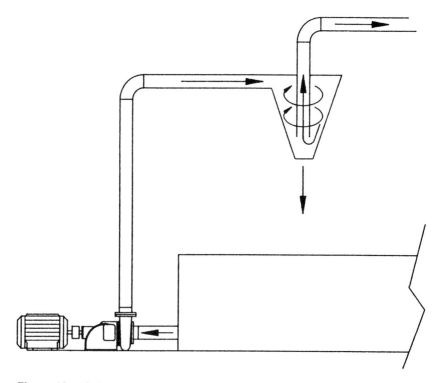

Figure 13 Hydrocyclone system. (Courtesy Clarmatic Industries, Inc.)

separation may occur using a solution-type water miscible product containing particulate from cast iron grinding. Modification of the typical hydrocyclone has been done and has resulted in a dumbbell-shaped unit instead of the typical conical shape. Whatever the outside configuration, the separation parameters are the same.

Another device which can be added to primary separation tanks is the solids separating centrifuge. This unit is designed to remove particles by centrifugal force. A spinning bowl receives the particle-laden metalworking fluid. The force against the fluid and particles separates the particles by moving them to the side of the bowl. The bowl fills with particulate and eventually requires cleaning. Some centrifugal solid separators have liners to facilitate easier cleaning. It is customary to use one of these units on small flows and/or small dirt load systems. These conditions are usually found in individual machines and not large central systems. There is another type of centrifugal separator which will remove particles; however, it is generally applied to systems to solve another separation problem. These will be addressed later.

4. Magnetic Separators

The metalworking industry works a variety of metals, the properties of which are different. Some of the metals interact with a magnetic field. When such a metal is worked, a magnet can be used to remove fine particulate (see Fig. 14). The metalworking fluid is passed in close proximity to the permanent magnet system. As the fluid passes by the magnet, the particles which are magnetizable "stick" to the magnet. The magnet is generally rotating and brings the particles up and out of the liquid. As the magnet continues movement, a nonmagnetic blade removes the accumulated particles from the magnet. If operational changes are made to a system, such as intermittent magnet movement, the particles accumulated can form a "cake" on the magnet. Nonmagnetic particulate can be trapped in this "cake" as the metalworking fluid passes through it. In this manner, the fluid can be cleaned

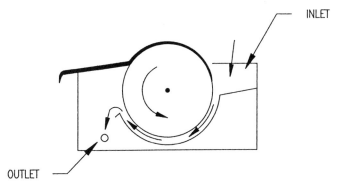

Figure 14 Magnetic system. (Courtesy Clarmatic Industries, Inc.)

of both magnetic and nonmagnetic particulate. The difficulty with this system is the flow rates attainable and the cleanliness obtainable. The application of magnetic separation systems is generally used on systems requiring less than 250 gal (946 l)/min. Another area of use for magnetic systems is in the primary separation of metal fines prior to positive filtration. This application allows for bulk separation of solids to take place before finer filtration. The magnetic system is satisfactory for magnetizable particle removal, but has its drawback in gal/min serviceable and consistency of fines removed.

B. Filtration Systems

The actual separation of particulate by introducing a media or filter into the fluid stream constitutes filtration of the fluid. This filter needs to be supported in some sort of system hardware. A force is applied and the filter system is activated.

There are three driving forces for filtration systems: gravity, vacuum, and pressure. There are a number of different filter materials which can be selected. It is not the scope of this chapter to give a presentation of selection criteria, but it can generaly be said that the more dense the fabric in ounces per square yard, the finer the attainable filtration. This is valid in comparing fabric densities in the same family. There are a number of criteria used for the selection of filter materials. [5]

As with the selection of fabric, the selection of hardware also varies and can be discussed in two categories: those units that use disposable media and those using permanent media.

1. Disposable Media

Systems are produced in a number of configurations. These include bags, cartridges, rolled goods, chopped paper, and a host of material referred to as precoats (see Fig. 15). The operation can be manual or automatic. Bag filters are filter media in a bag form. The bag(s) may be suspended at the end of a pipe with a special fitting or enclosed in a housing. The driving force is pressure: less than 15 psi (103 kPa) for the end-of-pipe assembly and up to 150 psi (1034 kPa) for housing units. The bag is selected to retain chips and particulate but allow a flow of metalworking fluid through the system. When a pressure drop is noted across the bag, indicating the bag is plugged or there is a significant reduction in flow, the bag filter is manually changed. The bag filters can be purchased in various sizes. This provides some flexibility in the square foot of filter area available. A wide range of micron retention bags are available providing additional flexibility. Typically, a flow rate of 25 to 50 gal/min/ft^2 (1000 to 2000 l/min/m^2) can be obtained. However, type of dirt and viscosity of the fluid may modify those numbers downward. For selected applications: washers, individual machines, and recycling equipment, the bag filter has proven effective.

Cartridge filters are used to provide a positive afterfilter for other filtration

Filtration Systems 291

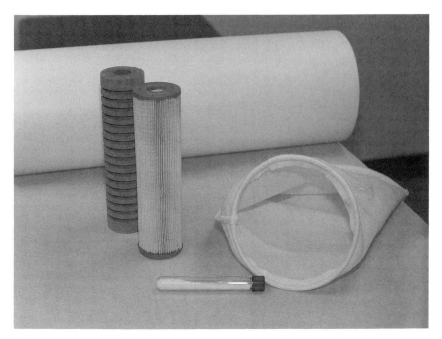

Figure 15 Forms of disposable media.

systems. These units consist of individual cartridges or clusters contained in multiple cartridge housings (see Fig. 16). The driving force for cartridge filtration is also pressure. Therefore, the housings for the cartridges are made for low (75 psi [517 kPa]) to high (300 psi [2068 kPa]) pressure. The flow rate for cartridge filtration is less in gallons per minute per square foot than bag filters. The application of these units is not generally on large flow rate systems but as "guard" filters to prevent spurious particulate from entering a close tolerance area. This close tolerance area could be a through-the-drill supply of coolant, a tapping operation, or application to a fine finish machine.

Disposable media on a roll provides the maximum flexibility because it can be used for gravity, vacuum, or pressure separations. The basic use of the rolled goods is in the gravity-driven pieces of equipment. In a gravity separation device, the metalworking fluid containing particulate flows or is pumped from the machine into an open container (see Fig. 17). In the bottom of the container is a perforated plate which supports the media (see Fig. 18). As the liquid level deepens, the head of liquid forces metalworking fluid through the media. The particulate is trapped on the filter media. As more and more particulate is collected on the filter media, a greater resistance to flow is encountered. The metalworking fluid becomes deeper and a cycle-inducing device is activated. This may be a float ball, limit switch,

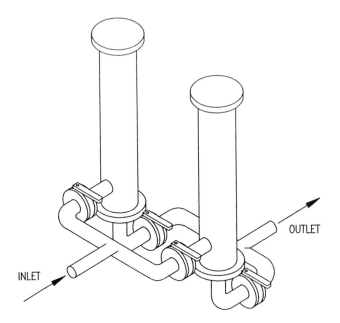

Figure 16 Cartridge filter system. (Courtesy Clarmatic Industries, Inc.)

conductivity probe, or other device. When the device is activated the filter media is indexed through the tank by means of a conveyor. This provides new filter media in the tank and results in improved flow rate. The liquid level drops and filtration continues until the cycle repeats itself. The indexing and movement of the media could be done manually, replacing the filter media or pulling the media through the system.

Vacuum systems are like gravity systems except there is a pump which is used to create a negative pressure under the media and its support structure (see Fig. 19). This pump can be an air pump used to evacuate a large volume of air from under a media and support system or more typically a pump to draw liquid through the filter media faster than gravity would normally allow. The use of pumps generally increases the time between indexes of the media. The liquid-drawing filter pump had been selected on its merit of being able to provide good vacuum characteristics. However, the concept of a single pump being used as the filter vacuum pump and machine supply pump has meant a compromise of characteristics. These characteristics are pressure and volume. As the vacuum increases, the flow and pressure decrease. It is, therefore, important to select a pump with a good net positive suction head. This system pump concept is a compromise because it requires good vacuum character as well as providing adequate pressure and volume for the machine and flushing systems. Typical vacuum conditions can reach 12 to

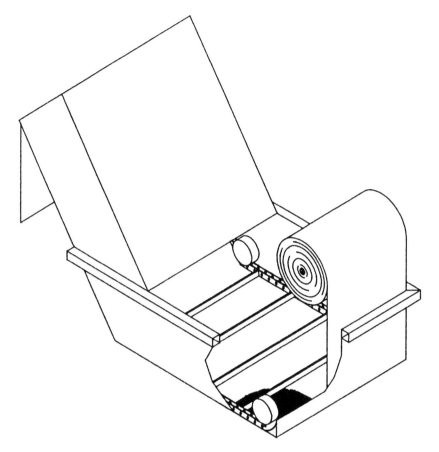

Figure 17 Gravity media system. (Courtesy Clarmatic Industries, Inc.)

Figure 18 Perforated plate support.

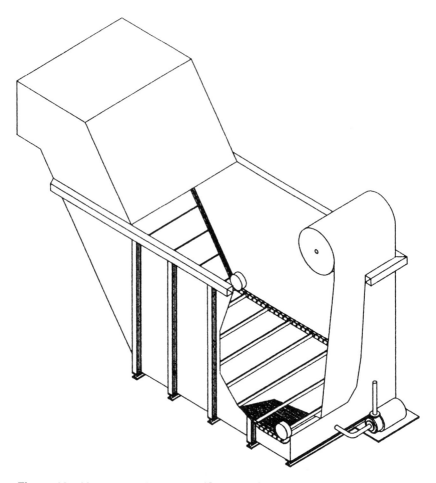

Figure 19 Vacuum media system. (Courtesy Clarmatic Industries, Inc.)

15 in. (41 to 51 kPa) of mercury. Beyond 15 in. (51 kPa) of mercury, justified benefit is low. Typical indexing methods for a vacuum media system include a mercury vacuum switch, differential pressure switch, or a timer. The first two index relative to the reduction in flow caused by the buildup of particulate on the filter media. This causes the vacuum pump to draw more vacuum and this is subsequently sensed by the switch. The timer is used on materials that do not cause a substantial decrease in flow rate or buildup in vacuum such as steel machining, aluminum machining, or grinding. Because these chips and particles form a porous pile or cake, indexing on vacuum may not be adequate to keep ahead of the chip loading in the system.

To determine the timer setting, a calculation is performed to determine the quantity of solid material removed. Based on an expansion factor, a volume of chips can be approximated from the solid stock removed (see Table 4). Knowing the capacity of the conveyor in the filtration system, the timer can be set for an interval which will prevent overloading of the system. This setting has to be done for the worst condition, i.e., the most chips produced at maximum production. At any other condition which would make less chips, the indexing will be more frequent than necessary.

When the vacuum system indexes, the filter media moves a short distance. Therefore the media is never fully replaced. This changes the gallons per square foot passing through the new area. If a system is purchased to provide 10 gal/min/ft^2, this condition only occurs at start up. At any other time during operation, the filter's moderate indexing minimizes the new area and causes the gallons per minute per square foot to increase. In fact, the only reason for large tanks with large filter areas is to provide physical volume of metalworking fluid.

Another system is the pressure-driven unit in which the filter media is supported on a metal or cloth belt, slotted, or perforated plate. The support and filter media are enclosed between two shells. The metalworking fluid and particulate are introduced into the cavity above the filter media and forced through the media. This process continues until the particulate builds a cake and the resistance causes an increase in pressure. When a preset pressure on a switch is reached, the filter supply pump is turned off, air is introduced into the cavity to remove the metalworking fluid and dry the particulate cake. After this process is complete, the two shells separate and the filter media is advanced entirely out of the system. The two shells close and the process begins again. The usual configuration is for the two shells to be in the horizontal position. Multiple horizontal shells can be used with multiple rolls of filter media. The pressure settings vary for each type. Indexing pressures

TABLE 4 Approximate Expansion Factors Based on Densities of Solids and Chips

	Density of solids[a]	Volume expansion factor	
Steel	490 (7850)	Grinding	5
		Machining	7
Cast iron	480 (7690)	Grinding	5
		Machining	5
Aluminum	170 (2720)	Grinding	5
		Machining	10
Brass	560 (8970)	Turnings	3

[a]In units of lb/ft^3 and (kg/m^3).

for two-shelled units are usually 7–10 psi (50–70 kPa) while multiple-shelled units can run to 35 psi (240 kPa). The limitation for any of these systems is the mechanism which keeps the shells closed. The closing mechanism used is air pressure in conjunction with air bags or cylinders. The advantage of the pressure-driven unit is that a reasonably dry cake can be discharged from the system. With pressure filtration, each time the filter indexes, a new complete filter media is introduced into the system.

The disposable media filter material comes in a variety of retention capabilities. As the filtration process proceeds through its cycle, the particulate being deposited on the media is building up. This buildup in particulate forms a "cake." In some cases this cake is actually necessary in order to attain fine particle filtration. The filter media acts only as an initial barrier upon which the cake can build. This process is referred to as depth filtration.

2. Permanent Media

Permanent media filters use the same driving forces for the filtration process as disposable media filters. However, the indexing cycle moves or removes only the cake from the media. This leaves the same media to be the support and initial barrier for the particulate in the next cycle. Permanent medias are made of wire mesh, woven fabric belts, or wedge wire screen (see Figs. 20–24). The primary difference is the backwash or blow down of the cake of particulate formed on the media. In the disposable media systems, the cake is carried out of the system with the filter

Figure 20 Wedge wire permanent media panels or drum.

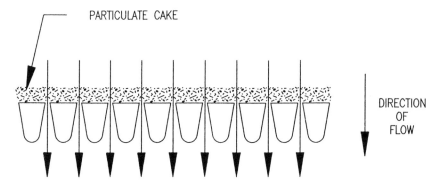

Figure 21 Wedge wire screen cross section. (Courtesy Clarmatic Industries, Inc.)

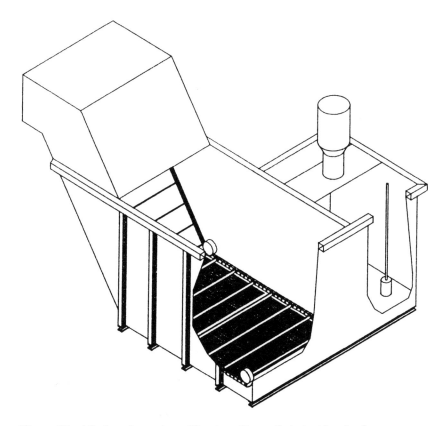

Figure 22 Wedge wire system. (Courtesy Clarmatic Industries, Inc.)

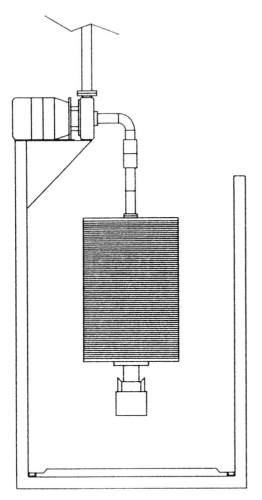

Figure 23 Vertical wedge wire drum system. (Courtesy Clarmatic Industries, Inc.)

media. In permanent media systems, the cake accumulated on the media can be blown off with air or metalworking fluid. The edge of a piece of metal, called a "doctor" blade, can also be used to remove the cake from the media. This essentially removes the particulate cake and provides a renewed area for reestablishing the cake. Generally, permanent media have a very open character. The percent open area and micron size of the openings is large in comparison to the small particles

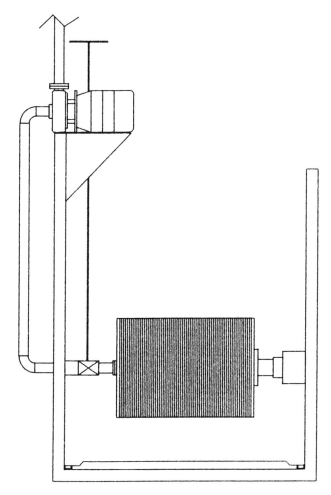

Figure 24 Horizontal wedge wire drum system. (Courtesy Clarmatic Industries, Inc.)

to be removed. It is necessary and a requirement for fine filtration that a cake be established and maintained for as long a period as possible. After index and before a new cake is established, migration of large particulate can occur. After the cake has been established there is an increasing improvement in the cleanliness of the filtrate. As with disposable media systems, the cake does the filtering in a permanent media system.

VI. ANCILLARY SYSTEMS

A. Extraneous Oil Removal Units

The metalworking fluid being recirculated in the filtration system is subject to varieties of outside contaminations. One of these is oil which is introduced from the machine tool hydraulic, way, and lubricating systems. This oil leaks into the fluid and becomes in varying degrees part of the recirculating metalworking fluid. Most units incorporated into a system for the removal of this oil act best on free oil, which will separate from the recirculating metalworking fluid given the time. These units pick up this free oil by a wetting of oil-loving material in the form of belts, ropes, and coalescer media. These removal vehicles utilize the affinity of polypropylene or stainless steel to surface coat with free oil. As these vehicles become wetted with oil they are constantly having oil removed from them by the natural separation process of gravity (coalescer) or a blade squeegee device. The free oil removed travels down a chute into a container for disposal or reuse. One difficulty with these devices is the need for cleaning because of the oil-wetted fine particulate which floats with and is removed by the vehicles. The coalescer media and squeegee devices need cleaning or they will become plugged. The coalescer-type units have an advantage over the ropes and belts because they can be set up to remove free oil and not a large quantity of metalworking fluid.

Other removal units available work on the difference in specific gravity between the oil and metalworking fluid. These devices are centrifugal and were originally designed for the removal of one fluid from another. The separation process takes place under varying amounts of relative centrifugal gravities. These can be low or high depending on the equipment purchased. The low-speed units do not subject the metalworking fluid to high gravity forces and therefore separate the free oil with less retention time than would be required under normal gravity conditions. The higher-speed units can subject the metalworking fluid to very high forces. These forces can be high enough to separate not only the free oil but also the metalworking product itself. Beside the possible deleterious effect on the fluid, these units tend to require more maintenance and a higher degree of technical support. All of the units will tend to remove some material which can form in the metalworking fluid other than free oil. These consist of invert emulsions, soap scum, hard-water precipitates, and fine particulate.

B. Metalworking Fluid Makeup Units

The very use of the metalworking fluid will require an addition of either water, to compensate for evaporative loss, or metalworking fluid concentrate, to replenish various ingredients lost. The additions to meet these requirements can be done manually by adding water or concentrate as needed based on an analytical determination. This has proven an effective method for smaller clarification systems servicing one or a few machines. In systems of large capacity, the need for addition can be high and alternative addition techniques are used. These techniques rely on a control means to determine if water or premixed water and metalworking fluid should be added. Additions are not made because of on-line analytical tests but rather the indication of reduction in overall system volume. Because of this method of replenishing the system, only water can be added with adjustments made later in the concentration, or a premixed metalworking fluid concentrate and water can be added. The latter is the preferred method because it adds a new amount of concentrate each time volume replenishment is required. This type of addition also means that there is less fluctuation in the metalworking fluid concentration from day to day. One unit available works on a venturi principle. As water passes over a fixed orifice plate a certain amount of fluid concentrate is drawn into the water stream. This mix is discharged into the system. Because of changes in water pressure and flow, clogged orifices, and different concentrations required at different seasons of the year, this unit may not provide uniform and consistent concentration deliveries as required.

A few units use pumps to pump the concentrate into a water stream. One unit employs a water-actuated proportioning pump which when supplied with water, draws up a quantity of concentrate into a separate chamber. When the pump continues to function, the concentrate is mixed with the water stream. This unit has a variable screw adjustment which can be changed to give different concentrate deliveries to the water stream. The concentration in this type of unit varies depending on the stroke of the water-actuated pump. For more truly premixed metalworking fluid, a mix chamber is needed on the downstream side of the pump. Another pump system includes an electrically operated pump—either centrifugal, piston, or tube—to deliver concentrate into a flowing stream of water. Concentration adjustments are made either by changing the feed or stroke on the pump, or opening or closing a valve limiting or increasing the flow of concentrate. These systems also require a mix chamber in order to provide more uniform premix additions. A variety of makeup units are also available which measure the amount of water added to a system and add a preset amount of concentrate. Usually these additions are not premixed but added as two separate streams of liquid. This unit relies on the filter system turbulence to mix the metalworking fluid. In a few cases pumps have been placed on systems and set to add an amount of metalworking fluid concentrate to the system each and every day with no regard to variations in the requirements of water additions. This addition technique is used based on the assumption that concentrate replenishment is required uniformly each day and evaporative loss of water is consistent each day.

C. Temperature Control

Certain applications require that the machine, tooling, and workpiece maintain a relatively uniform temperature. These applications include fine tolerance work, such as honing and mirror finish grinding. However, where dimensional stability is important, some machining systems are being considered for temperature control. The temperature of the fluid is selected based on the ambient room temperature plus or minus two or three degrees, or a compromise which is economically justifiable. The heat input is determined by the horsepower of the machines, peripheral equipment, and the work done. Other considerations are the evaporative loss of water, room temperature, water added to the system, and temperature of the parts. Once a temperature has been determined and the necessary calculations made, the selection of a cooling system can be pursued. Two different types of cooling can be used: evaporative cooling from an outside water-cooling tower and a mechanical chiller. Each of these choices has a number of different operational parameters which contribute to their particular advantages and disadvantages. Selection is made by working closely with those trained in this field.

D. Alarms and Controls

The controls available on most systems have changed as the technology and electronic gear have changed. The changes have occurred in the electronic hardware and the information retrieval which can be interfaced with other systems plant wide, but this has not changed the functional requirements of the system. This function is to produce consistently "clean" metalworking fluid for use at the machine tool at adequate volume and pressure. Indications of trouble in the operation of the filtration system have always required observation or physical interaction. It is necessary to provide sensors and alarms on filtration systems to monitor their continued mechanical performance.

VII. CONCLUSION

For all of the options available in the filtration of metalworking fluid, the best option and primary goal is to provide a mechanically sound, continuously functioning system which will deliver acceptably clean fluid at the tool–workpiece interface. This can be done by reviewing the particulate produced in the metalworking operation, setting standards to be met, reviewing methodologies for accomplishing these standards, and giving attention to the primary goal. Filtration systems can contain a plethora of devices and controls. If they do not add to the primary goal, they are a nicety and not a necessity. Discern what the overall economics and cost of the system are and put the emphasis on "clean" metalworking fluid.

ACKNOWLEDGMENTS

I wish to thank my partner Merlin P. Hoodlebrink for his support and understanding, Charles E. Aring for drawing the representations of filter units in conjunction with Robert J. Fox of Clarmatic Industries, Inc. and Sandra D. Weber of Brandt & Associates, Inc. for preparing the table and typing the text.

APPENDIX: OTHER SOURCES OF INFORMATION

P. J. C. Gough, *Swarf and Machine Tools*, Hutchinson & Company, London (1970).

M. Opachak, *Industrial Fluids—Controls, Concerns and Costs*, Society of Manufacturing Engineers, Dearborn, MI (1982).

D. B. Purchas, *Solid/Liquid Separation Technology*, Uplands, Croydon, England (1981).

R. K. Springborn, *Cutting and Grinding Fluids: Selection and Application*, American Society of Tool and Manufacturing Engineers, Dearborn, MI (1967).

R. H. Warring, *Filters and Filtration Handbook*, Gulf, Houston, TX (1981).

D. Zintak, *Improving Production with Coolants and Lubricants*, Society of Manufacturing Engineers, Dearborn, MI (1982).

REFERENCES

1. B. L. Nehls, *J. Am. Soc. Lubr. Eng.* 179–183 (1976).
2. J. J. Joseph, *Coolant Filtration*, Joseph Marketing, East Syracuse, New York, p. 15–18 (1987).
3. R. H. Brandt, *Lubr. Eng.* 254–257 (1972).
4. R. S. Marano, G. S. Cole, and K. R. Carduner, *Lubr. Eng.* 376–382 (1991).
5. P. Z. Chrys, *Man* 42–48 (1991).

11
Metalworking Fluid Management and Troubleshooting

GREGORY J. FOLTZ
Cincinnati Milacron
Cincinnati, Ohio

I. INTRODUCTION

The metal removal process consists of three variables: (1) the machine tool, (2) the cutting tool or the grinding wheel, and (3) the metalworking fluid. Each of these is significant and important in producing any part. There are a number of aspects pertinent to each variable and their interaction that must be understood in order to make the metal removal process occur [1,2]. While the purpose of this chapter is to discuss the control and management of the metalworking fluid variable, it is also important to understand the machine tool and the cutting tool or grinding wheel.

Many of the variable aspects of the machine tool include the setup, feed and speed rates, metal removal rate, alignment, drive systems, and vibration. The age of the machine and how it has been maintained will also influence performance. The type of cutting tool (HSS, carbide, ceramic, etc.) or grinding wheel (silicon carbide, aluminum oxide, CBN, diamond, resin bond, vitrified bond, etc.) is very important. The sharpening or regrinding of the tools and the truing and dressing of the grinding wheels can affect performance.

All of these variables can be optimized, but if the proper fluid is not used, poor performance can result. It is also important to remember that when troubleshooting a system, these variables, as well as the fluid, should be considered. In most cases, a well-organized maintenance/service plan exists for each machine tool and plant engineers can precisely define the number of parts per tool or pieces per wheel dress. Yet selection and maintenance of the metalworking fluid, the third important variable, is not well understood.

The main functions of a metalworking fluid are to control heat and provide lubricity [3–5]. It must also flush away the chips and protect the machines and workpieces from corrosion. When these functions are integrated into the machine tool and cutting tool/grinding wheel framework, an efficient metal removal system will result. It is, therefore, very important that this fluid variable be properly selected, controlled, and maintained, in order to achieve top performance. This fluid management program then becomes an integral part of a plant's operation as the importance of the metalworking fluid to the entire process becomes understood.

When a metalworking fluid management program [6] is in place, the fluctuation in this variable (correct fluid, concentration, pH, dirt volume, tramp oil, etc.) is reduced and more consistent quality parts can be produced. Finish, size, and geometry problems are eliminated. Productivity can be increased as machine and tool/wheel performance are optimized with the proper fluid. The plant's working environment is improved as offensive odors, irritating mists, skin problems, and dirty machines are controlled. The bottom line is improved costs.

Properly controlled fluids do not need to be dumped as often [7]. This eliminates costs associated with machine downtime, disposal, and new fluid purchase. As efficiency of operations are improved, the cost of producing each part will drop. More parts can be produced and fewer wheels or cutting tools are required. The advantages of a metalworking fluid management program and the ability to troubleshoot any problems that may occur are therefore key elements in a plant's operation.

II. FLUID SELECTION PROCESS

A fluid management program begins with the selection of the proper fluid for the job. There are four categories of fluids [8,9]: straight oils, soluble oils, semisynthetics, and synthetics. The performance of these products can range from light duty to very heavy duty operations. Metalworking fluids will have different performance properties depending on their chemical composition. This can be affected by the oil levels, amount of chemical lubricants and extreme-pressure additives, cleanliness properties, biocide levels, and a variety of other factors. The selection criteria that follow are designed to define the requirements for a particular job.

A. Selection Criteria

In order to achieve optimum performance, the correct fluid must be selected and based on a review of the variables of the entire operation [10]. These include the following.

1. Size of Shop

For a small shop with a few machines doing a variety of work on a variety of metals, a very general purpose product is selected to minimize the number of products required. For a large plant producing large quantities of the same part, a product very specific to the needs of that operation can be selected.

2. Type of Machines

It is important to consider the age and design of a machine tool before selecting a product. Some machines, especially older models, were designed so that the metalworking fluid also serves as the lubricating fluid for the moving parts and gears. In that case, a fluid with a high degree of physical lubricity will be required. The seals on the machines must also be inspected to ensure that they are designed to be used in a water environment. If not, it may be necessary to use a straight oil-type product.

3. Severity of Operations [11]

The severity of the operation will dictate the lubricity requirements of the fluid. Two types of fluid lubricity exist, chemical and physical, so that it is not always necessary to use an oil-containing product to achieve good machining/grinding characteristics. Stock removal rates and feeds and speeds, together with finish requirements, must be considered. Metalworking operations can be divided according to their severity: light duty (surface grinding cast iron), moderate duty (turning, milling steels), heavy duty (centerless grinding, sawing steels), and extremely heavy duty (form and thread grinding, broaching). If a series of operations are to be performed with one fluid, it is necessary to select the most critical operation, because in most cases it will dictate the fluid selection.

4. Materials

The kind of material being worked (cast iron, steel, aluminum, titanium, copper, glass, carbide, plastics, etc.) is very important in fluid selection [12,13]. The corrosion control and/or staining properties of some fluids may not be compatible with all materials. Some fluids are formulated specifically for certain metals. The hardness and machinability of the material must also be considered.

5. Quality of Water

Since water is the main component (90–95%) of any water-based metalworking fluid mix, its quality can be an important factor in performance of the fluid [14].

Water quality is covered in greater detail later. Water hardness greater than 200 ppm can produce mix stability problems with many emulsion-type products. Water with a high chloride or sulfate level (greater than 150 ppm) can promote corrosion and/or rancidity. On the other hand, soft water (less than 80 ppm hardness) can lead to foam with many products. It is important to know the water quality before selecting a product.

6. Type of Filtration [15,16]

Individual machine sumps or central systems each make different demands on a fluid. The type of filtration used, e.g., settling or some type of positive filtration-using media (paper, cloth, or wire screens) or a separator such as a centrifuge or cyclone, will also affect the fluid selection process. Settling systems obviously require fluids with good settling characteristics. Media filters require fluids capable of passing through the media without clogging. Separators require products that are sufficiently stable to undergo the demands of this process. Filtration is described in greater detail in Chapter 10.

7. Contamination

Contamination has a drastic effect on the life and performance of a metalworking fluid. Lubricating oils, way lubes, hydraulic oils, rust inhibitors, floor cleaners, and heat treat solutions are some of the contaminants often found. Different fluids have different mechanisms for handling these contaminants, especially the oils. Some may be emulsified and others rejected. While most cutting fluids can handle some contamination, the greater the amount, the shorter the fluid life and the more erratic the performance.

8. Storage and Control Conditions

Where and how a fluid is stored prior to use can affect its performance. Many products will freeze if stored outside or in unheated warehouses during winter conditions. Other products, if stored outside under the hot sun, will be degraded. The compatibility of the fluid with the plant's mixing conditions must be considered. For water-based fluids, the concentration control procedures must be considered. If they are very lax or nonexistent, then a product with a very wide operating range should be selected.

9. Freedom from Side Effects

In some grinding operations, the use of a very transparent fluid is desirable. At some plants, a particular product color or odor may be requested. Certainly fluids should be free from misting and dermatitis problems. They should all be safe and pleasant to use. The fluid should not leave an objectionable residue or cause problems with the paint on machine tools.

10. Ease of Disposal/Recycling [17–19]

In many factories, the most critical element in the selection of a metalworking fluid is its compatibility with the waste treatment process. If the fluid cannot be effectively and economically treated, then any performance advantages are negated. Plants will typically have a very specific waste treatment test that a product must pass before it can be considered for testing. With the advent of more in-plant recycling systems, products are also judged on their ability to be effectively recycled through the plant's existing or planned treatment system. In recycling operations, it is frequently necessary to standardize an entire plant on just one product, in order to have the recycling system work. This must be considered in fluid selection, see Chapters 12 and 13.

11. Chemical Restrictions

Because of concerns over the health and safety aspects of a particular chemical or because of some environmental or disposal issue, certain plants may restrict the use of chemicals that may be found in some formulations. Certain industries, such as aerospace and nuclear power plant component manufacturers, have restrictions on halogen compounds. It is necessary to obtain not only a list of the restricted chemicals, but also the allowable limits for them. In some cases, trace amounts of these materials may appear as an impurity in a formulation, and may not be present in a sufficient quantity to restrict the use of the product.

12. Performance vs. Cost

The objective of any fluid is to achieve maximum performance at minimum cost. In calculating cost it is necessary to consider all the factors and not just the cost of the product. Considerations include used fluid disposal costs, downtime for cleaning, lost production, machine cleaning, recharging costs, tool life, tool resharpening costs, etc. Small improvements in tool life may be difficult to measure, but consider the fluid's sump life, additive costs, shop cleanliness, operator acceptance, and other related factors that contribute to the cost of the fluid in use. When considering cost it is also important to use the common denominator of "mix gallon cost," the cost of the product per gallon multiplied by its recommended dilution ratio. If a product costs less ($6.00/gal vs. $8.00/gal) but is used at a stronger concentration (10% vs. 5%), then the actual cost comparison for a mix gallon is $0.60 vs. $0.40. The product that costs more is actually less expensive to use, without considering the factors noted earlier.

B. Supplier Evaluation

Using these selection criteria, the requirements for any metalworking job can be very well defined. However, before a particular product can be selected, many other parameters must be considered. There are over 300 suppliers of metalworking fluids in the United States offering a wide range of products. It is necessary to evaluate the suppliers and determine how their products, business practices, and philosophies compare with your needs. In this analysis, we assume that the major goal is not necessarily to find the lowest cost product, but to find the product that is most cost effective in terms of performance.

1. Quality

W. E. Deming stresses [20] independence from mass inspection of incoming goods or finished materials. Statistical evidence that quality is built into the finished product should be required of all suppliers. Many users of metalworking fluids have their own quality standards that must be met and these are well defined for the supplier. Suppliers may have their own programs of quality control and quality assurance. The ability to produce consistent lots of quality material is essential when the production process is so dependent on the metalworking fluid. A visit to a supplier, with a review of processes and procedures for quality, may be beneficial.

2. Delivery

The ability of a supplier to quickly deliver material and the ability of the user to maintain a minimal inventory are becoming more critical in the metalworking fluid industry. Supplier location involving both production and warehouse facilities should be evaluated. For large fluid users, there is a growing trend to be less dependent on material supplied in drums and more dependent on material supplied in refillable totes or in bulk.

3. Health/Safety Testing [21,22]

The users of any chemical need to be assured that the products are reasonably safe. This can be accomplished in a number of ways. A review of the MSDS supplied by the manufacturer is the primary method. This should include information on any hazards associated with the product as well as information on the safe use of the product. Some users may also require product composition information in order to make their own evaluation of the product's safety. In addition to the MSDS information, some suppliers can provide additional testing information based on specific evaluations regarding oral, dermal, inhalation, or eye exposure.

4. Service

Metalworking fluids are used in a very dynamic working environment. The selection of the proper fluid may require some assistance. The methods to control the fluid and care for it in use must be explained to the user. Questions relating to safe use of the product must be addressed. When a problem arises, laboratory evaluation may be needed to resolve any questions. All of these issues can be addressed if the supplier has a good service program. The user should investigate the supplier's capabilities in these areas before making a product selection.

5. Performance Data/Laboratory Testing

When a metalworking fluid is developed, the supplier will have run many laboratory screening tests in areas such as corrosion control, lubricity, oil emulsification, rancidity control, and foam. Some of this testing may be done based on the supplier's procedures and other testing according to industry standards or guidelines (ASTM) [23]. This data should be reviewed in terms of the customer's desired performance and selection criteria.

6. Case Histories

In judging how a product will perform in a particular application, it is frequently very helpful to review any case history data that may be available from the supplier. It is important to compare these applications in terms of setup, work material, water quality, filtration, flow rates, etc. If a particular product is successfully used in a certain application, there is a higher level of confidence in the new application.

C. Product Evaluation

A metalworking fluid user can define an answer to the various selection criteria and most suppliers can furnish the information requested in an evaluation. If additional information is needed in the selection process, it is typically gained by a laboratory evaluation and/or an in-plant testing program.

1. Laboratory Testing

Many plants will have a set of initial laboratory screening criteria [24–28] that a product must pass before it can move any further. Tests such as lubricity, corrosion control, and rancidity control are some of the many performance procedures used to screen metalworking fluids. Also, chemical tests may be run to develop a product profile, waste treatment compatibility, and background information on the product's quality. Several tests are typically chosen that are known to be key to the success of the product in a particular operation. For example, on cast iron machining applications, a corrosion test using cast iron chips is a typical laboratory evaluation.

2. In-Plant Testing [29]

The true measure of any product's performance is a test on the actual application. An individual machine or a small central system with several isolated machines may be used. This is the best method to simulate all of the variables that will be encountered in a normal use situation. It is important in this type of testing to "qualify" the entire process with the existing product or standard. Define a measurable set of performance criteria that are to be evaluated, i.e., tool life, parts per dress of the wheel, machine cleanliness, sump life, and product odor. Then set up a system for measuring this data, along with key product specifications such as concentration, pH, bacteria counts, etc. Typically it will take at least eight weeks, maybe more, to develop a sufficient data base by which products and their performance can be judged. Work with the supplier on these tests, because even though an objective is to hold certain variables constant, improved product performance may be achieved by altering some variables in combination with the test fluid. For example, modifying the coolant application with a synthetic to obtain a better flow to the cut zone, [30] may show improved performance over a soluble oil-type fluid. The most important item on in-plant testing is to establish measurable parameters before the testing begins. In this way the actual performance of various products can accurately be compared and judgments made on product selection.

With information on selection criteria, supplier evaluation, and product evaluation, the proper fluid can be selected for any application. This is the first step in fluid management. It is now necessary to control and maintain that fluid in the work environment to achieve optimum long-term performance.

III. WATER

Water is the major ingredient in a water-soluble metalworking fluid mix. It may amount to as much as 90–99% of the mix as used. Therefore, the importance of water quality to product performance cannot be ignored [31,32].

Corrosion, residue, scum, rancidity, foam, excess concentrate usage, or almost any metalworking fluid performance problem can be caused by the quality of the water used in making the mix. Untreated water always contains impurities. Even rainwater is not pure. Some impurities have no apparent effect on metalworking fluid. Others may affect it drastically. By reacting or combining with metalworking fluid ingredients, impurities can change performance characteristics. Therefore, water treatment is sometimes necessary to obtain the full benefits of water-soluble metalworking fluids.

A. Water Quality

Water quality varies with the source. Water may or may not contain dissolved minerals, dissolved gases, organic matter, microorganisms, or combinations of these impurities causing deterioration of metalworking fluid performance. The amount of dissolved minerals, for example, in lake or river water (surface water) depends on whether the source is near mineral deposits. Typically, lake water is of a consistent quality, while river water varies with weather conditions. Well water (ground water), since it seeps through minerals in the earth, tends to contain more dissolved minerals than either lake or river water. Surface water, however, is likely to contain a higher number of microorganisms (bacteria and mold) and thus need treatment. Typical water hardness throughout the United States is shown in Fig. 1.

Some metalworking plants use well water and have detailed information on

Management and Troubleshooting

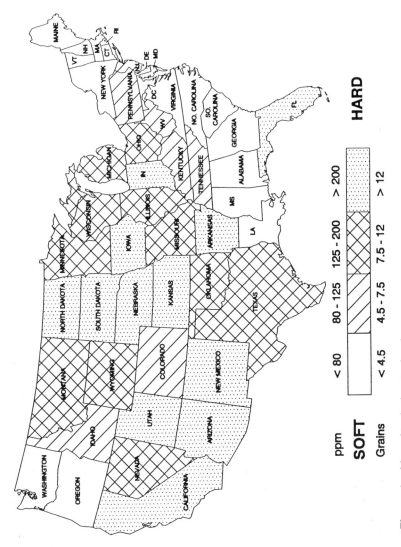

Figure 1 Map of typical water hardness.

its composition. Most, however, use water supplied by a municipal water works, which maintains daily or weekly analyses of the water. To estimate the effect of water on a metalworking fluid mix, measurement of the following provide sufficient data in most cases:

Total hardness as calcium carbonate	Chlorides
Alkalinity "P" as calcium carbonate	Phosphate
Alkalinity "M" as calcium carbonate	Sulfates
pH	

1. Total Hardness

Of the water analysis results, total hardness has perhaps the greatest effect on the metalworking fluid mix. Hardness comes from dissolved minerals, usually calcium and magnesium ions and is reported in parts per million (ppm) and expressed as an equivalent amount of calcium carbonate ($CaCO_3$). Hardness may also be expressed in terms of "grains," with one grain equal to 17 ppm hardness. The ideal water hardness for making a metalworking fluid mix ranges from 80 to 125 ppm. The term "soft" is used for water if it has a total hardness of less than 100 ppm and the term "hard" if total hardness exceeds 200 ppm. Test kits and test strips are available from many manufacturers for testing water quality.

a. Soft Water When the water has a total hardness of less than 80 ppm, the metalworking fluid may foam—especially in applications where there is agitation. Foam causes problems when it overflows the reservoir, the machine, the return trenches, etc. Foam may also interfere with settling-type separators (since it suspends swarf and prevents settling), obscure the workpiece, and diminish the cooling capacity of a water-based metalworking fluid. Soluble oil and semisynthetic products typically foam more readily in soft water than synthetics.

After a metalworking fluid is exposed to chips, dirt, and tramp oil for a few days, foam tends to dissipate. If it must be eliminated immediately, inspect the system for physical conditions that contribute to excessive foam. Fluid flowing through sharp turns or "waterfalls," high-pressure nozzles, or malfunctioning pumps could be responsible. If physical causes cannot be eliminated, foam depressants, chemical water hardeners, antifoam, or oil are useful to decrease the foam.

b. Hard Water Hard water, when combined with some water-soluble metalworking fluids, promotes the formation of insoluble soaps. The dissolved minerals in the water combine with anionic emulsifiers in the metalworking fluid concentrate to form these insoluble compounds that appear as a scum in the mix. Such scum coats the sides of the reservoir, clogs the pipes and filters, covers machines with a sticky residue, and may cause sticking gages.

Because soluble oils typically have the least hard-water stability, hard water has a more obviously detrimental effect on them. Separation of the mix is apparent in severe cases, and is characterized by an oil layer rising to the top of a fresh mix.

Semisynthetics and synthetic metalworking fluids may not be visibly affected by water hardness. Some are formulated with good hard-water tolerance. However, dissolved minerals react with ingredients other than emulsifiers. In these reactions, the metalworking fluid ingredients change or are tied up and, consequently, the product never attains peak performance.

Dissolved mineral content increases in a metalworking fluid mix with use. After a 30-day period, dissolved solids in the mix can increase three to five times the original amount. This results from the "boiler effect" that exists in a metalworking fluid reservoir. That is, water evaporates and leaves dissolved minerals behind. Then, makeup (usually 3–10% per day) introduces more with each addition, and dissolved minerals continue to accumulate. Therefore, even with water that has very low dissolved mineral content initially, dissolved minerals can build up rapidly and cause problems.

2. pH

pH is an expression that is used to indicate whether a substance is acidic, neutral, or alkaline. A pH of 7 is neutral, between 0 and 7 is acidic, while 7 to 14 is alkaline (basic). Water in the United States normally varies from 6.4 to 8.9 in pH, depending on the area and source of water. The buffering ability of a metalworking fluid is far greater than that of any clean water supply. pH adjustments to the water are rarely needed.

3. Alkalinity

Two kinds of alkalinity exist in water: "P" alkalinity and "M" alkalinity. P alkalinity is the measure of the carbonate ion (CO_3^{2-}) content and is expressed in ppm calculated as calcium carbonate. This is sometimes referred to as *permanent alkalinity* and, as such, is not changed by boiling as is the M alkalinity. M alkalinity is the measure of both the carbonate ion content (P alkalinity) and the bicarbonate ion (HCO_3^-) content. This value is also expressed in ppm, calculated as calcium carbonate. It is referred to as *total alkalinity* and *temporary alkalinity*. This is because its value can be lowered to that of P alkalinity by boiling.

Metalworking fluids typically perform best when the pH is between 8.8 and 9.5. They require a certain amount of alkalinity for good cleaning action and corrosion and rancidity control. If pH and total alkalinity become too high, however, pitting and staining of nonferrous metals may occur. Skin irritation is another possible problem. Currently, there appears to be no satisfactory treatment for alkaline water, so careful product selection is critical.

4. Chloride

When chloride ion (Cl⁻) content is high (above 50 ppm) in the water used to make metalworking fluid mixes, it is more difficult for the product to prevent rust. Richer concentrations of the metalworking fluid mix may sometimes counteract the effect of chlorides. In other cases, excessive chloride ions must be removed from the water prior to use by demineralizing.

5. Sulfate

Sulfate ions (SO_4^{2-}) also affect the ability of a metalworking fluid to prevent rust, though not as much as chloride ions. In addition, they can promote the growth of bacteria. If sulfate ion content exceeds 100 ppm, richer concentrations of the metalworking fluid mix may improve corrosion and rancidity control.

6. Phosphate

Phosphate ions (PO_4^{3-} and others) contribute to total alkalinity and stimulate bacterial growth, leading to problems of skin irritation and rancidity, respectively. If phosphate ions are found in the mix water, they should be removed by demineralization to prevent these problems.

B. Water Treatment

There are two processes that are commonly used in treating hard water: water softening and demineralization.

1. Water Softening

In this process, the water passes through a zeolite softener. The softener exchanges calcium and magnesium ions (positively charged ions which are largely responsible for hardness) for sodium ions. In effect, water that was rich in calcium and magnesium ions becomes rich in sodium ions. The total amount of dissolved minerals has not decreased, but sodium ions do not promote the formation of hard-water soaps. Corrosive, aggressive negative ions are not removed by the zeolite and can continue to build up in the metalworking fluid mix, leading to corrosion problems or salty deposits. Thus, the use of "softened" water is not recommended with water-soluble metalworking fluids.

2. Demineralization

Deionizers or reverse osmosis units are used to demineralize water. Deionizers *remove* dissolved minerals. This is done selectively or completely, depending on the type and number of resin beds through which the water passes.

It is not necessary to obtain pure water for metalworking fluid mixes. A hardness level of 80–125 ppm is suitable. Usually a two-bed resin deionizer produces water of sufficiently high quality, as opposed to a more expensive mixed-bed deionizer needed to obtain pure water.

Reverse osmosis removes dissolved minerals by forcing water through a semipermeable membrane under high pressure. Typically, this process removes 90 to 95% of the dissolved minerals.

3. Choice of Water Treatment

The chemistry of the water as determined by a water analysis, water quantity needs, water quality requirements, and economics (capital and operating costs) are considerations in selecting suitable water treatment. Softening of hard water eliminates the scum that forms in some metalworking fluid mixes, but increases the possibility of rust problems. Typically, deionizers are lower in capital costs than reverse osmosis units, but higher in operating costs. Deionizers can provide higher-quality water; however, resin beds must be regenerated frequently. If not regenerated frequently, water quality deteriorates and the resin beds also serve as an excellent environment for massive growth of bacteria. Reverse osmosis units do not require regeneration, but do require membrane replacement in time, depending on the water quality fed into the units. Pretreatment systems, prior to either the deionizing or reverse osmosis unit usually lengthen resin or membrane life.

With either method of demineralization, foam can be a problem when initially charging a metalworking fluid system. To avoid foam, the initial charge could be made with untreated water (except in cases where dissolved mineral content is excessive) and subsequent makeup could be mixed with the demineralized water. Chips, grinding grit, and debris eventually will add impurities to the initial charge, but the amount is not significant when compared to using untreated water for daily makeup.

Many metalworking fluid users treat poor-quality water before using it in fluid mixes. The benefits vary, depending on the water quality before treatment and the type of metalworking fluid that is used. In one case, composition of a fluid user's city water varied widely in dissolved minerals content because of frequent changes in processing by the municipal water works. After passing this water through a mixed-bed deionizer, consistent quality water with zero hardness was obtained. The cost of demineralization roughly equaled the amount saved in reduced usage of soluble oil concentrate. In addition, filter media consumption was reduced, while fluid filtration improved significantly. Demineralized water has also decreased additive usage and a corresponding incidence of skin irritation. Likewise, the amount of residue on machines was less, and, what was present was more fluid. This user concluded that the benefits of using demineralized water were well worth the investment. Also, the water is now of consistent quality, which eliminates one major variable when looking for the source of any metalworking fluid performance problem.

IV. METALWORKING FLUID CONTROLS

Metalworking fluids, specifically the water-soluble types, are all formulated to operate within a certain range of conditions in areas such as concentration, pH, dirt levels, tramp oil, bacteria, and mold. When fluid conditions fall out of this range, in one or more of these areas, performance problems can develop. It is therefore necessary to have a set of tests run on some regular basis on the fluid mix to keep it within these operating conditions. Some general purpose-type products may have very wide operating ranges in several areas so that they can withstand the abuse of being used in an environment with limited control. The performance of other products may require that they be controlled in certain areas (i.e., concentration) very close to their listed specification.

The importance of controlling a metalworking fluid has been understood for many years and continues to grow [33–35]. Working with the supplier and understanding the needs of a particular operation will usually dictate the frequency and degree of the control required. In some plants, where the fluid is very critical to the operation, such as aluminum can production, checks on concentration and pH are made every 4 h. In other manufacturing plants, such as automotive components or bearings, large central systems are used and checks are typically made once a day. It is much more difficult to control a plant where many individual tanks are utilized. The tanks may be checked once a week, or the fluid may be controlled by simply monitoring the output of a premix unit or a recycling system. In this section, some of the typical tests used to control water-soluble metalworking fluid mixes, in use, will be described.

A. Concentration

Water-soluble metalworking fluids are typically formulated to operate in a concentration range of 3–6%, although concentrations up to 10% are not uncommon for many heavy duty applications. Concentration is *the* most important variable to control. Concentration is not an absolute value but rather a determination of a value for an unknown mix based on values obtained from a known mix. There are certain inaccuracies, variables, and interferences in any method. This must be considered when evaluating the data.

In some products, several methods for measuring the concentration may exist, depending on the different components of the product. One method may measure alkalinity, another the anionic components, and another the nonionics. Initially all the methods may agree, but as the fluid mix ages and as contaminants are introduced, different values may be obtained. While this does not necessarily imply a problem, it may certainly lead to some concern. In some cases, an additive package may be necessary to rebalance the product. With other products, this may be considered a normal aging process.

Metalworking fluid suppliers formulate their products so that, at the correct

Management and Troubleshooting

operating range (see Fig. 2), the proper level of the chemicals needed for performance (lubricants, rust inhibitors, biocides, etc.) are present. To achieve this performance, the fluid must therefore be maintained within this range. This is done by means of some concentration control procedure. Various methods are available.

1. Refractometer

This is an optical instrument that measures the refractive index of a metalworking fluid. The refractometer reading, obtained from a scale of numbers in the unit, is then converted into a concentration value via a factor or graph made from taking readings on known mixes of various concentrations. A 0% concentration would have a 0 refractometer reading. Depending on the model, the refractometer should always be zeroed with water before taking a reading. Generally, synthetic fluids have very small values for refractometer readings (1–3). Small differences in scale readings become rather large differences in concentration. Soluble oils have rather

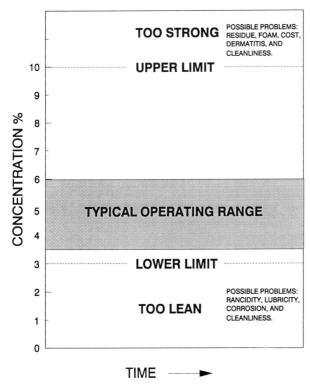

Figure 2 Metalworking fluid typical operating range.

high refractometer readings, and in many cases, the reading will correspond directly to the concentration.

On a refractometer, the reading is taken where there is a distinction between two colors, typically black or dark blue and white. As metalworking fluids, in use, pick up more contamination, especially tramp oil, the distinction between these two colors becomes much less clearly defined. The ability to obtain an accurate reading becomes much more difficult.

Overall, refractometers are a fairly quick method to check concentration and are certainly sufficient for many operations, however, the inherent inaccuracies must always be kept in mind.

2. Titrations

Chemical titration methods can be established to measure certain components or groups of components in any fluid mix. These would include measuring the alkalinity, the anionic content, the nonionic level, or the sulfonates. Some companies have made these titrations into small kits which can easily be used by anyone in a manufacturing plant. It is important to establish known controls for any of these methods. The interferences from contaminants or the change in a titration value due to the aging of the mix should also be established in order to give more accurate results. It is also possible to utilize specific ion electrodes and automatic titrators to run these types of concentration checks.

3. Instrumental

Instrumental methods typically allow for a very specific measurement of one compound. Instruments used include gas chromatography (GC), atomic absorption (AA), high-pressure liquid chromatography (HPLC), and fourier transform infrared (FTIR) [36]. These can be quite sophisticated and involved methods, requiring expensive instrumentation and lengthy sample preparation. To that extent, they are more frequently found in the laboratories of the metalworking fluid supplier and not the customer. One or more key components are chosen to track, sometimes with the assumption that other ingredients will stay in relative balance. In other cases, a component may be measured to detect a depletion or a buildup. In cases of depletion, an additive may be used to restore this component.

B. pH

A pH measurement determines the degree of acidity or alkalinity of a metalworking fluid mix. Metalworking fluids are typically formulated and buffered to operate in a pH range of 8.5 to 9.5. This is somewhat of a compromise. If the pH ran higher, the fluid would provide excellent ferrous corrosion control, but may have problems with mildness and nonferrous corrosion protection. A lower pH would be good for mildness and nonferrous corrosion control but may cause problems with rancidity control and ferrous corrosion protection.

pH is also a good, quick indicator of the condition of the fluid. A pH below

Management and Troubleshooting

8.5 is typically a result of bacterial activity. This can affect mix stability, ferrous corrosion control, and microbial control. Additives can be used to increase the pH of a mix. A pH greater than 9.5 is generally the result of some form of alkaline contamination, and will affect the mildness of the fluid. It is very difficult to correct a high pH, short of dumping the system.

pH values on a fluid mix can be obtained by using pH paper, which is dipped into the mix and observed for a color change, or by using a pH meter. Many models are available, ranging from inexpensive handheld units to rather elaborate laboratory bench models. When using any meter to check the pH of a metalworking fluid, it is important to ensure that the meter has been standardized with the appropriate buffers and also that the electrode(s) is clean. The oils found in most used fluid samples can quickly foul pH electrodes causing inaccurate readings. Cleaning with isopropyl alcohol eliminates this problem.

C. Dirt Level

Dirt or total suspended solids (TSS) in a metalworking fluid mix includes metal chips and grinding wheel grit. Recirculating dirt, whether it is a large quantity of small particles or just one or two large particles, can affect part finishes [37], lead to dirty machines, and clog coolant supply lines. Recirculating metal fines can also lead to rust problems if they deposit on parts.

Dirt or TSS measurement is typically an indication of the effectiveness of the filtration system and/or the settling and chip agglomeration properties of the fluid. To obtain a representative sample for testing, it is important to sample the fluid from the clean coolant nozzle and to be aware of the current indexing cycle of the filter system.

Many procedures exist for determining dirt load in a fluid. A simple method is to centrifuge a sample in a calibrated centrifuge cone. Other methods involve filtering a sample through a specific size filter paper, drying, perhaps ashing, and then weighing. Typically, dirt volumes in excess of 500 ppm or 20 μm in size can lead to problems. Each operation will dictate the type of filtration required. Working with suppliers of both the metalworking fluid and the filtration system is the best method to address concerns in this area, especially if an adjustment to the filtration system is required. There are chemical additives available to assist in chip settling and filtration.

D. Oil Level

Almost every metalworking fluid contains oil. It is either an ingredient, the product oil in a soluble oil, or one of the major contaminants in the form of tramp oil. In both cases, it is useful to know the level of the oil present. Product oil can give an indication of the concentration and the tramp oil of the amount of contamination.

Tramp oil can be in two forms, *free* or *emulsified*. Free oil is that oil which is not emulsified and basically floats on the top of the mix. Emulsified tramp oil is nonproduct oil which is either chemically or mechanically emulsified into the product. Free oil can generally be removed by skimmers or belts, while emulsified tramp is much more difficult to remove, even with a centrifuge or a coalescer.

The sources of tramp oil can be hydraulic leaks, way or gear lube leaks, or from the forced lubrication systems that are found on many machines. The type of oil used can also make a difference in its emulsification or rejection properties with a particular metalworking fluid. Some oils are formulated to emulsify themselves into any water systems, while others have better rejection capabilities. The same goes for metalworking fluids. Depending on the formulation, they may be designed to emulsify a certain level of oil or to completely reject it.

Free oil is measured by centrifuging a sample of mix or simply by allowing it to stand for several hours and then reading the amount of oil that is floating on the surface. Total oil is determined by completely splitting a mix with sulfuric acid and then reading the total oil. If a concentration has been determined by some alternate method or if a "clean" sample at the same concentration has been subjected to this same break, then the product oil can be calculated. Subtracting this product oil from the total oil will give the total tramp. Subtracting the free oil from the total tramp will indicate the emulsified tramp oil.

Generally, high tramp oil levels will affect a product's cleanliness, filterability, mildness, corrosion, and rancidity control. Many mechanical methods exist for minimizing the leaks and removing the oil [38].

E. Bacteria and Mold Levels

Metalworking fluids do not exist in a sterile environment and can develop certain levels of organism growth [39]. The water environment of most fluids is conducive to biological growth. Fluids are formulated to handle bacteria and mold in different ways. Some products contain bactericides and fungicides. Other products are formulated with ingredients that will not support biological growth. Still others are formulated so that, while bacteria and mold may grow, no offensive odors or performance problems develop. Regardless of the type of product, knowing the level of microbial growth and, in most cases, being able to control it is a very useful tool [40]. Bacteria and mold levels can be determined in several ways.

1. Plate Counts

Specific agars for bacteria and mold are prepared. The fluid mix is appropriately diluted and then introduced to the plates along with the agar. The plates are incubated for 48 h at approximately 36°C. After that time the colonies are counted.

2. Sticks

Several companies have developed very specific test systems for determining bacteria and mold counts in metalworking fluids, using "sticks" or "paddles" which have already been prepared with various agars. These sticks are immersed in a fluid sample for 2–3 s, put into a plastic container, and incubated at 25–30°C for 24–48 h. They are then "read" by comparing their appearance to a chart which relates this appearance to a specific number of organisms.

Most suppliers set limits of about 10^5 as the maximum bacteria levels and 0 as the maximum mold level. If counts exceed these levels, some form of treatment may be needed [41].

3. Dissolved Oxygen

Another method to get an indication, but no specific counts, on biological activity is with a dissolved oxygen (DO) check. When a metalworking fluid mix is exposed to the air or is pumped about in air, it will absorb a certain amount of oxygen. At 68°F, a circulated fluid mix will dissolve about 9 ppm oxygen. When aerobic bacteria grow, they use some of the oxygen and also excrete certain gases which drive some of the remaining oxygen out of the mix. Using this phenomena, the DO of a mix can be measured to get a relative indication of biological problems. This determination can be based on one DO reading. A value of less than 3 ppm would indicate a problem. Alternatively, an initial reading can be taken, followed by a reading after the mix has been allowed to sit for 2 h. A difference of 2–3 ppm usually indicates a problem. DO is a good method for a quick, on-site determination of any biological problems.

F. Conductivity

Another metalworking fluid parameter frequently measured for an indication of the fluid's condition is conductivity. The unit of conductivity is the microSieman (μS). Conductivity can be measured on any type of commercially available conductivity meters. A typical 5% metalworking fluid mix in tap water will have a conductivity of about 1500 μS. Conductivity can be altered by the mix concentration, buildup of water hardness, buildup of chloride or sulfate from the water, mix temperature, dissolved metals, and just about any other contaminant. Since so many variables can affect conductivity, a single reading is of little value. Observing any trends in these conductivity readings over a period of time may be useful in assessing mix condition and aging, as well as helping in problem-solving for residues or unstable mixes. Relate the conductivity values to the concentration values to look for any indications of contamination.

V. CARE AND MAINTENANCE OF THE FLUID

Prolonging the life of the metalworking fluid and optimizing its performance are very dependent on the control of the metalworking fluid system [42–44]. This control is as important as the selection of the proper fluid and includes maintenance of the mechanical components as well as the metalworking fluid. The problems that beset metalworking fluids in central system applications are the same as those in individual machines, only the magnitude is greater. A program to accomplish this control should include the following steps.

1. Assign the responsibility for control. If a coordinated program is not established to control the system, it will result in no control. One department or one individual should be responsible for checking fluid concentration and other specified parameters and for making any additions of water, concentrate, or additives to the system. These additions should be recorded for future reference. This person or department will be more mindful of additions, know the reason for making them, and not use concentrate or additive additions as the only means to resolve a production problem. When a control program is not utilized, frequent system dumps and excess usage, resulting in increased costs, can easily occur since no one really knows the status of the system.

2. Clean the system thoroughly before charging with a fresh mix. Dirt and oil can accumulate in relatively stagnant pockets or quiet areas in the central systems or individual machine. If not removed, such accumulations not only cause dirt recirculation in a fresh charge, but also provide an instant inoculation of bacteria to the fresh mix.

3. Maintain the concentration of the metalworking fluid at the dilution recommended for the particular operation. Dilutions are indicated on the label and in the product literature.

Many plants run daily concentration checks on central systems. Individual machines are usually checked on a less frequent basis. As mentioned earlier, the fluid concentration can be checked with a refractometer, a mini-titration kit specific to that product, a laboratory titration procedure, or an instrumental method. Concentration can be controlled by use of premixed fluid or a proportioning system. Reviewing this concentration information can indicate trends and possible problems long before they show up on the production line. Lean concentrations can lead to rust, rancidity, poor tool life, lack of lubricity, and other problems. Maintaining a stronger than recommended concentration can result in foam, skin irritation, residue, increased costs, and other problems.

The fluid mix is lost from the system by both evaporation and carry-off or splashing. Depending on the type of operation, type of fluid, and part configuration and handling, the amount of mix lost by either of these means can vary. By evaporation, only water is lost. By splashing or carry-off, both water and fluid concentrate are lost. Therefore, each time water is added to the system, metalworking fluid concentrate should also be added at a ratio that has been selected to maintain the proper dilution in the system. This will keep product components in their proper balance and minimize any selective depletion of these components. For grinding operations, the premixed fluid is usually a leaner dilution than for machining. This is because grinding typically loses more water to evaporation.

Management and Troubleshooting

4. Keep the metalworking fluid free of chips and grit. This is a major factor in fluid life. Recirculating dirt can lead to unsightly buildup on the machines, plugged coolant lines, poor finish in grinding [45], and tool wear in machining. Chip buildup in reservoirs can drastically reduce the volume of the system and deplete product ingredients. Positive filters with some type of disposable media do a better job of removing small fines than settling tanks. On individual machines, regular cleanouts of the reservoir or sump should be utilized to keep this buildup under control. The use of fluid recycling could be a cost efficient option.

5. As mentioned earlier, the quality of water needed to make a metalworking fluid mix is a very important factor in performance. Remember, the ideal hardness of water for making a metalworking fluid mix ranges from 80 to 125 ppm.

6. Aerate the metalworking fluid mix by keeping it circulated. This circulation prevents the growth of anaerobic bacteria that cause offensive odors. Many central systems continually circulate even when production is not running, others utilize timers to circulate the fluid for a short time on a set schedule during any nonproduction hours or days. In individual machines, an air hose can be used to bubble air through the mix while the machine is not operating. Atmospheric oxygen is detrimental to the growth of odor-producing anaerobes. During circulation, oxygen continually enters the metalworking fluid, but, when a system is shutdown, this cannot occur.

7. Provide good chip flushing at the machines and in the trenches. If chips do not reach the filter, they deplete certain constituents of the metalworking fluid and furnish an excellent breeding ground for bacteria. It is essential that the chips reach the filter in order that they might effectively be removed. Trenches, return lines, system capacity, retention time, flow rates, and other design parameters must all be adequately sized to provide this good filtration. Washdown nozzles may need to be installed on the machines or in the trenches to keep the metalworking fluid moving back to the sump or filter. Check that these nozzles are set at flow rates sufficient to keep the chips moving but not excessive, which could result in foaming.

8. Employ good housekeeping practices. Foreign matter that is allowed to accumulate in a metalworking fluid has a drastic effect on its life and performance. While a good high-quality metalworking fluid is formulated to cope with a certain amount of contamination, the greater the amount of contamination, the shorter the life and the more erratic the performance of the fluid. Avoid using reservoirs as a "garbage" disposal. Cigarette butts, food scraps, sputum, and candy wrappers, for example, inoculate the metalworking fluid with bacteria and furnish food for growth. Do not dump floor-cleaning solutions into the reservoir. Many contain chemicals, such as phosphates, which may contribute to skin irritation, promote the growth of odor-producing microorganisms, or cause the product to foam.

9. Remove extraneous tramp oils. Minimize the leakage of oils into the system through proper maintenance of seals and lubrication systems. If excess

quantities of oils leak into the system, the metalworking fluid performance can be reduced. High oil levels can extract oil-soluble lubricants, emulsifiers, and microbicides from the fluid. Lubricating and hydraulic oils contain food for bacteria. They may also blanket the surface of the fluid, excluding air, and thereby provide ideal conditions for the growth of odor-producing bacteria. If allowed to build up, extraneous oil causes smoking, reduces the cooling action of the fluid, increases residue around the machine area, and makes the machines look dirty. Oil-removing devices such as skimmers, coalescers, oil wheels, or centrifuges can be used to prevent oil buildup.

Using this program, it is possible to achieve improved production and long, trouble-free metalworking fluid life in central systems and individual machines.

VI. METALWORKING FLUID TROUBLESHOOTING

Problems with metalworking fluids can be related to a number of causes, many of which are not inherently fluid problems. Improper machine setup, coolant application, or product selection can all lead to problems. It is certainly more complex to solve the problems in central systems compared to individual machines. A much larger volume of fluid, more production, and many more operators are involved. In some cases, there may be a combination of many factors causing the problem.

Logical thinking is the first step in any problem-solving effort. Get the facts, analyze them, plan a course of action, and implement the plan. For an individual machine, the solution may be a simple dump and recharge. For a central system, the problem and the resolution will usually be much more complicated. In these situations, service help from the supplier via a phone call or personal visit is a likely course of action.

The problems most commonly attributed to metalworking fluids include: corrosion, rancidity or objectionable odors, excessive foam, insufficient lubricity resulting in poor tool life or an unsatisfactory finish on the part, objectionable residue or a dirt buildup on the machine, and safety concerns such as dermatitis or eye, nose, or throat irritation. In this section, many of the possible causes and the corresponding remedies have been listed.

Management and Troubleshooting

A. Corrosion of the Work or the Machine

POSSIBLE CAUSES	CORRESPONDING REMEDIES
The concentration of the metalworking fluid mix may be too low.	Make a concentration analysis by one of the described methods and adjust the mix to the recommended concentration. Determine and correct the cause of the low concentration (mixing errors, water leaks, hard water, etc.) For a central system mix, it may also be necessary to check and adjust the pH to the recommended standard.
The recommended concentration range of the fluid may be too low for this application.	Increase the concentration of the mix by ½ to 2% increments, depending on the product being used, to find the optimum concentration range.
The buildup of ions from the water supply (total hardness, chloride, or sulfate) may be too high for the product or the current concentration.	If possible, conduct laboratory testing to determine water quality and any buildup in the fluid. In some cases, increasing the concentration by ½ to 2% may provide control for a while. If the buildup is excessive, a system dump may be required. In areas of poor quality water, consider a water treatment system or products with optimum corrosion control.
The metalworking fluid tank may be full of chips or swarf, contaminated by tramp oil or other contaminants.	Clean the fluid, if possible, using filtration or recycling equipment. If it is not possible to adequately remove the contaminants, dump and recharge the system with fresh metalworking fluid. Excessive dirt can "plate out" on parts and lead to rust. Certain settling or filter aid additives may help.
Parts that are still wet with metalworking fluid may be touching other ferrous materials or dissimilar metals.	Avoid metal-to-metal contact in stacking parts after any metal removal operations. Use plastic coated wire baskets rather than metal tote pans. Dry the parts before prolonged storage. Use vapor barrier material between parts during handling and storage. The use of a rust preventive spray or dip may be needed for extended storage.
Hot, humid conditions may accelerate rust problems by slowing the drying action.	Increase the concentration of the mix. Improve plant ventilation. During severe weather conditions, it may be necessary to use a water displacing rust preventive or an additive to the fluid mix.
Fumes from acidic materials may be saturating the area.	Improve plant ventilation. Use fans to direct the fumes outside or provide some type of covering for parts and machines.

B. Rancidity or Objectionable Odor [24]

POSSIBLE CAUSES	CORRESPONDING REMEDIES
The concentration of the metalworking fluid mix may be too low.	Make a concentration analysis by one of the described methods and adjust the mix to the recommended concentration. Determine and correct the cause of the low concentration (mixing errors, water leaks, hard water, etc.) For a central system mix, it may also be necessary to check and adjust the pH to the recommended standard.
The recommended concentration range of the fluid may be too low for this application.	Increase the concentration of the mix by ½ to 2% increments, depending on the product being used, to find the optimum concentration range.
The fluid tank may be full of chips or grinding swarf, contaminated by tramp oil leakage, or other contaminants, such as food scraps.	Clean the fluid where possible using filtration and/or oil removal equipment. On small tanks, it may be advisable to dump and recharge the fluid. In certain conditions of rancidity, the use of biocidal additives to treat for specific bacteria and/or mold problems is the recommended treatment. Use only as recommended by your supplier. It is best to eliminate the source of the contamination.
A high dirt content indicates inefficient filtration. This could be due to an incorrect filter setting, a change in a setting, or defective or improper media.	Repair defective media. Restore filter to original settings and adjustments. Experiment to find more effective adjustments and settings. Increase retention time. Contact the filter manufacturer. Investigate the possibility of obtaining a higher percentage of the large swarf on the media in order to build a better filter cake. If necessary, thoroughly clean the system according to recommended procedures. Recharge with a fresh mixture. It may be advisable to use additives to assist in obtaining a usable mixture without cleaning the system. Contact the supplier for recommendations.
Extreme conditions of contamination, excessive dirt load, or both, may require a change in operational procedures.	Aerate the cutting fluid and increase filtration time by running the entire system up to 24 hours per day and on weekends, if necessary.
The system may be contaminated from the old cutting fluid or construction debris.	Thoroughly clean the system according to recommended procedures. Recharge with a fresh mixture.
The sulfate content of the water may be too high for this specific product.	Have a water analysis made. If the sulfate content is over 150 ppm, use a higher concentration of metalworking fluid, change to a product that is more compatible with this condition, or use treated water.

Management and Troubleshooting

POSSIBLE CAUSES	CORRESPONDING REMEDIES
Excessive amounts of lubricating oils may be leaking into the system. These oils often contain sulfur or phosphorus, which are ideal foods for bacteria.	Change to more compatible oil. Prevent leakage into the system. If this is not possible, consider installing oil removal equipment such as oil skimmers or centrifuges.

C. Excessive Foam

POSSIBLE CAUSES	CORRESPONDING REMEDIES
The concentration of the cutting fluid mixture may be too high.	Make a concentration analysis and adjust to the recommended concentration. Determine and correct the cause of the mixture being too high. Most frequently, this is a human error or mechanical problems with metering devices.
The recommended concentration range may be too high for this application.	Decrease the concentration of the mixture by 0.5-2.0% increments, depending on the fluid being used, to find the optimum concentration. CAUTION: If the concentration is too low, other problems (rust, rancidity, etc.) may develop.
The level of the cutting fluid in the reservoir may be low, causing air to be drawn into the pump.	Fill the reservoir to the normal operating level with water and concentrate at the recommended concentration.
A crack in the pump housing or intake piping may be allowing air into the system.	Inspect the pump and piping system. Repair or replace defective units.
High outlet pressures, high fluid velocities, sharp corners in the return system, or excessive waterfalls may create high agitation.	Locate any of these foam-producing conditions and reduce or eliminate where it is possible.
The water may be too soft to use with this specific product.	Have a water analysis made. If the total hardness is less than 50 ppm, change to a product that is more compatible with soft water.
The system may be contaminated from some external source such as indiscriminate disposal of floor cleaners, washing compounds, etc.	Determine and eliminate the source of indiscriminate dumping of other shop materials into the cutting fluid reservoir. It may be advisable to add antifoaming additives to assist in returning the fluid to a normal condition. Contact the supplier for recommendations. If the system is highly contaminated, thoroughly clean according to the recommended method. Recharge with a fresh mixture.

D. Unsatisfactory Surface Finish or Burn on Parts from a Grinding Operation

POSSIBLE CAUSES	CORRESPONDING REMEDIES
The concentration of the cutting fluid mixture may be too low.	Make a concentration analysis and adjust the mixture to the recommended concentration. Determine and correct the cause of the low concentration (e.g., mixing errors, water leakage, recirculated grit, hard water, etc.).
The recommended concentration may be too low for these specific conditions.	Increase the concentration of the mixture by 0.5 - 2.0% increments, depending on the fluid being used, to find the optimum concentration.
The flow of the cutting fluid may be inadequate or it may not be reaching the metal removal area.	Increase the volume and readjust the nozzle so that a maximum amount of fluid reaches the metal removal area. Foam or entrained air may be getting into the cut zone in place of fluid. Follow the remedies for reducing foam.
The cutting fluid tank may be full of chips or grinding swarf; contaminated by oil leakage or by other matter.	Drain and thoroughly clean the reservoir and cutting fluid piping system according to recommended procedures. Recharge with a fresh product.
The grinding wheel may be incorrect for this application.	Determine if the wheel is acting too hard or too soft. Change wheel grade accordingly.
The water may be too hard to use with this specific product.	Have a water analysis made. If the total hardness is over 200 ppm, change to a product that is more compatible with hard water or use treated water.

Management and Troubleshooting

E. Cutting Tool or Grinding Wheel Life Is Not Satisfactory

POSSIBLE CAUSES	CORRESPONDING REMEDIES
The concentration of the cutting fluid mixture may be too low.	Make a concentration analysis and adjust the mixture to recommended concentration. Determine and correct the cause of the low concentration (e.g., mixing errors, water leakage, recirculated grit, hard water, etc.).
The recommended concentration range of the fluid may be too low for this application.	Increase the concentration by 0.5 - 2.0% increments, depending on the product used, to find the optimum concentration.
The flow of the cutting fluid may be inadequate or it may not be reaching the metal removal area.	Increase the volume of fluid being used and readjust the nozzle so that the maximum amount of fluid reaches the metal removal area.
The cutter or tool design may be incorrect for this application.	Analyze the tool geometry in relation to the application. Consult the tool engineering specialist. Change the geometry to obtain improved chip formation.
A high dirt content indicates inefficient filtration. This could be due to incorrect filter setting, a change in a setting, or improper media.	Repair defective media. Restore filter to original settings and adjustments. Experiment to find more effective adjustments. Increase retention time in settling systems. Contact the filter manufacturer. Investigate the possibility of obtaining a higher percentage of the large swarf on the media in order to build a better filter cake. If necessary, thoroughly clean the system according to recommended procedures. Recharge with a fresh mixture. It may be advisable to use additives to assist in obtaining a usable mixture without cleaning the system. Contact the supplier for recommendations.

F. Skin Irritation [46]

POSSIBLE CAUSES	CORRESPONDING REMEDIES
Regardless of the cause, skin irritation is a medical problem and should be treated immediately.	Have the worker report immediately to properly trained medical personnel. Although the ailment may be unrelated to the cutting fluid, it should be investigated and treated by a competent person.
The concentration of the cutting fluid mixture may be too high.	Make a concentration analysis and adjust to the recommended concentration. Determine and correct the cause of the mixture being too high. Most frequently, this is a human error or mechanical problems with metering devices.
The recommended concentration range may be too high for this application.	Decrease the concentration of the mixture by 0.5 - 2.0% increments, depending on the product used, to find the optimum concentration. CAUTION: If the concentration is too low, other problems (i.e., rust, rancidity, etc.) may develop.
The soap in the washrooms may be too harsh and irritating.	Change to a mild, but equally effective cleaning agent.
The operator's hands may be immersed continually in the metalworking fluid.	Encourage the use of waterproof barrier creams or protective gloves. Use material handling devices where feasible.
The operator may be coming in contact with harsh irritating chemicals outside, or even inside the company.[47]	Determine if the operator has any activities where he might come in contact with such chemicals (i.e., solvents used in painting, cleaners, and solvents used in automotive repair work, etc.). Substitute the irritating products for ones which will not affect the operator's skin.
The operator may be subject to skin irritation because of poor hygienic conditions.	Encourage washing frequently, wearing freshly laundered work cloths and using protective gloves, aprons, boots, etc., especially if there are excessive splash conditions.

POSSIBLE CAUSES	CORRESPONDING REMEDIES
The sump or reservoir may be contaminated from some external source such as indiscriminate disposal of floor cleaners, washing compounds, construction debris, etc.	Determine and eliminate the source of indiscriminate dumping of other shop materials into the cutting fluid reservoir. Improve hygienic practices. If the system is excessively contaminated, thoroughly clean according to the recommended method. Recharge with a fresh product.

G. Eye, Nose, or Throat Irritation

POSSIBLE CAUSES	CORRESPONDING REMEDIES
Regardless of the cause, these are medical problems and should be treated immediately.	Have the worker report immediately to trained medical personnel. Although the ailment may be unrelated to the cutting fluid, it should be investigated and treated by a competent person.
The concentration of the cutting fluid mixture may be too high.	Make a concentration analysis and adjust the mixture to the recommended concentration. Determine and correct the cause of the high concentration (e.g., mixing errors).
The recommended concentration may be too high for the specific conditions.	Decrease the concentration of the mixture by 0.5 - 2.0% increments, depending on the fluid being used, to find the optimum concentration. CAUTION: If the concentration becomes too low, other problems (rust, rancidity, etc.) may develop.
There may be irritating fumes coming from some other operation in the plant or outside the plant.	Investigate ventilation conditions of heat treating or plating areas, and the plant in general. Improve unsatisfactory conditions with use of fans until permanent changes can be made. Investigate possible sources outside the plant and take corrective action if required.
There may be excessive splashing or misting the cutting fluid.[48]	Reposition the guards on the machine to contain the splash or mist. Grind chip breakers into cutting tools. Encourage the use of safety goggles or glasses.

H. Objectionable Residue [49,50]

POSSIBLE CAUSES	CORRESPONDING REMEDIES
The concentration of the cutting fluid mixture may be too high.	Make a concentration analysis and adjust to the recommended concentration. Determine and correct the cause of the mixture being too high. Most frequently, this is a human error or mechanical problems with metering devices.
The recommended concentration range may be too high for this application.	Decrease the concentration of the mixture by 0.5 - 2.0% increments, depending on the fluid being used, to find the optimum concentration. CAUTION: If the concentration becomes too low, other problems (rust, rancidity, etc.) may develop.
The cutting fluid reservoir may be full of chips or grinding swarf; contaminated by oil leakage and food remnants; or contaminated by other matter.	Drain and thoroughly clean the reservoir and cutting fluid piping system according to recommended procedures. Recharge with a fresh product.
There may be excessive misting conditions due to inefficient guards.	Design and place the guards, shields, etc., so that misting (especially from grinding operations) is confined to the immediate area of the cut.
The system may be contaminated from some external source such as oil leaks from the machine tools.	Locate and repair all oil leaks. Remove extraneous oil by means of oil skimmers or a centrifuge. If necessary, clean the system thoroughly according to recommended procedures. Recharge the system with a fresh product mixture.
The water may be too hard to use with this specific product.	Have a water analysis made. If the total hardness is over 200 ppm, change to a product that is more compatible with hard water or use treated water.

VII. CONTRACT FLUID MANAGEMENT

Managing metalworking fluids can be a very involved and time-consuming process, especially as the products become more complex and the control techniques more sophisticated. As more plants recognize the value of properly maintaining their fluids from a productivity and a waste minimization standpoint, they realize that a certain degree of expertise is required to accomplish this. For that reason, the concept of contract fluid management is growing in popularity [51–55]. Under this plan the metalworking fluid supplier or another chemical vendor has complete responsibility for the usage and control of the chemicals within the plant [56]. An on-site person(s) will run regular checks on the fluid parameters, calculate additions, recommend additives, recommend system dumps, and participate in all in-plant activities (meetings, training programs, etc.) that are related to these chemicals. In essence, a partnership is formed between the supplier and the user.

For an agreed upon monthly fee over the course of the contract (typically 2–3 years), the user has the benefit of the expertise of the supplier to manage and control all aspects of the chemical usage. The supplier, with better control programs, can improve fluid performance, resulting in better quality and productivity for the user. In some cases it is possible to reduce fluid usage and disposal, resulting in even greater cost benefits.

REFERENCES

1. E. S. Nachtman and S. Kalpakjian, *Lubricants and Lubrication in Metalworking Operations*, Marcel Dekker, New York, pp. 1–61 (1985).
2. E. R. Booser, ed., *CRC Handbook of Lubrication*, CRC Press, Boca Raton, FL, pp. 335–356 (1984).
3. R. K. Springborn, *Cutting and Grinding Fluids: Selection and Application*, American Society of Tool and Manufacturing Engineers, Dearborn, MI, pp. 5–30 (1967).
4. D. C. Zintak, ed., *Improving Production with Coolants and Lubricants*, Society of Manufacturing Engineers, Dearborn, MI, pp. 8–49 (1982).
5. M. E. Merchant, Fundamentals of cutting fluid action, *Lubr. Eng.*, August 1950.
6. R. M. Dick and G. J. Foltz, How to maintain your coolant system, *Mach. Tool Bluebook 30*, January 1988.
7. J. Ivaska, Green management, *Cutting Tool Eng. 39*, October 1991.
8. E. S. Nachtman and S. Kalpakjian, *Lubricants and Lubrication in Metalworking Operations*, Marcel Dekker, New York, pp. 63–105 (1985).
9. *Waste Minimization and Wastewater Treatment of Metalworking Fluids*, Independent Lubricant Manufacturers Association, Alexandria, VA, pp. 2–4 (1990).
10. E. R. Booser, ed., *CRC Handbook of Lubrication*, CRC Press, Boca Raton, FL, pp. 361–365 (1984).
11. Lubrication in metalworking, *Tribol. Int.* 171–175, June 1982.
12. R. K. Springborn, *Cutting and Grinding Fluids: Selection and Application*, American Society of Tool and Manufacturing Engineers, Dearborn, MI, pp. 53–65 (1967).

13. *Machining Data Handbook*, 3rd Edition, Volume Two, compiled by the Technical Staff of the Machinability Data Center, Metcut Research Associates, Cincinnati, OH, pp. 16/17–16/96 (1980).
14. C. Yang, The effects of water hardness on the lubricity of a semi-synthetic cutting fluid, *Lubr. Eng. 133*, March 1979.
15. J. Joseph, *Coolant Filtration*, Joseph Marketing, East Syracuse, NY (1987).
16. M. Opachak, ed., *Industrial Fluids: Controls, Concerns, and Costs*, Society of Manufacturing Engineers, Dearborn, MI, pp. 25–104 (1982).
17. D. C. Zintak, ed., *Improving Production with Coolants and Lubricants*, Society of Manufacturing Engineers, Dearborn, MI, pp. 167–212 (1982).
18. *Waste Minimization and Wastewater Treatment of Metalworking Fluids*, Independent Lubricant Manufacturers Association, Alexandria, VA, pp. 15–159 (1990).
19. J. C. Childers, Metalworking fluids—A geographical industry analysis, *Lubr. Eng. 542*, September 1989.
20. W. E. Deming, *Quality, Productivity, and Competitive Position*, Massachusetts Institute of Technology, pp. 267–311 (1982).
21. E. S. Nachtman and S. Kalpakjian, *Lubricants and Lubrication in Metalworking Operations*, Marcel Dekker, New York, pp. 215–222 (1985).
22. *Waste Minimization and Wastewater Treatment of Metalworking Fluids*, Independent Lubricant Manufacturers Association, Alexandria, VA, pp. 26–30 (1990).
23. American Society for Testing Materials, *Book of ASTM Standards*, Section 5, Vol. 05.01, 05.02, and 05.03 (1987).
24. E. S. Nachtman and S. Kalpakjian, *Lubricants and Lubrication in Metalworking Operations*, Marcel Dekker, New York, pp. 107–116, 133–156 (1985).
25. R. K. Springborn, *Cutting and Grinding Fluids: Selection and Application*, American Society of Tool and Manufacturing Engineers, Dearborn, MI, pp. 83–114 (1967).
26. M. D. Smith and J. E. Lieser, Laboratory evaluation and control of metalworking fluids, *SME Technical Paper*, MR73-120, Society of Manufacturing Engineers, Dearborn, MI (1973).
27. E. O. Bennett, The biological testing of cutting fluids, *Lubr. Eng. 128*, March 1974.
28. H. R. Leep and S. J. Kelleher, Effects of cutting conditions on performance of a synthetic cutting fluid, *Lubr. Eng. 111*, February 1990.
29. J. E. Clock, What coolant selection taught us, *Mod. Mach. Shop 86*, November 1986.
30. E. R. Booser, ed., *CRC Handbook of Lubrication*, CRC Press, Boca Raton, FL, pp. 366–368 (1984).
31. D. C. Zintak, ed., *Improving Production with Coolants and Lubricants*, Society of Manufacturing Engineers, Dearborn, MI, pp. 167–171 (1982).
32. M. Opachak, ed., *Industrial Fluids: Controls, Concerns, and Costs*, Society of Manufacturing Engineers, Dearborn, MI, pp. 232–236 (1982).
33. W. M. Coursey, The application, control and disposal of cutting fluids, *Lubr. Eng. 200*, May 1969.
34. W. A. Sluhan, Coolant management: Rx for ending coolant headaches, *Carbide Tool J 21*, March–April 1985.
35. M. Opachak, ed., *Industrial Fluids: Controls, Concerns, and Costs*, Society of Manufacturing Engineers, Dearborn, MI, pp. 75–81 (1982).

36. R. E. Johnston, M. Fayer, and S. DeSimone, Multicomponent analysis of a metalworking fluid by fourier transform infrared spectroscopy, *Lubr. Eng. 775*, September 1988.
37. R. S. Marano, G. S. Cole, and K. R. Carduner, Particulate in cutting fluids: Analysis and implications in machining performance, *Lubr. Eng. 376*, May 1991.
38. M. Opachak, ed. *Industrial Fluids: Controls, Concerns, and Costs*, Society of Manufacturing Engineers, Dearborn, MI, pp. 70–74 (1982).
39. E. O. Bennett, The biology of metalworking fluids, *Lubr. Eng. 227*, July 1972.
40. E. R. Booser, ed., *CRC Handbook of Lubrication*, CRC Press, Boca Raton, FL, pp. 371–378 (1984).
41. *Waste Minimization and Wastewater Treatment of Metalworking Fluids*, Independent Lubricant Manufacturers Association, Alexandria, VA, pp. 31–46 (1990).
42. T. D. Howes, H. K. Toenschoff, and W. Heuer, "Environmental aspects of grinding fluids," CIRP Grinding STC Keynote Paper, August 1991.
43. G. Skells, Fluid management skills, *Cutting Tool Eng. 52*, October 1990.
44. M. Opachak, ed., *Industrial Fluids: Controls, Concerns, and Costs*, Society of Manufacturing Engineers, Dearborn, MI, pp. 65–82 (1982).
45. W. M. Needelman, F. A. Fiumano, and J. A. Masters, Controlling grinding coolant contamination in an automotive plant, *Lubr. Eng. 479*, August 1989.
46. E. O. Bennett, Dermatitis in the metalworking industry, *ASLE Special Publication SP-11*, American Society of Lubrication Engineers, Park Ridge, IL, (1983).
47. E. O. Bennett, Stop metal dermatitis before it starts, *Manuf. Eng. 36*, August 1991.
48. M. Opachak, ed. *Industrial Fluids: Controls, Concerns, and Costs*, Society of Manufacturing Engineers, Dearborn, MI, pp. 7–9 (1982).
49. D. R. Bell, N. L. Matthews, and M. Zabik, Out, damned smut!, *Mach. Tool BlueBook 37*, January 1988.
50. T. J. Drozda, Those troublesome cutting fluid residues, *Manuf. Eng. 55*, January 1983.
51. Coolant management improves the cash flow, *Metalworking Prod. 39*, November 1990.
52. L. C. Archibald and C. Bowes, Who knows the real cost of poor fluid management?, *Eur. Mach. 39*, January–February 1991.
53. Jobbing out fluids management, *Am Mach. 45*, January 1991.
54. J. C. Childers, Chemical contract management panel discussion, *Lubr. Eng. 22*, January 1992.
55. J. Joseph, Keeping fluids clean in GM-Spring Hill, *Am. Mach. 42*, October 1991.
56. Keeping your cool, *Manuf. Eng. 52*, June 1991.

12
Recycling of Metalworking Fluids

RAYMOND M. DICK
Cincinnati Milacron
Cincinnati, Ohio

I. INTRODUCTION

In 1990, there were approximately 81 million gallons of metalworking fluids manufactured in the United States [1]. Of all the metal removal fluids consumed in the United States, 90% was used by the fabricated metal, transportation equipment, and machinery industries [2]. The vast majority of metalworking fluids consumed are found in individual machines rather than central fluid systems. According to the 1989 *American Machinist* Metalworking Survey, published in November 1989, there are 1,870,753 metalworking machines in the United States. Industry experts at Henry Filter Systems in Bowling Green, OH, estimate that there are approximately 5000 central systems in the United States. While some machines may use straight oils or no fluids at all, the majority of metalworking machines use water-based fluids.

Based upon the total volume of lubricants used in the United States, the disposal of these fluids, and oily wastewater in general, has become a much more important issue during the past 20 years. The metalworking industry has reevaluated its use of lubricants and chemicals as a result of environmental, health and safety, and productivity reasons.

Since the early 1970s, there have been laws enacted to protect surface and groundwater quality. These laws have directly impacted the treatment and disposal methods of oil–water emulsions. These laws include the Federal Water Pollution Control Act (or Clean Water Act), the Clean Water Act Amendments, the Safe Drinking Water Act, and the Resource Conservation and Recovery Act [3]. The end result of this legislation is that stricter methods are required for the proper treatment and disposal of metalworking fluids. Contract hauling costs for oily wastewater have increased by as much as ten to 20 times the cost in the 1970s. Landfilling of oily wastes and sludges containing liquids has been banned. Stricter sewer discharge standards require more effective wastewater treatment methods.

Health and safety issues concerning the use of chemicals in the workplace have also become prominent. Metalworking fluids are carefully studied for operator health and safety characteristics because of the close contact operators have with the fluids. However, these clean fluids become contaminated during use with metal chips, fines, grinding wheel solids, various lubricating oils and greases, cleaners, solvents, bacteria, and other materials either purposely or accidently discharged into the sump. These contaminants and lack of proper fluid controls are largely responsible for frequent fluid discharge or dumps.

For the reasons mentioned earlier, the use and control of metalworking fluids have become an important issue for metalworking plants. The purpose of this chapter is to discuss the subject of metalworking fluid recycling as it relates to management and equipment designed to extend fluid life. Some information found elsewhere in this book is repeated here in order to give the reader a comprehensive overview of all aspects of this important subject.

II. BACKGROUND AND HISTORY OF FLUID RECYCLING

In the early 1970s and before, metalworking fluids were considered consumable products, designed for a relatively short life prior to disposal. Many of these fluids were discharged directly to the sewer or contract-hauled to landfills for less than ten cents per gallon.

More recently, the cost of fluid disposal has rapidly increased and, along with liability concerns, there has been a growing importance placed on fluid management. Fluid recycling systems and management techniques have rapidly gained importance as plants seek better fluid life and reduced disposal costs.

Today, greater importance is placed on understanding and managing the metalworking process. The metalworking process consists of the machine, operator, tool or wheel, fluid, and workpiece. Additional process variables may include the water quality, filter systems, and machine variables (such as lubricant systems, sump design, and workpiece handling). In terms of relative cost, the fluid costs have been low, which, in the past, resulted in poor management and frequent disposal.

With today's industry emphasis on productivity, waste minimization, and cost control, many plants have installed fluid recycling equipment. The goal is to optimize fluid performance, reduce oily wastewater volume, and reduce fluid concentrate and disposal costs.

As environmental regulations become stricter, more and more emphasis will be placed on fluid recycling. This chapter identifies the basics of fluid management, as well as fluid recycling technologies and equipment.

III. BASICS OF FLUID MANAGEMENT

For many years, metalworking fluids have been used to increase the productivity of metalworking operations. Oil-in-water emulsions provide benefits of lubricity, reduction of friction, and reduction of heat in the metalworking process. In the past, these fluids have been relatively inexpensive to purchase, use, and dispose. More recently, fluid costs have rapidly increased, primarily due to rising disposal costs. Many metalworking plants are investigating alternative processes to better manage the use and disposal of metalworking fluids.

Since the early 1940s, when the initial research was completed on the metal removal process, water-based fluids have been used to improve metalworking operations [4]. Straight oils have been replaced in many applications due to safety and health reasons. The major problems with straight oils in metalworking plants are fire hazards, slippery floors, and general housekeeping difficulties. There are serious health concerns with breathing oil mist [5]. Compared to straight oils, water-based fluids provide improved machining and grinding characteristics, part finishes, tool life, wheel life, and allow for higher speeds in many operations [6]. In addition, they are cleaner and safer to use.

As metalworking plants seek improvements in productivity and economics, metalworking fluids must be evaluated. Along with improvements in machine tool technology, tooling, materials, and automation, there are improvements possible with the use of metalworking fluids.

To improve overall management and control of fluids, each plant must evaluate the following fluid use areas:

Water quality
Fluid selection
Fluid controls
Contaminant removal systems

A. Water Quality

The quality of the water mixed with metalworking fluid concentrates is very important to the performance of these fluids. Fluid life, tool life, part finish, foam characteristics, product residue, part or machine corrosion, mix stability, and concentrate usage are all affected by water quality [7].

During normal fluid use, evaporation and carry-off losses require daily additions of fluid make-up. This process increases the quantity of total dissolved solids (TDS) in the fluid. Figure 1 depicts the theoretical increase in total dissolved solids because of evaporation losses and 10% daily make-up. The higher the initial TDS of the water source, the more rapid the TDS increase over time. As certain dissolved solids increase in quantity, problems will develop with the fluids. For instance, mineral and hardness salts, particularly chlorides and sulfates, contribute to corrosion at a level of approximately 100 ppm. Sulfates also promote the growth of sulfate-reducing (anaerobic) bacteria in fluids and create a "rotten egg" odor [7].

It is important to have a water analysis completed on the plant's water. The fluid manufacturer may recommend treated water if the dissolved solids, hardness, minerals, or metals are at a high enough level to cause metalworking fluid application problems. The type of treated water used may be deionized water, distilled water, or reverse osmosis treated water. Water softening will remove the calcium and magnesium hardness ions, however, it can contribute to the corrosiveness of a metalworking fluid since sodium chloride and sodium sulfate are more corrosive than the hardness minerals [8]. A water-hardness comparison is found in Table 1 [8].

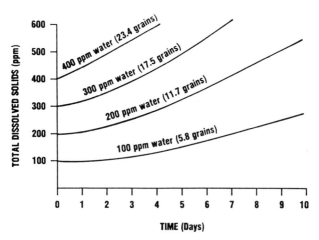

Figure 1 Increase in total dissolved solids with 10% daily make-up TDS vs. time.

TABLE 1 Water-Hardness Comparison

CLASSIFICATION	PARTS per MILLION	GRAINS per U.S. GAL
Very soft water	Less than 17	Less than 1
Soft water	17-52	1-3
Medium hard water	52-105	3-6
Hard Water	105-210	6-12
Very hard water	More than 210	Greater than 12

B. Fluid Selection

Because of the large variety of fluids available, it can be a difficult and time-consuming process to find the best fluid for a given operation. The fluid selected will have the greatest impact on a plant's fluid management program. Recommendations from the supplier will be helpful in narrowing the field of fluids from which to choose.

To understand the basic classes of fluids available, refer to Table 2. The four major classes of fluids are straight oil, soluble oil, semisynthetic, and synthetic [8]. The water-based fluid concentrates are typically mixed with water at ratios from 1:10 to 1:50, depending on the specific application and fluid type.

With a knowledge of fluid types and assistance from suppliers, it is important to evaluate the following fluid-related parameters:

Performance: as indicated by laboratory screening tests and field testing [8].

TABLE 2 General Classes of Metalworking Fluids

CLASS	% PETROLEUM OIL IN CONCENTRATE	APPEARANCE OF FRESH FLUID
Synthetic	0	Transparent to opaque
Semisynthetic	2 - 30	Transparent, Translucent, or opaque
Soluble Oil	60 - 90	Opaque
Straight Oil	100	Transparent

Health and safety: per material safety data sheets and other available data from suppliers [9].
Waste treatment: using screening tests, field tests, and supplier assistance.
Quality standards of fluid: with ability of supplier to provide consistent quality, using statistical process control methods.
Technical service support: for proper controls and troubleshooting of fluids with supplier's assistance.
Cost: using overall cost of fluid including purchase price, labor (mixing, transporting, and controlling), machine downtime (for charging and disposing of fluid), waste treatment, and operator safety and health.
Delivery: with supplier providing "just-in-time" delivery.

Each application must be thoroughly evaluated to understand the fluid requirements. For instance, it is important to know the specifics of the operation:

Type of machining, grinding, etc.,
Central system or individual machine sumps,
Tooling and setup,
Type of metals,
Type of water,
Part requirements (tolerances, finish, and rust protection),
Machine requirements (lubrication, seals, paint, cleanliness, and visibility of work area).

While laboratory or screening tests are useful and necessary, the most beneficial information is available from production machines.

In most plants, personnel from manufacturing, maintenance, safety, purchasing, the laboratory, and wastewater treatment areas will have input on the fluid selection process. Ideally, one person or group will have the authority and responsibility to select the best *overall* fluid. It is best if only one fluid at one concentration can be used in a plant for purposes of management and recycling. If this is not possible, then it is very important to minimize the number of different fluids used in a plant.

C. Fluid Controls

Given the selection of the best fluid for a specific application, it is important to identify the needed controls to maintain optimum fluid performance. Every fluid has a range of parameters within which it is designed to operate. For example, concentration, pH, and contamination level (oil, dirt, and bacteria) are parameters that ideally are controlled for each water-based fluid. The supplier of the fluid must be able to identify the parameters and ranges to be used in controlling the fluid.

The majority of fluid problems arise due to improper concentration, when the fluid mix becomes too rich or too lean. Evaporation and carry-off losses, approximately 10% per day, alter the fluid's mix ratio.

The concentration will need to be checked on a frequent basis, preferably daily, but at a minimum of once per week. Concentration check methods are discussed in Chapters 7 and 11.

Similarly, it is necessary to check fluid pH on a frequent basis and maintain it in the recommended range. One common occurrence with fluids is that bacteria generate acid by-products, which lower the fluid's pH. If this situation occurs, it will be necessary to control the bacteria level and readjust the pH of the fluid.

After the initial charge of a fresh mix of fluid into a machine reservoir (or sump), this fluid becomes contaminated with oil, dirt, metals, bacteria, and other materials as a result of its use. For optimum fluid performance and life, fluid contaminants must be controlled. These contaminants can be minimized with good maintenance and housekeeping programs. With many machines, lubricating oils and greases cannot be isolated from the fluid. These contaminants, as well as metal contaminants, are an expected by-product of the machining, grinding, or other metalworking operations. Many of the contaminants that cause fluids to be disposed of frequently are foreign materials, such as floor sweepings, cleaners, solvents, dirt, tobacco, and food. With the goal of improved fluid management, education, and revised plant practices will be required to improve housekeeping and sanitation of the fluids.

Contaminant levels of oil, dirt, and bacteria can be monitored to determine how the quantities are changing over time. In many cases, corrective action can be taken to remove the oil, filter solids, or control bacteria to prevent disposal of the fluid. With any product there is a finite life of the fluid. The decision to dispose of the fluid is usually based on an oily and dirty appearance or foul odor. However, by monitoring fluid parameters on a routine basis, the fluid can be better controlled, leading to improved fluid performance and an extended useful life.

D. Contaminant Removal Systems

In many metalworking operations, contaminant removal systems are used to enable the machine to provide a certain finish, tolerance, production rate, etc., on a part. Contaminant removal systems are also used to maintain the fluid in a clean condition to minimize disposal frequency. Two general classes of contaminant removal systems are those for central systems and those for individual machines. Flexible manufacturing systems (or cells) may employ either central system or individual machine reservoirs. Contaminant removal systems are becoming more and more necessary for plants interested in better fluid management and control.

1. Central Systems

A central system is a large reservoir which supplies fluid to several individual machine tools. The central systems can range in size from a few hundred gallons to over 100,000 gal. Where identical or similar operations are performed on many individual machines, a central system is used to supply one fluid to all the machines. One major advantage of the central system is that it has a contaminant removal system for solids and, in some cases, oil to maintain the fluid in a clean condition. Also, since only one fluid is used, a daily fluid sample will provide a control system for monitoring concentration, pH, and contamination levels. With proper fluid controls and management techniques, the typical central system fluid will have a life of one to three years.

Table 3 is a list of the general types of contaminant removal equipment used on central systems [10]. Most of the systems employ some type of filtration to remove solids (metal chips, grinding swarf, and dirt). They can be as simple as settling and dragout systems or more advanced, such as the positive filters. Equipment such as a centrifuge or coalescer may be added to the central system to control tramp oil.

2. Individual Machines

There are a wide variety of contaminant removal systems for individual machines. Table 4 is a list of some of the more commonly used systems [10]. One of the most difficult aspects of controlling fluids in individual machines is that many plants do not monitor these fluids because of the number of samples and tests required. In addition to a large number of individual sumps, there may be different fluids and different concentrations that make the control task more difficult. However, it is recommended that daily checks are made of the concentration, since individual

TABLE 3 Contaminant Removal Equipment for Central Systems

EQUIPMENT	REMOVES		
	OIL	DIRT	BACTERIA
Settling/Dragout		X	
Multiple Weir		X	
Flotation	X		
Positive Filters			
Gravity		X	
Pressure		X	
Vacuum		X	
Centrifuge	X	X	
Cyclone		X	
Coalescer	X	X	
Pasteurization			X

TABLE 4 Contaminant Removal Equipment for Individual Machine Tools

	REMOVES		
MEDIA-BASED SYSTEMS	OIL	DIRT	BACTERIA
Filtration		X	
Pressure		X	
Vacuum		X	
Gravity		X	
NATURAL FORCE SYSTEMS			
Settling/Gravity		X	
Oil Skimmers	X		
Coalescers	X	X	
Aeration	X		
MECHANICAL SEPARATION SYSTEMS			
Cyclones		X	
Centrifuges	X	X	
Magnetic Separator		X	
OTHER			
Pasteurization			X

machine fluids can have a rapid change in fluid concentration, even in one day. A simple refractometer test is adequate for the daily checks; however, the chemical titration is a more accurate test and recommended for long-term control.

For many individual machines, contaminant removal systems are provided to handle one particular contaminant. For instance, a grinder may have a combination dragout/paper media filter to keep the fluid clean. A milling machine may have a dragout system/chip conveyor to remove the metal chips. However, few individual machines have contaminant control equipment to control all types of fluid contamination, such as oil, dirt, and bacteria.

Because of the difficult control situation, many plants are seeking better methods to control fluids in individual machine sumps. Typically, there is economic justification in seeking improved methods since the fluids in individual machines may be disposed of as frequently as once a week.

3. Fluid Recycling

An effective method to extend fluid life for individual machine tools is the use of batch treatment fluid recycling systems capable of removing contaminants such as tramp oil, dirt, and bacteria; and to readjust the fluid concentration before the fluid is returned to the individual machine. The fluid from each machine is treated with the batch treatment equipment on a frequent basis to minimize the contaminants.

Though there are several types of systems on the market, each plant must determine the feasibility of a fluid recycling system for its own purposes. A plant survey is recommended as the first step to identify the number of machines, sump capacities, frequency of disposal, and reason(s) for disposal. Also, data on fluid concentrate cost and gallons purchased as well as cost of waste treatment (or contract hauling) are important. This data is used to determine the economics of fluid recycling for a particular plant. An example of a fluid survey questionnaire is found in Table 5. Based on a thorough evaluation of the current fluid practices and proposed fluid management changes, a study must be completed to select the optimum fluid recycling system. Examples of equipment selection criteria are:

> Good economics (capital, operating, maintenance, and energy costs),
> Effective removal of contaminants such as oil, dirt, and bacteria,
> Make-up system to add concentrate or water to recycled fluid,
> Simple operation, low maintenance,
> Durable, quality equipment,
> Minimal floor space requirement,
> Warranty protection,
> Spare parts and service available from manufacturer.

E. Management Controls

In addition to water quality, fluid selection practices, fluid controls, and contaminant removal equipment, management controls are an important part of fluid longevity.

The fluid-use survey previously mentioned is used to identify particular machines or fluids that have high disposal frequencies. In addition, by talking to operators about the specific problems in a department or at a machine, we may learn of particular obstacles to fluid management or recycling. For example, there may not be a "standard practice" for the disposal of cleaners or solvents and these materials may simply be discharged to the metalworking fluid reservoir. These contaminants will directly influence the fluid performance and will make it impossible to recycle the cleaner or the metalworking fluid.

Another obstacle to fluid recycling in many plants is the actual sump design or the machine layout, which prohibits easy access to the fluid. Many sumps are poorly designed, especially in the base of a machine or simply in an inaccessible area, where the fluid can be trapped. Since the cleanout job becomes messy and time consuming, it is seldom completed. Another problem is that machines are placed in close proximity so that the sump cannot be reached with a sump cleaner.

In some cases, it may be necessary to redesign the machine sump or improve the machine layout to minimize the sump cleanout problems. Ideally, the sump is readily accessible to see, smell, sample, and service (add make-up, clean, etc.) the fluid.

Many plants find that equipment changes are necessary to ensure better

TABLE 5 Metalworking Fluid Management Questionnaire

FLUID SURVEY

PLANT: NUMBER OF MACHINE TOOLS:

LOCATION: AVERAGE SUMP SIZE:

MANUFACTURER OF: TOTAL GALLONS OF FLUID PURCHASED / YR.:

OPERATIONS: AVERAGE FLUID CONCENTRATE COST / GAL:

METALS: COST OF WASTE DISPOSAL / GAL:

FLUID(S) USED: LABOR COST / HR.

PLANT SURVEY

DATE: _____

MACHINE	DEPARTMENT	FLUID	FLUID CONCENTRATION	SUMP CAPACITY	DISPOSAL FREQ.?/YR.	GALLONS DISPOSED/ YEAR	REASON FOR DISPOSAL (OIL, DIRT, BACT., ETC.)
1.							
2.							
3.							
4.							
5.							
6.							
7.							
8.							
etc.							

control of the fluid. This may include the addition of spray hoses to manually flush machines, spray nozzles to minimize stagnant areas, or guarding to prevent carry-off.

The "people management" of fluids is very important as well. It is important that operators and plant personnel keep the fluid clean by obeying good housekeeping and hygiene practices. For example, food, drinks, tobacco, cleaners, solvents, paper, rags, floor-drying compounds, and dirt must be discarded properly and not put into metalworking fluids. Education and training is necessary for all plant personnel if better fluid management is to occur.

It is very helpful to document the fluid condition through simple test methods as discussed earlier (pH and concentration), and observe trends to predict fluid failure. A fluid log at the individual machine is helpful for the operator to complete daily fluid checks and observe fluid changes. If the fluid reaches an "out of control" condition, for example, the concentration is too low, then corrective action can be taken. See Table 6 for an example of a fluid log.

For a batch treatment recycling process, it is also very helpful to have a fluid recycling schedule. Every machine can then be cleaned out on a regular basis. Table 7 is an example of a machine cleanout schedule; every machine is cleaned on a one-, two-, or three-week cycle. It is necessary to process used fluids on a frequent basis to avoid contamination problems that overwhelm the fluid and require disposal. While there is no simple rule in terms of a recycling schedule, in many situations once a month is the typical minimum frequency.

Table 8 lists the typical factors that determine the required recycling frequency to extend fluid life. The fluid type and water quality have the greatest impact on fluid cleanliness, where some fluids tend to reject oil and dirt better than others. Typically, synthetic and semisynthetic fluids reject tramp oil better than soluble oils and tend to stay in a cleaner condition.

The type of operation and metals used will define the amount of contamination that the fluid receives. For example, grinding cast iron will place a large amount of graphite fines into a fluid, which are difficult to remove, and therefore will reduce the fluid life. Tramp oil leakage from the machine is an important source of fluid contamination, which can quickly overwhelm the fluid and require disposal.

The throughput of the machine and its run time will impact the fluid condition. An idle machine may, in fact, cause more problems with fluid rancidity than a continually pumped and aerated fluid.

Many machines are equipped with filters and dragouts to keep the fluid clean. This can greatly extend fluid life. Continuous filtration or routine sump cleanouts will have an important benefit on fluid life.

Obviously, the better the in-plant control of the fluid, the less frequent the need for disposal. Simple tests such as pH and concentration can greatly extend the fluid life. Maintaining the fluid volume at a full level in the sump is important to

TABLE 6 Fluid Log

DAILY COOLANT LOG

Machine # _____ Coolant in use _____

(Possible Format For Small Plants) Capacity _____ Coolant supplier _____

Date/Time	Concen-tration	pH	VISUAL CHECKS					Samples Taken for Analysis	ADDITIONS			Remarks
			Rust	Tramp Oil	Machine Build-ups	Rancidity	Color		Coolant Concentrate	Water	Other Additives	

Source: Handbook of Coolant Maintenance and Recycling for Small Users, Kalcon, Monlan Corporation, Kalamazoo, MI.

TABLE 7 Fluid Recycling Schedule

MACHINE NUMBER	SIZE (GAL)	RECYCLE EVERY WEEKS	WEEK #1 M T W H F	WEEK #2 M T W H F	WEEK #3 M T W H F	WEEK #4 M T W H F	WEEK #5 M T W H F	WEEK #6 M T W H F
00061	40	1	X	X	X	X	X	X
00049	50	1	. X . . .	. X . . .	. X . . .	. X . . .	. X . . .	. X . . .
00983	40	1	X	X	X	X	X	X . . . .
02564	25	1	. . X . .	. . X . .	. . X . .	. . X . .	. . X . .	. . X . .
02565	40	2	 X		 X		 X	
00017	15	2		. . X . .		. . X . .		. . X . .
00018	5	2	. . . X .		. . . X .		. . . X .	
00019	30	2		X		X		X . . . .
00026	80	2	. . . X .		. . . X .		. . . X .	
00043	20	2		. . . X X		. . . X X		. . . X X
00044	60	2		. . . X .		. . . X .		. . . X .
00053	10	3	 X			 X		
00054	10	3		. . X . .			. . X . .	
00058	80	3			X			X . . . .
00061	30	3		. . . X .			. . . X .	
00983	15	3				. . X . .		
00020	5	3			. . . X .			. . . X .
00021	10	3		. . X X .			. . X X .	
00023	30	3	X			X		
00027	20	3	X			X		
00030	10	3	X			X	X	
00043	5	3	X			X	X	
00068	10	3	X			X		
00056	50	3	X			X		
00069	5	3	X			X		

TOTAL GALLONS								
MONDAY			85	80	80	85	80	80
TUESDAY			80	80	80	80	80	80
WEDNESDAY			85	75	75	85	80	75
THURSDAY			80	80	80	70	85	70
FRIDAY			80	80	80	80	80	80

TABLE 8 Recycling Frequency Determination Factors

1. FLUID TYPE & WATER QUALITY
   Resistance to oil emulsification
   Resistance to bacteria /mold

2. CONTAMINATION
   Tramp Oil Leakage
   Solids

3. MACHINE USAGE
   Metals/Solids Loading
   Idle Fluid

4. MACHINE FILTRATION
   Sump Design/Access
   Cleanliness of Sump & Fluid

5. CONTROL
   Monitoring of Concentration, pH, Volume

6. AGE

fluid performance. Finally, age of the fluid will contribute to fluid failure because of oil emulsification, mineral buildup, metals accumulation, and product component depletion.

In addition to the various fluid management techniques we have discussed, one of the most important is simply to have the proper personnel operate and manage the program.

Ideally, one person or a small number of people should have the authority to operate the fluid management program. As the responsibility is passed to numerous people, less control of the fluids takes place. For example, a small group responsible for the fluid condition can make sure fluids are tested and properly adjusted. If each operator is responsible for a particular sump, then the tendency is to have fluids frequently discharged rather than worry about fluid testing and filtration.

In many plants, a "coolant committee" is set up to manage fluids. It may include representatives from purchasing, engineering, production, the laboratory, and waste treatment departments. Each group provides different priorities in terms of fluid selection and use, but as a team the best fluid management practices can be employed.

IV. FLUID RECYCLING TECHNOLOGIES

A. Filtration

As previously discussed, the individual sump environment creates many problems for extended fluid life. While many machines are equipped with individual machine filters, routine cleanout and maintenance are critical to long-term cleanliness and performance of the fluid. It is not unusual to find an improperly maintained paper media filter, dragout/conveyor system, or hydrocyclone, which results in fluid problems.

Many individual machines do not have any filtration at all. The sump may include a series of baffles and weirs to trap solids and oils. These machines are more susceptible to recirculating fines, bacteria growth, and fluid problems.

Filtration of metalworking fluid is critical for reasons of part finish, tool life, bacteria control, heat transfer, lubricity properties, etc. Critical operations, such as grinding, may have continuous filtration to minimize contaminant accumulation. Proper metalworking fluid filtration is probably the single largest problem with fluid performance. Typically, the combination of improper filter application (or lack of any filtration) and poor maintenance leads to premature fluid failures.

For these reasons, many companies have opted for a fluid recycling program that can help improve fluid cleanliness. A set recycling schedule is developed and equipment such as high efficiency sump cleaners are used to routinely clean individual machine sumps. For many plants, this is the best way to guarantee routine cleaning, which is necessary to keep ahead of fluid contamination problems. For more information on fluid filtration see Chapter 10 or Ref. 10.

There is no single absolute level of filtration required for optimum fluid performance. Each process must be evaluated to define the degree of filtration required. For many plants, a level of 200 ppm of total suspended solids or less is commonly found in a used fluid. Though some recirculating fines are always present, the goal should be to minimize these, especially in the critical processes. Central systems offer the best solution for continuous filtration. Where there are numerous machines on similar operations (for example, grinding crankshafts), central systems offer the best method of fluid control. The fluid is monitored on a daily basis to maintain it "in control."

B. Oil Removal

Various types of oils and greases are used for machine tool lubrication, such as hydraulic fluids, gear oils, way oils, etc. Many of these fluids either drain or are washed back to the metalworking fluid sump.

Depending on the fluid type (synthetic, semisynthetic, or soluble oil), these lubricating oils and greases can become emulsified with the metalworking fluid or can simply float on the fluid surface, as "free" oil. In either case, "tramp" or "extraneous" oil is a leading cause of fluid failure.

The typical mechanism of failure is due to a free oil layer inhibiting oxygen transfer to the metalworking fluid. This causes anaerobic bacteria to flourish and the well-known side effects (pH is lowered due to acidic by-products of bacteria growth and hydrogen sulfide gas is produced, giving the "rotten egg" odor). Oil components may also act as food sources for the organisms. Therefore, it becomes important to control tramp oil leaks into the fluid. Once the fluid is contaminated, remove as much oil as possible without harming the product (metalworking fluid emulsion).

Commonly found oil removal devices are skimmers (disk, rope, or belt), coalescers, and centrifuges. Also, there are numerous types of filters and oil sorbent materials used to help remove oil.

The separation of tramp oil, or the rise rate of oil droplets, is dependent on several factors, including droplet size, specific gravity, and temperature [11]. This relationship is expressed by Stokes Law as follows:

$$V_r = \left(\frac{g}{18\mu}\right)(S_w - S_o) D^2$$

where

V_r = velocity of the rise rate of oil droplets
g = acceleration due to gravity (981 cm/s)
S_w = specific gravity of water (metalworking fluid)
S_o = specific gravity of oil
D = diameter of oil droplet
μ = viscosity of water (metalworking fluid) at specified temperature

In a used metalworking fluid, we define three states of oil. These are chemically emulsified oil, mechanically emulsified oil, and free oil. Figure 2 depicts the relative size of these oil states. For the most part, we assume the fluid emulsion is stable and therefore, we are more concerned with the ever-changing state of the free and mechanically emulsified oil. It is important to understand that the degree of emulsification will impact the separation efficiency of any oil removal equipment. Therefore, the degree of fluid turbulence, type of metalworking fluid, fluid temperature, and mineral content of the fluid will affect the separation of "tramp" or extraneous oil from the product.

In nearly all cases, our goal is simply to remove free oil and loosely emulsified oil (such as mechanically emulsified oil) from the used metalworking fluid. We do not want to remove the product oil or other components that would harm the product performance.

As with solids filtration, there is no universally approved standard for the amount of acceptable tramp oil. In many fluids, a small amount of tramp oil may

FREE OIL
 > 150 microns

MECHANICALLY EMULSIFIED OIL
 > 20 < 149 microns

CHEMICALLY EMULSIFIED OIL
 < 20 microns

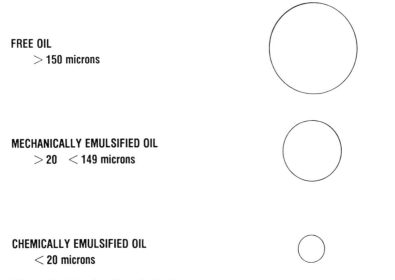

Figure 2 Metalworking fluid oil states.

actually have some benefits, such as higher lubricity, decreased foam, and softer residue. However, in most cases the level of free oil and loosely emulsified oil needs to be controlled to a level of 0.5% or less for optimum fluid performance and life. It does not make sense to target chemically emulsified oil, since if this form of oil is removed, we remove valuable product components.

To reduce the tendency for a buildup of chemically emulsified oil, it is necessary to minimize oil contamination, select a product type (such as synthetics and semisynthetics) that is less susceptible to tramp oil emulsification, use high-quality demineralized water, and use oil removal equipment to reduce the level of tramp oil.

As opposed to solids filtration, where central systems offer a distinct advantage to fluid management, the central system typically makes tramp oil removal more difficult. The high pumping recirculation rates with central systems increase the tendency for oil emulsification. However, as mentioned previously, the biggest problem is not the highly emulsified oil, but the free floating and loosely emulsified oil. Routine sump cleanouts and oil filtration are required to keep ahead of oil contamination for individual machine sumps. The two major types of equipment employed for this purpose are the coalescer and centrifuge.

C. Bacteria Control

The most common reason for fluid disposal is rancidity or fluid odor. Typically, the type of chemicals found in metalworking fluids are good food sources for bacteria. The tramp oils and fines also foster bacteria growth due to nutrients and growth sites. Therefore, most metalworking fluid formulations include a bactericide or fungicide to counteract bacteria and/or mold growth in used fluids. Some newer fluid formulations feature the use of chemicals that are "biostatic" or "biostable," which reduce the tendency for bacteria growth and odor development without the need for a biocide.

Whatever the fluid formulation, fluid cleanliness is a key to extended fluid life and performance. As discussed previously, the fluid type and water quality have a major impact on the tendency for bacteria to grow. Typically, the higher oil-containing products have more bacteria growth problems, where the synthetic-type products have a greater tendency for mold growth. Other common problems leading to increased bacteria growth are improper fluid concentration, lack of pH control, higher mineral content (especially phosphates and sulfates), and reduced fluid volume.

Bacteria use surface and film attachment to reproduce. The ideal sump is easily accessible for frequent cleanouts to prevent the buildup of fines and bacteria colonies. In addition to filtration, pasteurization is employed to control bacteria growth. By raising the fluid temperature to 165°F for at least 15 s, heat-sensitive bacteria will be killed [12]. Other treatments such as ultrafiltration and radiation have been used to remove and kill bacteria in used metalworking fluids, however, those techniques are not widely used for this purpose.

D. Chemical Additives

Metalworking fluids are consumable products which eventually will require disposal. Up to this point, the primary techniques to extend fluid life mentioned were fluid management and contaminant removal methods. However, if the fluid chemicals are depleted, certain additive "packages" may be used to refortify the product.

As the fluid age increases, there is a greater chance for product losses, which can lead to instability and product failure. The most common chemical additives are caustic soda (to raise pH) and biocide. In certain cases, dyes, odorants, and surfactants are added to compensate for product changes or losses. Other additives, such as antifoams or chelating agents are useful for certain fluid problems. In most cases, additives are used for central systems or fluid recycling programs where there is a greater emphasis on extended fluid life.

V. FLUID RECYCLING EQUIPMENT

There are numerous approaches to fluid recycling, including the use of existing or "add-on" filtration equipment to either individual machines or central systems. Batch and continuous systems can be used to supplement existing machine filtration. In addition, many companies now offer a fluid recycling service, where portable recycling equipment is used to process fluids as needed.

The importance of individual machine sump maintenance cannot be overstated. Even with a batch treatment system or a fluid recycling service, many of the fluid spoilage conditions occur as a result of poor sump maintenance.

Sections III and IV reviewed the overall fluid management basics and fluid recycling technologies. This section will discuss available systems to recycle metalworking fluids. For the most part, these systems target the individual machine-type manufacturing facility, which has the greatest need in terms of fluid management. The two primary technologies used for fluid recycling systems are the coalescer and the centrifuge. In each case, these technologies are primarily targeted to separate the tramp oil from the used metalworking fluids. However, to some extent the coalescer and centrifuge will remove solids but this may become a maintenance problem unless prefiltration is used. As discussed previously, Stokes Law defines the separation of oil droplets from the used metalworking fluid. Figure 3 reveals the impact of increasing oil droplet diameter, specific gravity differential, and temperature on the oil droplet rise rate.

Coalescence uses the property of oil attraction to polypropylene media (or oleophilic, "oil-loving" materials) for removal of tramp oil. Figure 4 shows a typical configuration for a coalescer, the vertical tube. These tubes are used in the oil removal tank of a batch treatment recycling system.

The centrifuge uses a series of plates, or a disk stack, spinning at a high rate

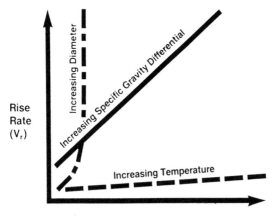

Figure 3 How variations in oil parameters and application conditions affect an oil droplet's rate of rise [11].

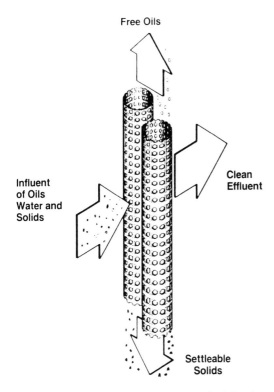

Figure 4 Vertical tube coalescing: principle of operation [11].

to physically separate materials of differing specific gravities, that is, oil, water, and solids. The advantages of a centrifuge, typically, are its high throughput rate (about 2 gpm) and the centrifugal force, which provides oil separation. One disadvantage can be the amount of maintenance time for certain applications where solids and greases require frequent bowl and disk stack cleaning. Also, the centrifuge has a higher incidence of long-term repairs because of its nature of processing abrasive fluids at high process rates. Figure 5 is a schematic of a typical centrifuge.

One newer technique, ultrafiltration (UF), is used on certain fluids to remove contaminants yet reuse the effluent. The UF process seems particularly suited to certain solution or synthetic products where the chemicals pass through the membrane while the contaminants, such as oil, dirt, and certain bacteria are contained. [13,14] This technique provides an excellent quality effluent where practically all the oil and particulates are removed. Because of a very small pore size, the UF process is not suitable to recycle semisynthetics or soluble oil-type emulsions.

Figure 6 depicts a typical batch treatment fluid recycling system. The typical procedure is to follow a machine clean-out schedule (for example, cleaning each

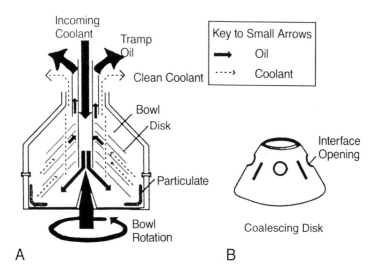

Figure 5 A. Schematic of a centrifuge commonly used for tramp oil removal. This centrifuge will remove some fine particulate if it is more dense than the fluid. The arrows show the passage of coolant through the unit. B. Drawing of a typical coalescing disk used to aid in the separation of the tramp oil from the coolant. The opening in the surface of the disk should be placed at the interface of the tramp oil and coolant where separation occurs. (Courtesy Henry Filtration Co., Bowling Green, OH.)

machine once per month) by removing all the fluids and solids from the sump using a high efficiency sump cleaner. The filtered fluid is then processed through the recycling system, where the clean fluid is tested for concentration and then adjusted. New make-up is added and the reclaimed and refortified fluid is returned to the clean sump. In practice, a split tank sump cleaner is used to clean out the sump using the "dirty fluid" compartment and then recycled fluid from the "clean fluid" tank is immediately returned to the sump. That is, the sump cleaner draws the clean fluid from the recycling unit and has this fluid available for a recharge once the individual machine is cleaned out.

A typical plant using a batch treatment recycling process uses one fluid and one concentration to simplify the clean-out and exchange program. Otherwise, sump cleaners and recycling systems must be cleaned out prior to a fluid change to eliminate cross contamination.

Figure 7 is a more detailed view of a batch treatment unit using coalescer/pasteurization/filtration techniques. A combination of filtration, coalescence, and pasteurization has proven to be an effective and economical method to recycle fluids from individual machines. The equipment may be used in a wide variety of

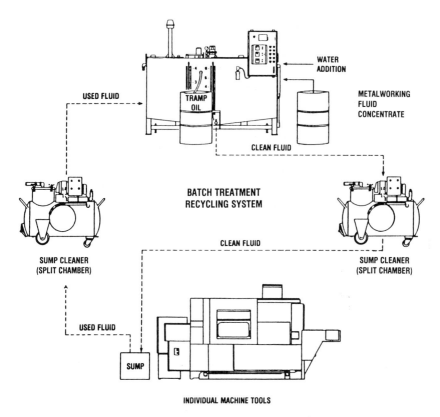

Figure 6 Metalworking fluid management equipment. (Courtesy Cincinnati Milacron, Cincinnati, OH.)

plants and applications to improve fluid management of individual machine fluids. The advantages of such a batch treatment module are:

> Excellent removal of contaminants: reduces free oil to 0.5% or less, controls suspended solids levels to 0.1% or less, controls bacteria levels to 100,000 counts/ml.
> Low maintenance, 1 h/week to maintain unit.
> Low operator labor, 1 h/8 h of operation.
> Simple operation.
> Minimal floor space.
> Low costs, total operating and energy costs typically less than 15 cents/gal.
> Durable pump, valves, heater, coalescer media.

The batch treatment module consists of three compartments—the used fluid tank, the oil removal tank (or coalescer tank), and the clean fluid and make-up tank.

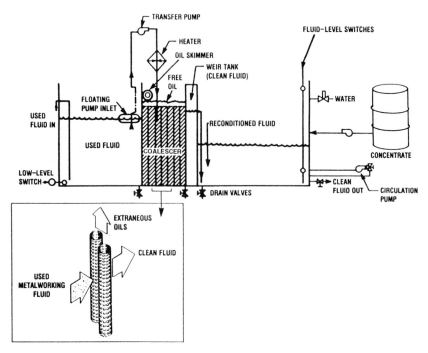

Figure 7 Metalworking fluid treatment with filtration/pasteurization/coalescence. (Courtesy Cincinnati Milacron, Cincinnati, OH.)

The used fluid is pumped from the surface of the used fluid tank through the transfer pump, heater, and coalescer. The fluid temperature is raised to 73.8°C (165°F) to lower bacteria counts as a result of pasteurization. Also, by raising the fluid temperature, improved extraneous oil separation results in the coalescer tank.

The fluid next passes through the coalescer media, where the media attracts and separates extraneous oil from the metalworking fluid. Coalescence uses the principle of Stokes Law, which states that as the diameter of the oil droplet doubles, the oil droplet rise rate increases by a factor of four. The coalescer media is made of polypropylene, which attracts oil to it in preference to water. The oil separates to the top of the tank and is removed by an oil skimmer. The clean fluid overflows into the clean fluid tank, where the fluid concentration is checked and fluid make-up is added. Typically, 50% by volume of new fluid make-up is added to the clean tank to make up for fluid losses (evaporation, carry-off, and restoring depleted fluid components). The combination of clean recycled fluid and new make-up fluid provides a fluid that can now be returned to the machines. The fluid is returned to the machine with a sump cleaner (clean side) or through a return drop line system.

The key to success of the fluid recycling program is scheduling the recycling to minimize contaminants and eliminate the frequent disposal of fluids.

1. Plant Application

As an example of a metalworking fluid management application, a company in northeastern United States has been successfully recycling fluids since April, 1985. This company manufactures braking systems for transit vehicles. Prior to fluid management, their greatest need was improved manufacturing productivity. A flexible manufacturing cell was installed consisting of 12 machines, including horizontal and vertical machining centers, and lathes. This company manufactures over 50 different parts using the various metals of cast and ductile iron, aluminum, and steel. The plant uses one fluid, a soluble oil, for all the operations.

The fluids are recycled approximately every one to two weeks, with an average of about 500 gal/week recycled. The individual machines are cleaned out using a sump cleaner with a capacity of 90 gal in each compartment. One compartment has filtration for the used fluid, the other compartment is for the clean, recycled fluid.

The fluid management program has provided a typical fluid life of over one year, and the disposal costs have been nearly eliminated. The fluid concentrate purchases have been reduced from approximately 24 drums/year to 12 drums/year. The fluid management program has provided benefits of extended fluid life, reduced fluid costs, and improved fluid performance.

Typically, fluids from individual machines will need to be recycled at least once a month. While the simplest schedule for fluid recycling is a set frequency (monthly, weekly, etc.), many plants may want to be more specific as to the exact point at which fluids must be recycled or disposed. In addition, limits based upon laboratory tests can be set up for each fluid with the help of the fluid supplier to indicate at what point the fluid needs recycled or disposed. For instance, if the pH value decreases, extraneous oil value increases or bacteria quantity increases, and action will be required. Certain fluids and operations can tolerate more contamination, which is the reason specific limits need to be set up by the fluid supplier and the metalworking plant.

2. Wastewater Treatment and Disposal

With an effective fluid management program in place, a metalworking plant can reasonably expect to lower its fluid concentrate purchase costs by 30 to 60%, depending on current fluid management practices at the plant. Reduced disposal volume will vary from 50 to 80% for most plants. However, even with management improvements, daily fluid make-up requirements may represent 20 to 30% of the concentrate purchase costs for most plants. Even the best fluids, used with good fluid management, will need to be disposed eventually. Typical reasons for disposal are:

High bacteria or mold counts, resulting in breakdown of product.
Excessive contaminants (oil, dirt, etc.)
Excessive buildup of dissolved minerals and metals.
Selective depletion of product components.
Mechanical breakdown on the machine or central system which requires pumping out fluid.

The major waste treatment and disposal options available are:

Contract hauling
Chemical treatment
Ultrafiltration
Incineration
Evaporation

The selection of the treatment method for disposal will depend on factors such as volume of wastewater generated, composition of wastewater, classification of hazardous versus nonhazardous, availability and cost of contract hauling, and whether the plant has access to a sewer system.

Ideally, with a careful fluid management program, the wastewater volume generated will be minimal. The high cost of contract hauling in many areas has resulted in plants opting for in-plant controls and equipment to eliminate the cost of contract hauling. The technology is available to waste treat fluids, using several stages and types of equipment, to provide a water clean enough for reuse for cooling water, parts washer, or metalworking fluid make-up. This technology enables plants to approach the goal of "zero discharge." Even with advanced wastewater treatment equipment, there are waste by-products that must be disposed. Once again, the assistance of the fluid supplier will be helpful in selecting the optimum waste treatment or disposal method.

VI. SUMMARY

The economics of fluid use is changing rapidly, since improved fluid productivity, environmental safety and health, and proper disposal are major needs. Purchase price, labor, machine downtime, productivity (quality, production rates, scrap, tool or wheel life), operator safety and health, and disposal costs are all part of the overall fluid use cost.

As metalworking fluid needs change and costs increase, it is increasingly important for plants to implement a fluid management program. The areas of fluid selection, water quality, fluid controls, contaminant removal equipment, wastewater disposal, and overall economics must be evaluated. Through careful fluid management, metalworking plants can generate substantial improvements in fluid performance and economics.

REFERENCES

1. *Report on the Volume of Lubricants Manufactured in the United States*, Independent Lubricant Manufacturers (1990). Presented to the Independent Lubricant Manufacturers Association 1991 Annual Meeting, Sept. 28–Oct. 1, by E. Cleves, Interlube Corporation, p. 2.
2. "Metalworking fluid trends 1991," speech by K. E. Rich, Lubrizol Corporation, November 1, 1991.
3. J. L. Leiter and R. A. Fastenau, *Waste Minimization and Wastewater Treatment of Metalworking Fluids*, Kelly, R. Dick, and Dacko, eds., Independent Lubricant Manufacturers Association, pp. 8–10 (1990).
4. G. Schaffer, The AM award M. Eugene Merchant, *Am. Mach. 124*(12):90–97 (1980).
5. W. E. Lucke, Cutting fluid oil mist in the shop, SME Tech. Paper MR78-266 (1978).
6. K. N. Bennett, Iron Age's guide to metalcutting fluids, *Iron Age*, pp. 18–26 (November 1984).
7. R. K. Springborn, ed., *Cutting and Grinding Fluids: Selection and Application*, ASTM, Dearborn, MI, pp. 102–104 (1967).
8. T. J. Drozda and C. Wick, eds., *Tool and Manufacturing Engineers Handbook*, Vol. 1, McGraw-Hill, New York, pp. 25–26, 29, 361–369 (1983).
9. B. H. Pinkelton, The OSHA hazard communication standard, *ASLE 43*(4):236–243 (1987).
10. J. J. Joseph, *Coolant Filtration*, Joseph Marketing, East Syracuse, NY, pp. 27–28, 37–44 (1985).
11. "A guide to understanding the treatment of oily wastewater," AFL Industries, Form 800138, pp. 3, 7.
12. E. C. Hill and R. Elsmore, Pasteurization of oils and emulsions, *Biodeterioration 5*: 469 (1983).
13. R. O. Sköld, Field testing of a model water based metalworking fluid designed for continuous recycling using ultrafiltration, *J. Soc. Tribol. Lubr. Eng. 47*(8):653–659 (1991).
14. R. O. Sköld and S. M. Mahdi, Ultrafiltration for the recycling of a model water based metalworking fluid: Process design considerations, *Lubr. Eng. 47*(8):686–690 (1991).

13

Waste Treatment

PAUL M. SUTTON
P. M. Sutton & Associates, Inc.
Bethel, Connecticut

PRAKASH N. MISHRA
General Motors Corporation
Warren, Michigan

I. INTRODUCTION

Environmental consciousness in manufacturing operations is becoming increasingly important and is receiving greater attention at all levels. In the past, the environmental impact of industrial practices was not considered a significant factor in technical decision making and manufacturing plant practices. This has changed as a result of public opinion and government regulation, prompting industry to take into account the environmental impact of manufacturing. Also, cost benefits have been realized from environmentally sound manufacturing practices such as waste reduction and recycling and reuse of materials. As a result, industrial decision making is increasingly accommodating environmental concerns.

An environmental problem of major concern is the proper disposal of unwanted waste oils and the treatment of wastewaters containing oils from manufacturing operations. There are thousands of sources of oily wastewater. The largest volumes result from metalworking, food processing, and vehicle cleaning operations but those operations account for less than half of the total volume [1]. Oily wastewater is generated by virtually every major industry including paper products companies, glass manufacturers, tobacco companies, and of course oil refiners. The manufacturing plants within these industries range from small to large but there are a larger number of small plants generating oily wastewater volumes in the range from 2 to 189 m^3/day (500 to 50,000 gal/day) [2]. The use of metalworking fluids in manufacturing operations in the automotive industry typically results in wastewater flows ranging from 76 to 2839 m^3/day (20,000 to 740,000 gal/day).

Efforts must be made to reduce the discharge of used metalworking fluids from manufacturing plants in order to reduce the impact of metalworking fluids on the external environment. In this respect, maximizing the recycle and reuse of the metalworking fluids can be very cost effective and is commonly practiced. This can substantially reduce the amounts of used metalworking fluids that have to be discharged into the external environment. Apart from the discharge of used metalworking fluids, parts washing operations also lead to the discharge of wastewaters containing metalworking fluids.

Straight oils and soluble oil-in-water emulsions were the primary metalworking fluids used in the automotive and other manufacturing industries until the late 1960s. The hydrocarbons contained in these products are normally refined paraffins or unsaturates with straight or branched carbon atoms. The carbon number is normally in the 10 to 20 range, representing a low viscosity to a heavy oil. A dispersant in the form of a surfactant is normally dissolved in the oil resulting in a white or gray-colored stable suspension when the solution is added to water. Viscosity modifiers, corrosion inhibitors, and biocides are often included in these fluid mixes.

The use of petroleum-based, nonwater-soluble metalworking fluids creates oil mist in the plant, presents a fire hazard, and poses a significant housekeeping burden. These factors and the oil crisis of the early 1970s led to the introduction of semisynthetic and synthetic metalworking fluids. Semisynthetics are chemically preformed microemulsions containing some oil, whereas synthetics by definition contain no oil and are true solutions of complex organics in water (i.e., water-soluble organics) [3]. The exact chemical composition of synthetic and semisynthetic metalworking fluids is held proprietary by the manufacturers of these materials. In general, they contain complex glycols, amines, amides, esters, fatty acids, and other organics. Material safety data sheets are often inadequate for assessing the impact on the waste treatment system. The main impetus behind the use of synthetics is the substantial productivity and tool life increases that have been possible [3]. Apart from that, the metalworking fluid has a longer life in the system. These factors alone can represent substantial cost savings. In addition, oil mist is virtually eliminated and housekeeping tasks tremendously reduced. In spite of these benefits, there are environmental concerns about the use of synthetics. From the waste treatment perspective, many synthetics are not compatible with the wastewater treatment unit operations installed on-site by industry and designed for dealing with petroleum-based free oils and oily emulsions. In addition to interfering with the performance and efficiency of these unit operations, the complex soluble organics making up the synthetic mixtures pass through the oily wastewater treatment system. The result is either that additional, on-site treatment steps are required or the treated wastewater discharged by the manufacturing plant will contain the complex organics, negatively impacting the downstream, municipal treatment plant operations.

Historically, manufacturing plants, when possible, have discharged their oily process wastewaters into the local sanitary sewer system believing that the municipal wastewater treatment plant at the end of the sewer line could handle the discharged materials [4]. Wastewaters containing large concentrations of oil and grease and complex organics can present several problems when received by the municipal treatment plant. The oil and grease can:

1. Present a fire and explosion hazard,
2. Impact on the mechanical operating equipment in the plant, clogging screens, fouling instrumentation equipment, and interfering with skimming operations, and
3. Inhibit biological treatment processes when present above a threshold concentration.

The complex organics:

1. Represent a significant oxygen demand to the treatment plant impacting its performance, and
2. Can be the source of toxic chemicals impacting on both the operation of the treatment plant and its effluent quality.

The deleterious effects that oily wastewaters can have on municipal treatment plants led in the past to the development of local pretreatment ordinances limiting the discharge of oil and grease to concentrations typically less than 100 mg/l. More recently, when the effects of discharges of toxic chemicals to the treatment plants were recognized, the federal regulatory body, the U.S. Environmental Protection Agency (EPA), issued regulations governing industrial discharges to municipal sewers. These "pretreatment regulations" limit industrial discharges of not only oil and grease, but heavy metals, acid and bases, and toxic organic chemicals. The regulations became effective on August 25, 1978.

The regulations developed by the EPA were of two types: categorical or uniform national standards applicable to industry groups, and standards created by local municipal authorities to serve their unique needs. Local limits differ widely, as illustrated in Table 1. The EPA relies on the state and local authorities to enforce

TABLE 1 Effluent Limitation for Several Western Cities (1983)[a]

	San Francisco, CA	Denver, CO	Phoenix, AZ	Los Angeles, CA
pH	6.0–9.5	6.0–10.0	5.0–9.5	5.5–11
Temp. (°F)	125	150	150	140
Oil and grease	100	50	100	600
COD	2500	—	—	
As		0.25	0.1	3
Cd		0.05	0.1	1.5
Cu		3.0	10.0	15
CN				
Total		1	2.0	10
Free		—	0.2	2
Sulfides	0.5	—	0.5	0.1
Pb		0.25	0.5	5
Ni		5.0	—	12
Ag		0.25	0.5	5
Cr (total)	5.0	0.45	0.5 (VI)	10
Zn		2.0	50	25
Ba		—	10.0	—
B		—	10.0	—
Mn		0.25	0.5	—
Hg		0.25	0.05	—
Se		0.05	0.1	—
Fe		15.0	—	—

[a]All results in mg/l except pH and temperature.
Source: Ref. 4.

compliance with both the local and categorical pretreatment regulations. Despite the pretreatment program's existence for over ten years, in 1989 the largest manufacturing facilities of all categories in the United States discharged at least 250 million kg (551 million lb) of toxic compounds into local municipal treatment plants, according to a 1991 report released by the U.S. Public Interest Research Group [5].

A thorough account of the evolution of the pretreatment regulations, the basis for them, the types of regulations (categorical and local), and the rationale for their promulgation is presented elsewhere [6].

The unit operations normally employed for treatment of wastewaters containing oil and grease and complex organics can be categorized as either physical, chemical, biological, or a combination of these process technologies. This chapter presents information concerning these technologies specifically as they are applied to treatment of wastewaters from metalworking and other manufacturing operations. Additional information is presented on the question of oil and grease characterization and measurement. Case histories will be used to illustrate the application of new treatment flowsheets for handling wastewaters generated from metalworking operations.

II. OIL AND GREASE CHARACTERIZATION AND MEASUREMENT

Oily wastes include greases as well as many types of oils. Grease is not a specific chemical compound, but a rather general group of semiliquid materials which may include fatty acids, soaps, fats, waxes, and other similar extractable materials [7]. Unlike some industrial oils, which represent precise chemical composition, greases are defined by the analytical method employed to separate them from the aqueous phase of a waste. A variety of tests can be used to determine oil concentrations in wastewater and sludges. Most tests involve extraction of the oil from water with a preferential solvent [8]. Solvents used include freon, hexane, petroleum ether, benzene ethyl ether, methylene chloride, and a variety of others.

Oil and grease can be characterized in three ways: by polarity, biodegradability, and physical characteristics. Polar grease and oils are normally derived from animal and vegetable materials, and are the characteristic form found in food-processing wastewaters. Nonpolar oils and greases are derived from petroleum or mineral sources. Generally, polar oils and greases are biodegradable, while nonpolar forms are considered bioresistant [7].

Five categories have been proposed to describe the physical forms of oil in wastewater [9]. They are:

1. *Free oil.* That which rises rapidly to the surface under quiescent conditions.
2. *Mechanical dispersions.* Fine droplets ranging in size from microns to a few millimeters in diameter, which are stabilized by electrical charges or other forces but not through the influence of surface active agents.
3. *Chemically stabilized emulsions.* Oil droplets similar to mechanical dispersions but with enhanced stability resulting from surface active agents at the oil/water interface.
4. *Dissolved or soluble oil.* Truly soluble species in the chemical sense plus very finely divided oil droplets (typically less than 5 μm diameter). This form generally defies removal by normal physical means.
5. *Oil wet solids.* Oil adhered to the surface of particulate material in the wastewater.

Although there are many variations in the analytical methods used to determine oil and grease, they all consist essentially of the following two steps:

1. Sample acidification and contact with the solvent, and
2. Separation of the solvent phase and quantifying of the oil and grease either gravimetrically following solvent evaporation or by infrared spectroscopy.

Table 2 provides a listing of published protocols and details of the available analytical methods are reviewed elsewhere [10,11]. Analytical results for oil and grease in wastewater and sludges can be expected to vary depending on the solvent used, the specifics of the extraction technique, and the final method used to quantify the oil and grease extracted. The American Society for Testing Materials recommended freon extraction followed by infrared analysis in its 1981 *Annual Book of Standards* [12].

Several components in petroleum wastewaters can cause varying degrees of interference in oil and grease analytical procedures [13]. Certain soluble organics, such as phenolics and short chain carboxylic acids, can interfere to the extent of indicating an oil and grease concentration level of 200% higher than the actual hydrocarbon concentration present in the wastewater [13]. These interfering compounds are extractable, soluble, nonhydrocarbon components and have also been shown to be present in the feed and effluent streams from metalworking wastewater treatment plants [14]. Modified analytical procedures are available to allow determination of "true" oil content of a wastewater stream [13].

TABLE 2 Listing of Published Analytical Methods for Determination of Oil and Grease

Method	Date	Type	Solvent
ASTM D1178-60	1967	Gravimetric	Methylene chloride
API 733-58	1958	Infrared	Carbon tetrachloride
CONCAWE 1/72	1972	Infrared	Carbon tetrachloride
EPA 00556	1976	Gravimetric	Freon 113
IMCO	1978	Infrared	Carbon tetrachloride
EPA 413.2/418.1	1978	Infrared/silica gel	Freon 113
ASTM D3921	1987	Infrared/silica gel	Freon 113
ASTM D4281	1987	Gravimetric/silica gel	Freon 113
PARCOM	1987	Infrared/fluorosil	Carbon tetrachloride/freon 11
APHA[a] 5520B/5520F	1990	Gravimetric/silica gel	Freon 113
APHA[a] 5520C/5520F	1990	Infrared/silica gel	Freon 113

[a]APHA: American Public Health Association.
Source: Ref. 11.

III. TREATMENT TECHNOLOGIES

The concentration of oil and grease in wastewater from metalworking operations varies significantly from plant to plant. The total oil and grease content is typically in the range of 10,000 to 150,000 mg/l [15]. The dissolved or soluble and emulsified oil content of these wastewaters normally varies from 100 to 5000 mg/l [16].

In the treatment of oily wastewaters, a primary level of treatment is used to separate the flotable or free and nonemulsified oils, and dampen the variation in or "equalize" the wastewater flow and concentration. Physical separation processes are normally employed to accomplish "free oil removal." In the secondary treatment phase, the emulsified oil is separated, normally either by breaking the oil–water emulsion chemically and then physically separating the oil and water phases, or by a straight physical separation process such as ultrafiltration. The secondary treatment phase normally achieves removal of both the emulsified oil and a large fraction of the dissolved or soluble oil.

In order to remove the complex, water-soluble or dissolved organics contained in metalworking fluid wastewater streams as a result of the use of synthetics, a tertiary treatment phase is normally required. A biological treatment process is often used to accomplish this step.

A. Primary Treatment

The physical separation processes employed for primary treatment take advantage of the difference in the specific gravity of oil and water. Gravity-type separators are the most common devices employed [17]. Wastewater equalization is often practiced as part of primary treatment in order to improve the performance of downstream processes.

1. Gravity Separators

In these devices the wastewater is held in a quiescent state allowing for gravity separation of the free oil which is then skimmed or pumped from the wastewater surface using appropriate devices. The gravity separator is normally equipped with provisions for removing solids that may settle to the bottom of the device. Where a significant fraction of the oil is adhered to the surface of settleable solids, gravity sedimentation rather than gravity flotation can result in appreciable reductions in the wastewater oil concentration (Table 3).

In theory, the performance of a gravity separator can be predicted by Stokes Law, but turbulence and short circuiting are common factors influencing the observed performance. The standard gravity separator tank used in refinery wastewater treatment is the API separator. The design of this device is based on standards published by the American Petroleum Institute [18]. The tank design is generally based on the volume of wastewater treated per cross-sectional area per unit time (e.g., m^3/m^2 day or gal/ft^2 min). In effect this parameter implies that longer

TABLE 3 Performance of Gravity Sedimentation in Removal of Suspended Solids Plus Oil and Grease

Industry	Suspended solids (mg/l)		Oil and grease (mg/l)	
	Influent	Effluent	Influent	Effluent
Adhesives and sealants	10,600	2260	2200	522
Copper foundry	52	20	30	6.2
Ferrous foundry	1500	64	14	2.7
Ink manufacture	1600	110	2400	260
Steel cold rolling	260	30	619	7
Steel hot forming	185	39	120	14
Leather tanning and finishing	3170	945	490	57
Paint manufacture	15,600	1400	2400	160

Source: Adapted from Ref. 17.

periods of wastewater retention allow better separation of the free or flotable oil from the water. The value of the design parameter depends on oil droplet size and its specific gravity. Small oil droplets require a longer retention time for separation and thus a larger tank area than larger droplets, which separate more quickly. The same is true for oil densities closer to water, which also require greater separation periods than very light oil [19].

Plate separators have been used to improve the gravity separation/oil removal process. The two common plate separators are the parallel plate interceptor (PPI) and the corrugated plate interceptor (CPI). These units differ from the API separator in that placement of included plates in the separation chamber results in a decrease in the distance an oil droplet has to move vertically before being removed. Coalescence of the small oil droplets into larger droplets can occur at the plate surface. The installation area required for the PPI and CPI units is normally less than 25% of the API separator area requirements. The smaller size can be a disadvantage. It has been reported that such units are unable to accommodate slugs of oil, where more time is required to ensure free oil separation [20]. Fouling of the plates has also been reported [20].

The oil–water emulsion leaving a gravity separator, designed according to API standards, has oil droplets less than 30 μm in diameter and an oil concentration of less than 200 mg/l according to one source [21]. Performance results reported from operation of various separators are highly variable with treated material or effluent oil values ranging from 10 to 115 mg/l (Tables 4 and 5). The effluent concentration will be affected by the oil loading to the separator and the feed or influent oil concentration. Gravity flotation of metalworking fluid wastewaters containing very high concentrations of free and emulsified oil (i.e., greater than

TABLE 4   Estimated Effluent Quality from Various Gravity Separators

Commercial gravity separator	Effluent oil concentration (mg/l)
Fram Akers Plate Separator	50–100
API Rectangular	50–75
Circular	50–75
Inland Steel Hydrogard	50–75
Shell Parallel Plate Interceptor	35–50
Shell Corrugated Plate Interceptor	35–50
Finger Plate Separator	35–50
Keene-GraviPak	20

Source: Adapted from Ref. 8.

1500 mg/l) resulted in effluent concentrations exceeding 500 mg/l on occasion [22]. It is reasonable to assume that properly designed and operated gravity separators will normally achieve a free oil removal performance in the range from 60 to 95% regardless of the oily waste source and concentration.

2. Wastewater Equalization

Industrial wastewaters fluctuate in flow and concentration with time, depending on the process and production cycle. The performance of secondary and tertiary phase treatment processes is normally improved when flow equalization is practiced. The performance of certain processes is also sensitive to variations in influent oil concentration. Depending on the extent of fluctuation, some form of wastewater equalization is often required prior to secondary treatment. Flow and concentration equalization can be accomplished by providing adequate holding tank capacity. It is desirable to provide more than one holding tank. The most common practice is

TABLE 5   Performance of Oil/Water Separators

Separator	Retention time (min)	Effluent oil concentration (mg/l)
Gravity tank	600	40–50
API separator	30	40–115
Parallel plate	30	25–70
Parallel plate	5	40–100
Corrugated plate	5	10–50

Source: Adapted from Ref. 8.

to have three holding tanks, each capable of holding one day of wastewater flow. Statistical analysis of flow data is necessary in determining the optimum holding capacity. Procedures are available for design of equalization facilities [23]. In the sequence of unit processes, equalization is normally preceded by free oil removal (i.e., gravity separation) if significant quantities of oil are present.

B. Secondary Treatment

Several different process technologies are utilized for breaking the oil–water emulsion that is left after the oily wastewater has passed through primary treatment. Emulsions can be broken by chemical, physical, and electrical methods.

Electrical methods are normally only employed when dealing with emulsions containing mainly oil with small quantities of water (i.e., water-in-oil emulsions). Electroflotation and electrocoagulation utilize electricity for the destabilization of water in oil. In electroflotation, small gas bubbles are formed through the electrolysis of water to oxygen and hydrogen gas [24]. In electrocoagulation, voltage is applied to a consumable electrode which then releases a metallic coagulant such as the ferrous ion [24].

Most oily wastewaters from metalworking operations consist primarily of water with lesser amounts of oil. Chemical and/or physical methods are the most popular methods for separating emulsified oils from metalworking fluid wastewaters (i.e., oil-in-water emulsions).

The oily floc following chemical demulsification is most often removed through air flotation, although sometimes clarification and/or filtration are used. In some cases, multiple steps of these physical unit processes are necessary to obtain the desired effluent quality. Sometimes, the chemical emulsion-breaking step needs to be repeated between multiple steps of dissolved air flotation.

Advances have been made in the use of membrane technologies such as ultrafiltration for recovery and reuse of spent emulsified metalworking fluids. Application of this physical process for treatment of metalworking fluid wastewaters has generally been limited to small flows (i.e., less than 76 m^3/day or 20,000 gal/day) because of high costs.

Evaporation has been used as a method of reducing the water content of spent metalworking fluids in small scale systems. The energy requirements and disposal costs of the concentrate are major barriers in using this approach on a large scale to deal with wastewaters containing spent metalworking fluids.

1. Physical Emulsion Breaking and Conventional Filtration

Physical emulsion-breaking processes include heating, centrifugation, and precoat filtration with the latter two being more common [25]. Centrifugation is normally applied to oily sludges or to small volume, oily wastewater streams [26]. In one treatment process example, mobile heating plus centrifugation has been applied to treat an emulsion from a metalworking plant with an oily wastewater volume of 38 m^3/week (10,000 gal/week) [26]. In the treatment scheme, magnesium chloride is added, the wastewater is heated to 95°C for 3 to 15 min, and centrifuged. Oil in the wastewater is reduced from 2000 to 4000 mg/l to less than 90 mg/l [26].

Oil removal by conventional filtration involves direct removal according to droplet size and induced coalescence [27]. Factors affecting performance include the type of wastewater [28] and the wastewater oil concentration, oil droplet size, suspended solids concentration, and the degree of flow variation [27,29]. Chemical pretreatment prior to filtration (i.e., addition of polyelectrolyte) will normally enhance oil removal [30]. Coalescence filtration is particularly suitable for separation of mechanical emulsions [31].

2. Ultrafiltration

Ultrafiltration is a pressure-driven membrane filtration process which uses molecular size pores to separate emulsions and macromolecules from a solution. It has been applied for oil recovery from wastewaters in many industries including adhesives and sealants, commercial laundries, synthetic rubber manufacturing, timber products processing, and metalworking operations. Unlike reverse osmosis, which provides separation down to the ionic level, ultrafiltration consists of a more open membrane and lower pressures are employed. Ultrafiltration membranes will reject solutes greater than approximately 0.001 µm in effective diameter. The size of a bacterial cell is typically greater than 0.5 µm. Ultrafiltration membranes cannot retain lower molecular weight soluble organic and inorganic compounds.

Determining the ultrafiltration membrane area requirement relies on specifying an operating liquid throughput or membrane flux. In treating oily emulsions, the flux normally depends on such factors as the concentration of oil and suspended solids in the feed stream, the membrane surface velocity, temperature, transmembrane pressure drop, surface fouling, and the extent of concentration polarization. Concentration polarization arises from the accumulation of solutes on the membrane surface. Solutes reach the membrane surface by convective transport of the solvent, a portion of which passes through the membrane. The rejected solutes often form a viscous gel layer on the membrane. This gel layer acts as a secondary membrane reducing the flux and often reducing the passage of low molecular weight solutes. Surface fouling is a result of the deposition of submicron particles on the surface as well as the accumulation of smaller solutes because of crystallization and precipitation. Modifying the membrane surface by chemical or physical methods can significantly improve its flux characteristics.

Figure 1 represents a simplified schematic of the flowsheet typically used for ultrafiltration of an oily wastewater. Hydrocarbon oil and grease concentrations of less than 10 mg/l have been achieved in the permeate or effluent from ultrafiltration systems treating wastewaters from General Motors (GM) automotive manufactur-

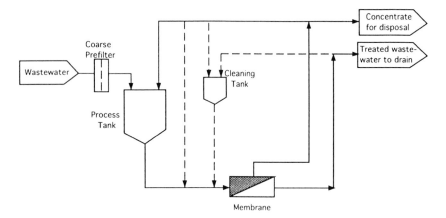

Figure 1 Simplified schematic of ultrafiltration flowsheet.

ing plants. The application of ultrafiltration for the treatment of metalworking fluid wastewaters will be further discussed in Sec. IV.

3. Chemical Emulsion Breaking

In order to break an oil–water emulsion chemically, the stabilizing factors must be neutralized to allow the emulsified droplets to coalesce [32]. The electrical charges on the emulsified droplet are neutralized by introducing an opposite charge through the addition of chemical emulsion breakers. The dielectric characteristics of water and oil cause emulsified oil droplets to carry negative charges, thus a cationic or positive charge emulsion breaker is required. Once the oil–water emulsion is broken, ideally two distinct layers are formed, an oil layer and a water layer. In actual practice, a scum or "rag" layer normally forms at the interface where solids and the neutralized emulsifier collect. The process usually consists of rapidly mixing the emulsion breaking chemicals with the wastewater followed by flocculation, and flotation or settling.

In breaking the emulsion, it is common to use sulfuric acid to lower the pH of the wastewater, followed by addition of an emulsion-breaking chemical such as alum and/or a polyelectrolyte. This emulsion-breaking process is usually conducted batchwise and involves the determination of the correct dosages of the treatment chemicals required for each batch of wastewater by laboratory analyses. The sulfuric acid converts the carboxyl ion in surfactants to carboxylic acids, allowing the oil droplets to agglomerate [32]. The addition of an inorganic coagulating chemical such as alum as an alternative to, or after sulfuric acid addition, aids in the agglomeration of the oil droplets. It has been claimed that organic demulsifiers such as polyamines are effective emulsion-breaking agents [32]. Lower dosages of the organics are often required and less chemical solids are produced. Alternative chemical demulsifying processes include [33]:

Addition of coagulating salts (e.g., aluminum, iron),
Addition of acids,
Addition of salts and heating the emulsion,
Addition of coagulating salts and treatment by electricity, and
Addition of acids plus organic cleaving agents.

Chemical treatment followed by gravity sedimentation is employed in a number of industries including those involving metalworking operations. One company reported treating oily wastewater from a ball and roller-bearing plant by coagulation with sodium carbonate, lime, and a polyelectrolyte flocculating agent to achieve oil and grease reduction from 302 mg/l to 28 mg/l—representing 90% removal in the treatment of up to 227 m^3/day (60,000 gal/day). Other industry results are presented in Table 6.

4. Air Flotation

Air flotation has been used for many years in the beneficiation of ores. Its first application in the wastewater treatment field was in the flotation of suspended solids, fibers, and other low-density solids. Flotation was also used for the thickening of activated sludge and flocculated chemical sludges. A detailed review of the process for the removal of oil from wastewaters is provided elsewhere [8].

TABLE 6 Oil and Grease Removal by Chemical Treatment Plus Gravity Sedimentation

Industry	Treatment chemical	Oil and grease (mg/l)		% Removal
		Influent	Effluent	
Paint manufacture	Sodium aluminate	1260	22	98
	Alum	1810	11	99
	Alum + lime	830	16	98
	Alum + lime + ferric chloride	393	91	77
	Alum + lime + polymer	980	22	98
	Alum + polymer	1700	880	48
	Alum + polymer	642	8	99
	Alum + polymer	1200	153	87
Commercial laundry	Alum + polymer	15	4	73
Steel pickling	Lime	3	1	66
Steel pickling	Lime + polymer	650	6	99
Steel pipe fabrication	Lime + polymer	5	4	20
Paint manufacture	Polymer	1100	22	98

Source: Ref. 27.

Air flotation is commonly employed as an alternative to sedimentation to separate the oil and water emulsion. The flotation process consists of four basic steps [8]:

1. Bubble generation in the oily wastewater,
2. Contact between the gas bubble and the oil droplet suspended in the water,
3. Attachment of the oil droplet to the gas bubble, and
4. Rise of the air/oil combination to the surface where the oil is skimmed off.

Although there are a number of different types of flotation systems, the two major commercial types most commonly used industrially are:

Dispersed or induced air flotation (IAF) in which air bubbles are introduced into the waste stream mechanically using high-speed impellers or a venturi nozzle, and

Dissolved air flotation (DAF) in which air is released.

DAF results from a pressurization/depressurization process in which very small gas bubbles are formed and rise to the suface with oil and suspended solids attached. A simplified schematic of a dissolved air flotation flowsheet is presented in Fig. 2. Typical performance results from the treatment of manufacturing plant wastewaters using DAF are presented in Table 7.

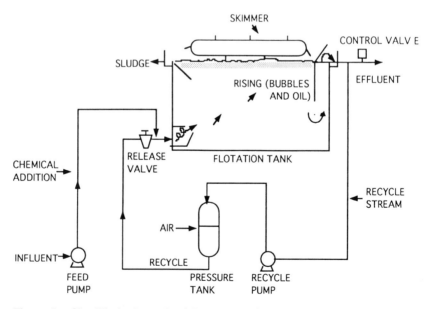

Figure 2 Simplified schematic of dissolved air flotation flowsheet. (From Ref. 8.)

TABLE 7 Air Flotation Performance Results in the Treatment of Metal-Bearing Wastewater

Author[a]	Oil and grease (mg/l)		% Removal	Chemicals utilized (mg/l)
	Influent	Effluent		
Barker et al.	1482	84	92	Alum, clay, polyelectrolyte
Ettelt[b]	819	14	98	2700 alum, 20 DOW A-23
Envirex[c]	587	5	99	100 alum, 1 polyelectrolyte
Ecodyne[d]	1170	75	94	1500 alum, 1 polyelectrolyte
	1620	8	99	Alum and polyelectrolyte

[a]Refer to Ref. 8 for citation details.
[b]Can-forming wastewater.
[c]Machine shop wastewater, manufacturer's literature.
[d]Manufacturer's literature.
Source: Adapted in part from Ref. 8.

Electro- and micro-flotation systems are newer concepts which are claimed to have a number of advantages over dispersed and dissolved gas flotation [8].

The key design variables controlling the treatment efficiency and performance of a flotation system are as follows [8]:

Gas input rate and volume of gas entrained per unit volume of liquid,
Bubble size distribution and degree of dispersion,
Surface properties of the suspended matter,
Hydraulic design of the flotation chamber,
Concentration and type of dissolved materials,
Concentration and type of suspended matter and oils,
Chemicals added,
Temperature, and
pH.

C. Tertiary Treatment

Biological and physical and/or chemical methods are used to remove the dissolved or water-soluble organics contained in metalworking fluid wastewaters. Reverse osmosis and activated carbon adsorption represent, respectively, potentially applicable physical and physical–chemical treatment processes.

Reverse osmosis membranes provide a barrier to the transfer of small molecular weight, dissolved organics and inorganics and thus are used to remove such contaminants as water-soluble organics, chlorides, and phosphates. Reverse osmosis membranes are easily fouled and as such the feed must be relatively free of oil and suspended solids.

Provided the dissolved organics remaining in the metalworking wastewater following secondary treatment are adsorbable onto activated carbon, this process is capable of achieving a high treatment performance. As in the case of reverse osmosis, pretreatment is critical to ensure little or no oil and that total suspended solids (TSS) are present in the feed to the activated carbon system.

Biological systems are the most popular method of treatment of metalworking fluid wastewaters containing water-soluble organics. Biological process reactors can be classified according to the nature of their biological growth. Those in which the active biomass is suspended as free organisms or microbial aggregates can be regarded as suspended growth reactors, whereas those in which growth occurs on or within a solid media can be termed supported attached growth reactors. Both suspended and supported growth reactors have been utilized to treat metalworking fluid wastewaters, following secondary treatment, for removal of the remaining five-day biochemical oxygen demand (BOD_5) and chemical oxygen demand (COD) [34–36], and for ammonium oxidation or nitrification [34].

A number of GM manufacturing facilities utilize significant quantities of synthetic metalworking fluids. During 1984 and 1985, the company completed extensive pilot plant studies at different manufacturing sites exploring various biological processes and process configurations for treatment of wastewater originating from existing oily wastewater treatment plants. On the basis of the results, the aerobic fluidized bed (AFB) process configuration was selected for full-scale implementation and subsequently the Oxitron system, a commercial embodiment of the AFB process configuration, was installed at four GM plant locations [34].

In the Oxitron system (Fig. 3), wastewater passes upward through a rectangular or circular reactor containing a bed of sand or granular activated carbon media at

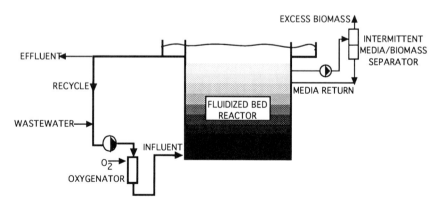

Figure 3 Oxitron aerobic fluidized bed process schematic. (From Ref. 37.)

Waste Treatment

a velocity sufficient to expand the bed, resulting in a fluidized state. Once fluidized, the media particles provide a vast surface area for biological growth, in part leading to the development of a biomass concentration approximately five to ten times greater than that normally maintained in conventional, activated sludge bioreactors.

The GM Delco Moraine, New Departure Hyatt (NDH) treatment plant flowsheet is typical of that employed by GM and consists of conventional primary and secondary treatment followed by tertiary treatment using the AFB process configuration (Fig. 4). In a performance evaluation completed in 1988, the two-stage Oxitron system achieved a median BOD_5 removal of 86%, together with essentially completed ammonium oxidation at a wastewater hydraulic retention time of less than 6 h [38].

IV. TECHNOLOGY APPLICATIONS AND CASE HISTORIES

The characteristic form of the oil and grease in the metalworking fluid wastewater (i.e., flotable or free oil, emulsified oil, soluble oil) and the volume of wastewater to be treated will dictate selection of the treatment flowsheet. Manufacturing operations generating large volumes of wastewater will often require primary, secondary, and tertiary treatment steps to achieve regulatory compliance. Two case histories are presented which fall into this category. A variety of smaller, preengineered and shop-fabricated treatment systems are available which are often applicable for handling the wastewater generated from small manufacturing operations. These

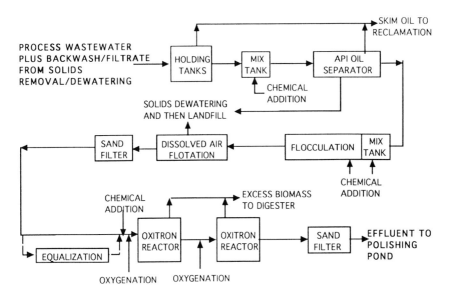

Figure 4 Industrial treatment at GM Delco Moraine, NDH plant. (From Ref. 38.)

smaller systems all employ operations which fall into the categories of primary, secondary, or tertiary oily wastewater treatment as described in the previous section.

For the small waste generator, belt and media skimmers are available for the removal of free oil from the surface of wastewater sumps and holding tanks. There are a large number of primary separator designs based on the characteristic that free oil floats in water and is attracted to a coalescing surface. These units are designed stage-wise, first for the removal of larger oil droplets and settleable solids, and subsequently for the removal of oil particles down to a few microns in diameter using a mesh of plastic or metal fibers.

Evaporators are often used to treat small quantities of oily wastewater. The unit consists of a holding tank that is heated to boiling either by gas, electricity, or steam heat. All types of water-based fluids may be treated by this method, which involves driving off water into the atmosphere to reduce the volume of the waste. These units are easy to operate, but their high-energy consumption and the disposal costs for the concentrated oily sludge preclude their application in a larger facility. Other potential problems include the possibility of fire or explosion and objectionable odors.

Small ultrafiltration package systems are increasingly popular for the removal of emulsified oil and a fraction of other more soluble components in metalworking fluids from metal cutting, grinding, and drawing operations. These systems can also be used to treat the fluids from alkaline and acid parts washing or cleaning baths. In this application the oils and particulate are removed from the bath while the permeate or the effluent from the ultrafilter represents a purified cleaner solution and can be returned to the bath. Package treatment systems are even available in which a biological reactor has been coupled to the ultrafiltration step providing both secondary and tertiary metalworking fluid treatment.

Two case histories have recently been reported involving large manufacturing facilities with flowsheets consisting of unit operations to provide primary through tertiary treatment of metalworking fluid wastewaters [39,40]. Unique to the flowsheets is the use of biological treatment and ultrafiltration stages.

A. TRW Plant

The TRW Steering and Suspension Division (TRW), in Rogersville, Tennessee, manufactures metal automotive parts during which they generate process wastewaters that contain free and emulsified oils, soluble organic materials, and low levels of metals such as copper, chromium, nickel, and zinc [39]. Synthetic fluids from machining, cutting, honing, and hydraulic applications constitute a majority of the measured oil in the process wastewater. The wastewaters were pretreated prior to discharge to Rogersville's municipal wastewater treatment plant.

In 1985, the Tennessee Department of Environment and Conservation along with the EPA mandated Rogersville to improve the efficiency and operation of their treatment plant and required the city to implement a pretreatment program for all its industrial users. The discharge from TRW was incapable of meeting the revised pretreatment standards for BOD, COD, oil, grease, and metals because of inadequacies in their pretreatment system. An engineering study was completed by TRW involving a technology screening step to identify alternative secondary and tertiary oily wastewater treatment options, laboratory treatability studies to develop process information for design and costing, and finally detailed design of a full-scale treatment system. Ultrafiltration followed by biological treatment in a suspended growth, sequencing batch reactor was the secondary–tertiary treatment scheme selected for full-scale implementation. A gravity separator was designed to accomplish free oil removal.

The full-scale TRW treatment system was designed for a process wastewater flow of 19 m^3/day (5000 gal/day). The system began operation in 1987. The full-scale system operating results during 1991 are presented in Table 8.

B. General Motors Plant

Until recently, industrial wastewaters generated at the GM Mansfield facility were treated in a conventional, physical–chemical oily wastewater treatment plant [40].

TABLE 8 TRW Full-Scale Treatment System Performance During 1991[a]

Parameter[b]	Maximum effluent concentration	Average effluent concentration	Pretreatment limit
BOD	350	158	350
COD	1048	370	600
TSS	505	172	350
Cadmium	<0.01	<0.01	0.050
Chromium, total	0.021	0.013	0.480
Copper	0.11	0.054	0.50
Cyanide	<0.01	<0.01	1.16
Lead	<0.05	<0.05	0.440
Mercury	0.0053	<0.005	0.05
Nickel	0.061	0.019	0.360
Phenol	0.131	0.068	0.80
Zinc	0.18	0.093	1.68
1,1,1-Trichloroethane	0.009	0.002	0.200
Surface Active Agents as MBAS	3.95	1.52	14
Total oil and grease	18	6.3	70

[a]All concentrations are expressed in mg/l.
[b]MBAS represents methylene blue active substances.
Source: Ref. 39.

The 35-year-old system required a high degree of operator attention and maintenance, and was prone to upsets in treatment performance. In June 1989 funding was approved to build a new treatment plant. The original appropriation was based on the use of a more conventional, physical–chemical treatment flowsheet with provisions made for the addition of a biological step in the future if required to meet more stringent, anticipated, effluent discharge regulations. In mid-1989, a GM corporate project team was established to review the design basis for the original appropriation request and to consider alternative wastewater management options. As a result of these efforts, a new treatment plant flowsheet was selected which combines physical and biological processes for the complete treatment of the manufacturing plant wastewaters (Fig. 5). Unique to the flowsheet is the application of the membrane biological reactor (MBR) system.

The MBR system was first developed in the 1970s for treatment of sanitary wastewater and consists of a suspended growth biological reactor combined with a membrane ultrafiltration unit process. The ultrafiltration step provides a positive means of liquid–solid separation preventing any loss of biological solids into the effluent and therefore allowing maintenance of a very high concentration of biomass in the suspended growth reactor. The MBR system is most attractive when applied in situations where long solids retention times (SRTs) are required, and physical retention and subsequent hydrolysis are critical to achieving biological degradation of pollutants (e.g., biological treatment of an oily wastewater). The

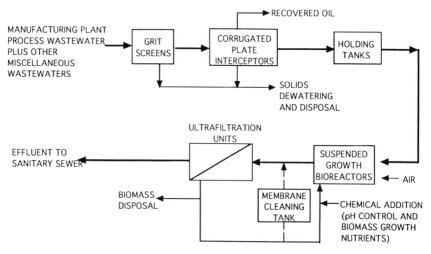

Figure 5 Simplified process schematic of the GM Mansfield industrial wastewater treatment plant. (From Ref. 40.)

ultrafiltration unit in the system retains oil and grease while it is biologically degraded in the aerobic reactor.

The GM Mansfield facility employs approximately 3500 people and is located in Mansfield, Ohio. The manufacturing plant fabricates hoods, fenders, floor pans, deck lids, and other sheet metal components. A wastewater characterization study, completed in 1989 following establishment of the corporate project team previously noted, indicated the manufacturing operations generated a maximum sustained wastewater flow of approximately 151 m^3/day (40,000 gal/day) [40].

In late 1989 and early 1990, the corporate project team identified and evaluated four alternative wastewater management options capable of meeting current regulatory requirements. In addition to evaluating three treatment flowsheets, off-site wastewater disposal was given consideration. The high operating costs, liability concerns, and other factors associated with this approach eliminated it during the evaluation. The three on-site treatment flowsheets evaluated were:

Flowsheet A: conventional physical–chemical treatment which consisted of flow equalization, grit and free oil separation, chemical coagulation, and dissolved air flotation.

Flowsheet B: ultrafiltration system treatment which included flow equalization, and grit and free oil separation prior to the ultrafiltration units, and

Flowsheet C: MBR system treatment which included flow equalization, and grit and free oil separation prior to the combined membrane and bioreactor units.

The flowsheets were evaluated against a number of quantitative and qualitative factors. The results of the evaluation are summarized in Table 9. Although the results indicated the ultrafiltration system option would represent the least-cost alternative (Table 9), it must be emphasized that this treatment flowsheet and the conventional physical–chemical alternative (Flowsheet A) did not include a biological treatment step. Comparative costs were developed previously for the MBR system flowsheet (Flowsheet C) and Flowsheet A plus add-on biological treatment, for the NDH plant. The MBR-based treatment scheme cost estimate was significantly lower than the physical–chemical plus add-on biological treatment system as built (Fig. 4).

The benefits realized by installing the MBR-based treatment flowsheet justified its selection for treatment of the GM Mansfield wastewater despite the lack of full-scale operating experience with this application of the MBR system. Credit was given to the fact that if the biological component of the treatment scheme did not perform in an acceptable fashion, modifications could be made to bypass the biological step, thus converting Flowsheet C to Flowsheet B.

In order to demonstrate the technical and economic advantages of applying

TABLE 9 GM Mansfield Wastewater Treatment Alternatives Evaluation Summary

Factor	Alternative treatment flowsheets		
	Flowsheet A: conventional physical-chemical (base case)	Flowsheet B: ultrafiltration system	Flowsheet C: MBR system
Simplicity of operation	Number of unit processes complicates operation	Simplest	Simpler
Chemical addition requirements	Significant	Lowest	Lower
Generation of residuals	Significant	Lower	Lowest
Effluent quality	Significant concentration of dissolved organics and suspended solids remaining	Better	Significantly better
Effluent recycle potential	A number of add-on unit operations required	Fewer add-on unit operations required	Highest recycle potential
Compatibility with manufacturing operations	Limits usage of synthetic metalworking fluids	Less compatible	Maximum compatibility
Capability of meeting future effluent quality requirements	Not capable	Not capable	Capable
Capital cost	Base case	Lowest	Lower
Operating cost	Base case	Lowest	Lower

Source: Ref. 40.

the MBR system to the treatment of oily wastewater and to develop information for design of full-scale systems, GM completed pilot plant studies on four different automotive manufacturing plant wastewaters at three different plant sites between September 1989 and October 1991 [40]. The results formed the basis for the design of the full-scale, demonstration MBR system for the Mansfield site. The MBR system was designed to handle 151 m^3/day (40,000 gal/day) of primary-treated wastewater with a mean COD of 5643 mg/l. The system consists of two 5.36 m (17.6 ft) diameter by 7.31 m (24 ft) high suspended growth reactors operating in parallel. Jet pod aerators provide oxygen, in the form of air, and mixing to the bioreactors. The bioreactors are coupled to three ultrafiltration units representing 195 m^2 (2100 ft^2) of total membrane area. Two ultrafiltration units are in use during normal operation.

Process start-up of the MBR system commenced in late September 1991. Results derived from routine plant monitoring and performance records over the three month period from February through April 1992 were used in order to assess the performance of the MBR system. The operating and performance results are stated in Table 10. The performance results indicated that although a high COD removal was achieved across the MBR system, a residual carbonaceous BOD_5 ($CBOD_5$) did appear in the effluent. This was thought to be due to a nutrient limitation. Although phosphoric acid was added as a biomass growth nutrient during the evaluation period, the effluent total phosphorus (TP) was less than 2 mg/l (Table 9). Nitrogen addition in the form of urea was added to the system beginning in early March. Although the mean effluent NH_3-N value stated was 4.5 mg/l, the result was noted as misleading as all but one value was less than the detection limit of the analytical procedure. The authors note that on that one day, the effluent $CBOD_5$ was less than 2 mg/l.

V. SUMMARY

An environmental problem of major concern is the proper disposal of unwanted waste oils and the treatment of wastewaters containing oils from manufacturing operations. A major source of oily wastewater results from metalworking operations. Straight oils and soluble oil-in-water emulsions were the primary metalworking fluids used in manufacturing industries until the late 1960s when semisynthetic and synthetic metalworking fluids were introduced.

The unit operations normally employed for treatment of wastewaters containing oil and grease and complex organics can be categorized as either physical, chemical, biological, or a combination of these process technologies. A primary level of treatment is used to separate the flotable or free and nonemulsified oils, and dampen the variation in or equalize the wastewater flow and concentration. Physical separation processes are normally employed to accomplish "free oil removal." In the secondary treatment phase, the emulsified oil is separated,

TABLE 10 GM Mansfield Full-Scale Demonstration MBR System Operating and Performance Results from February 1992–April 1992[a]

Operating and performance results	Value
Feed and reactor operating conditions	
Mean feed rate, m^3/day	116
Maximum daily rate, m^3/day	256
Minimum daily rate, m^3/day	65
Reactor HRT, days	2.26
Reactor total VS,[b] g/l	28.7
Reactor SRT, days	31
Performance results	
Feed COD, mg/l	14,184
Effluent values,[c] mg/l	
COD	799
$CBOD_5$	<41
TFOG	<18
HFOG	<9
NH_3-N	<4.5
TP	<1.8
NO_3-N plus NO_2-N	<1.1

[a]All values stated are means during operating period. HRT, hydraulic retention time; VS, volatile solids; SRT, solids retention time; TFOG, total fats, oils, and grease; and HFOG, hydrocarbon fats, oils, and grease.
[b]Samples taken alternately from reactors 1 and 2 in determining total VS.
[c]On certain days effluent $CBOD_5$, TFOG, HFOG, NH_3-N values were below detection limit. In those cases value was reported at the detection limit. All NO_3-N plus NO_2-N values were below detection limit of 1.1 mg/l total NO_3-N plus NO_2-N.
Source: Ref. 40.

normally either by breaking the oil–water emulsion chemically and then physically separating the oil and water phases, or by a straight physical separation process. The secondary treatment phase normally achieves removal of both the emulsified oil and a large fraction of the dissolved or soluble oil. In order to remove the complex, water-soluble or dissolved organics contained in metalworking fluid wastewater streams as a result of the use of synthetics, a tertiary treatment phase is normally required. A biological treatment process is often used to accomplish this step.

Two case histories have recently been reported involving flowsheets consisting of unit operations to provide primary through tertiary treatment of metalworking fluid wastewaters. Unique to the flowsheets is the use of biological treatment and ultrafiltration stages.

REFERENCES

1. N. L. Nemerow, *Liquid Waste of Industry: Theories, Practice and Treatment*, Addison–Wesley, Menlo Park, CA, pp. 435–439 (1971).
2. W. D. Alexander and P. L. Maul, Evaluation of a treatment system for spent machine coolants and oily wastewater, *Proceedings of the 36th Industrial Waste Conference*, Purdue University, Lafayette, IN, pp. 41–47 (1981).
3. P. M. Sutton, D. Kothari, P. N. Mishra, and L. Hachigian, Biological treatment of metalworking fluids: A new application for fluidized bed technology, *Proceedings of the Industrial Wastes Symposia, 58th Annual Conference*, Water Pollution Control Federation, Kansas City, MO, pp. 19–30 (1985).
4. G. F. Bennett and C. T. Philipp, *Industrial Wastewater Pretreatment, Water Conservation/Product Recovery/Pollution Abatement*, The University of Toledo, Toledo, OH (1986).
5. *Toxic Trick or Treatment: An Investigation of Toxic Discharges into the Nation's Sewers*, report prepared by the U.S. Public Interest Research Group, Washington, D.C. (1991).
6. G. F. Bennett, Impact of toxic chemicals on local wastewater treatment plant and the environment, *Environ. Geol. Water Sci*, *13*: 201 (1989).
7. J. W. Patterson, *Industrial Wastewater Treatment Technology*, 2nd ed., Butterworth, Stoneham, MA, pp. 273–302 (1985).
8. G. F. Bennett, The removal of oil from wastewater by air flotation: A review, *CRC Crit. Rev. Environ. Control*, *18*: 3, 189 (1988).
9. R. B. Tabakin, R. Trattner, and P. N. Cheremisinoff, Oil/water separations: The options available, Part 1 and Part 2, *Water Sew. Works*, July, p. 74 (1978), August, p. 72 (1978).
10. M. Kane, ed., *Manual of Sampling and Analytical Methods for Petroleum Hydrocarbons in Groundwater and Soil*, American Petroleum Institute, New York (1987).
11. K. Simms, Oil and grease analysis, *Wastewater Technology Centre Newsletter*, No. 20, pp. 6–7, Wastewater Technology Centre, Burlington, Ontario (1991).
12. *Annual Book of Standards, 31, Water*, Standard D-3921, American Society for Testing Materials, p. 654 (1981).
13. P. T. Sun, C. L. Price, J. C. Raia, and R. A. Balderas, Anomalies in oil and grease analyses of petroleum wastewaters and their implications, *Proceedings of the 42nd Industrial Waste Conference*, Purdue University, Lafayette, IN, pp. 151–162 (1987).
14. P. M. Sutton, P. N. Mishra, and P. M. Crawford, Combining biological and physical processes for complete treatment of oily wastewater, Proceedings of the 47th Industrial Waste Conference, Purdue University, Lafayette, IN, pp. 851–862 (1992).
15. J. J. Kulowiec, Techniques for removing oil and grease from industrial wastewater, *Poll. Eng.*, *11*: 49 (1979).
16. R. J. Brink, Operating costs of waste treatment at General Motors, *Proceedings of the 19th Industrial Waste Conference*, Purdue University, Lafayette, IN, pp. 12–16 (1964).

17. J. W. Patterson, *Industrial Wastewater Treatment Technology*, 2nd. ed., Butterworth, Stoneham, MA, p. 277 (1985).
18. *Manual on Disposal of Refinery Wastes*, 7th ed., American Petroleum Institute, New York (1963).
19. G. F. Bennett, Oil removal and recovery, presented at General Motors Water Pollution Control Management Course, GM Technical Center, Warren, MI, February (1989).
20. B. T. Davies and R. W. Vose, Custom design cut effluent treating costs: Case histories at Chevron USA Inc., *Proceedings of the 32nd Industrial Waste Conference*, Purdue University, Lafayette, IN, pp. 1035–1060 (1977).
21. N. D. Sylvester and J. J. Byeseda, Oil/water separation by induced-air flotation, *Soc. Pet. Eng. J. 20*:579 (1980).
22. R. W. Hare, P. M. Sutton, P. N. Mishra, and A. Janson, Membrane enhanced biological treatment of oily wastewater, presented at the 63rd Annual Water Pollution Control Federation Conference, Washington, D.C., October (1990).
23. W. W. Eckenfelder, Jr., *Industrial Water Pollution Control*, 2nd ed., McGraw–Hill, New York, pp. 40–48 (1989).
24. J. W. Patterson, *Industrial Wastewater Treatment Technology*, 2nd ed., Butterworth, Stoneham, MA, p. 290 (1985).
25. J. W. Patterson, *Industrial Wastewater Treatment Technology*, 2nd ed., Butterworth, Stoneham, MA, p. 283 (1985).
26. R. K. Chalmers, Treatment of wastes from metal finishing and engineering industries, in *Processes in Water Technology*, Vol. 1, Pergamon, New York (1972).
27. D. L. Ford and A. L. Elton, Removal of oil and grease from industrial wastewaters, *Chem. Eng.*, October 17, pp. 49–56 (1977).
28. W. Cawley, ed., *Treatability Manual, Vol. III, Technologies for Control/Removal of Pollutants*, US EPA 600-8-80-042-C, July (1980).
29. R. B. Tabakin, R. Trattner, and P. N. Cheremisinoff, Oil/water separations: The options available, Part 1 and Part 2, *Water Sew. Works*, July, pp. 74–77, August, pp. 72–75 (1978).
30. C. R. Symons, Treatment of cold mill wastewaters by ultrahigh-rate filtration, *J. Water Poll. Control Fed. 43*: 2280 (1971).
31. F. Berne and L. P. Aggarwal, Effluent treatment in the petroleum industry, *Effluent Water Treat. J. 14*: 26 (1977).
32. F. N. Kemmer and J. McCallion, eds., *The Nalco Water Handbook*, McGraw–Hill, New York, pp. 11-3–11-5 (1979).
33. A. Mertens, Treatment of water originating from metal industries, *Proceedings of the 22nd Industrial Waste Conference*, Purdue University, Lafayette, IN, pp. 908–925 (1967).
34. P. M. Sutton and P. N. Mishra, Fluidized bed biological wastewater treatment: Effects of scale-up on system performance, *Water Sci. Technol. 22*: 419 (1990).
35. B. R. Kin, M. J. Matz, and F. Lipari, Treatment of metal-cutting-fluids wastewater using an anaerobic GAC fluidized bed reactor, *J. Water Poll. Control Fed. 61*: 1430 (1989).
36. L. Polak, Biological treatability of industrial wastewater and waste machine tool coolants at John Deere Dubuque Works, *Proceedings of the 41st Industrial Waste Conference*, Purdue University, Lafayette, IN, pp. 123–131 (1986).

37. P. M. Sutton and P. N. Mishra, Biological fluidized beds for water and wastewater treatment, *Water Environ. Technol. 3*: 8, 52 (1991).
38. R. W. Hare, P. M. Sutton, P. N. Mishra, and K. F. Potochnik, Utilization of fluidized bed biological treatment at General Motors facilities: Pilot and full scale results, *Proceedings of the Industrial Wastes Symposia, 61st Annual Conference*, Water Pollution Control Federation, Dallas, TX, October (1988).
39. J. S. Dang, T. V. Clark, and D. V. Glenn, Treatment of synthetic hydraulic fluids by ultrafiltration and biological treatment—a case history, presented at the 47th Industrial Waste Conference, Purdue University, Lafayette, IN, May (1992).
40. M. D. Knoblock, P. M. Sutton, P. N. Mishra, K. Gupta, and A. Janson, Membrane biological reactor system for treatment of oily wastewaters: Pilot to full scale results, Proceedings of the Industrial Waste Symposia, 65th Annual Conference, Water Environment Federation, New Orleans, LA, pp. 13–24 (1992).

14

Contact Dermatitis and Metalworking Fluids

C. G. TOBY MATHIAS
Group Health Associates
Cincinnati, Ohio

I. INTRODUCTION

A. General Considerations

Occupational skin diseases account for a significant proportion of all occupational illnesses reported in the Bureau of Labor Statistics Annual Survey of Occupational Injuries and Illnesses. Machine tool industries are consistently found among Standard Industrial Classification categories with the highest incidence rates and numbers of cases of occupational skin diseases [1]. Contact dermatitis is an inflammatory condition induced by external contact of a substance or material with the skin surface. At least 90% of all occupationally acquired skin disorders are due to contact dermatitis [2]; most of these are irritant rather than allergic reactions. Metalworking fluids (MWFs) are important causes of occupational contact dermatitis, especially in machine tool industries, where they may be routinely blamed for any dermatitis that arises in association with exposure, whether or not such blame is ultimately justified. In at least one large epidemiological study of 286 machinists exposed to MWFs, the prevalence of contact dermatitis was 27%; only 2.8% of these were allergic reactions to ingredients of MWFs, the rest were irritant reactions [3]. A special form of contact dermatitis, contact folliculitis, induces an inflammatory reaction localized to hair follicles. Both contact dermatitis and folliculitis are best understood within the context of the normal structure and function of skin.

B. Structure and Function of Skin

The skin has two principal layers, the epidermis and the dermis. Each layer contains various structural elements which are important not only with regard to specific functions but also with regard to disease processes which may affect them.

The epidermis is a relatively thin layer compared to the dermis and ranges in total thickness from 100 to 200 µm. The outermost portion of the epidermis is the stratum corneum, a thin membrane 15 to 30 µm thick and composed of packed, cornified, dead cellular tissue with tight intracellular spaces filled with complex lipids. The stratum corneum constitutes the principal physical barrier against penetration of chemical substances and microorganisms into the living layers of skin and, ultimately, the body in general. The stratum corneum is thickest on the palms and soles, where it may be 75 to 100 µm thick. The bulk of the epidermal layer is composed of squamous cells (keratinocytes), which actively synthesize keratinous filaments, keratohyaline granules, and membrane-coating granules filled with complex lipids, all of which are ultimately destined to become the principal structural elements of the outermost protective stratum corneum. The innermost portion of the epidermis is a single cell layer of actively germinating cells, the basal cells. It takes approximately two weeks for newly generated squamous cells to mature and transform into the cornified cells of the stratum corneum, and an additional two weeks for the newly formed, cornified cells to desquamate into the environment.

Special pigment-producing cells, called melanocytes, are located along the basal cell layer of the epidermis. The pigment (melanin) protects the skin and body against the sun's harmful ultraviolet radiation and is responsible for tanning as well as racial differences in pigmentation. Melanin is packaged into granules within the melanocytes; under suitable ultraviolet light stimulation, these granules are transferred into surrounding keratinocyte cells. Melanocytes may be nonspecifically injured by severe inflammation from burns or chemical irritation, leaving residual postinflammatory changes of increased or decreased pigmentation. Melanocytes may also be selectively inhibited or destroyed by the toxic effects of some phenolic or catecholic chemicals which resemble the amino acid tyrosine, a precursor of melanin synthesis, leading to depigmentation of the skin.

Another specialized cell, the Langerhans cell, is also found along the basal and suprabasal layer of the epidermis. These cells participate in cutaneous immune surveillance and are responsible for selective antigen uptake in allergic contact dermatitis reactions. Following uptake, the antigen is presented to T-lymphocytes circulating through skin and regional lymph nodes, where further initiation and amplification of the immune response occurs.

The bulk of perceptible skin thickness is a deeper layer of connective tissue, the dermis, composed of fibrous proteins (e.g., collagen, elastin, and reticulin) embedded in an amorphous ground substance. The tensile strength and elasticity of the dermis provide protection from mechanical injury, while the permeability characteristics of the ground substance allow diffusion of nutrients from blood vessels to other cellular elements of the dermis and epidermis. Blood is distributed within dermal tissue through a highly developed and interconnected network of both superficial and deep vessels. A large volume of circulating blood may be brought close to the skin surface when the body temperature is elevated; dissipation of heat from blood vessels at the surface is an important component of the body's thermoregulatory reflexes. Dilatation of these vessels accounts for cutaneous erythema (redness) caused by various physical or chemical irritations to skin. Densely intertwined nerve tissue fibers also transverse the superficial dermis and function as sensory receptors.

Specialized appendage structures originate in the dermis but have ducts which traverse the epidermis to the skin surface. Sweat glands respond both to thermal stimulation and emotional stress. Although the primary function of sweat is evaporative cooling of the body surface when it is stressed by heat, secondary buffering against alkaline substances may be provided by lactic acid, amphoteric amines, and weak bases contained in sweat. Hair follicles are found on all cutaneous surfaces except the palms, soles, and mucous membranes. Hair grows cyclically with alternating periods of growth and quiescence. Follicles grow and rest independently of one another; some shedding occurs daily but is barely noticeable in a normal healthy state. While hair has some protective and insulating function in other mammals, this function is relatively unimportant in man. Instead, hair functions principally as a secondary sensory organ through stimulation of nerve endings in the richly innervated hair bulbs. Sebaceous glands are intimately associated with hair follicles and share the same duct opening to the skin surface. The oily content (sebum) of the sebaceous glands functions as a sexual attractant in some mammals, but its function in man is unclear. The fatty acids derived from sebum have some bacteriostatic and fungistatic properties and may have a limited function in this regard, once deposited on the skin surface.

II. PATHOGENESIS

A. Irritant Contact Dermatitis

The pathogenesis of contact dermatitis may involve either irritant or allergic mechanisms. Irritant contact dermatitis is an inflammatory reaction provoked by a direct, local toxic effect of MWF constituents or contaminants on epidermal cells; no immunological mechanisms are involved. Damaged epidermal keratinocytes release nonspecific chemical mediators causing dilation of dermal blood vessels (redness), leakage of fluids from blood vessels into the skin (swelling, small blisters, drainage of fluid from skin surface), and a variable cellular response (lymphocytes and polymorphonuclear cells). A strong irritant causes inflammation within minutes to hours following cutaneous exposure, sometimes called a chemical burn. Metalworking fluids, however, are only weak potential irritants and require frequent or prolonged skin contact for days to weeks before visible inflammation occurs. The constituents of MWFs responsible for irritant contact dermatitis have not been well defined. The potential to irritate skin is concentration dependent, and diluted MWFs are inherently less irritating than concentrates. Contamination of MWFs with lubricating oils, dirt, debris, and bacteria may increase irritant potential, presumably through alteration of pH and formation of more irritating breakdown products; some machinists anecdotally report that dirty or rancid MWFs are more irritating than clean, fresh MWFs. Prolonged entrapment of MWFs against skin by protective clothing, such as occurs when clothing becomes accidentally saturated or gloves are donned before washing hands contaminated with fluid, will also enhance the irritant potential. Finally, if the outermost protective stratum corneum is injured or damaged by some other mechanism, e.g., microscopic abrasions from swarf (metal chips) or an unrelated skin disorder, the resistance of skin to penetration by MWFs is lessened and the potential for irritation increased.

B. Allergic Contact Dermatitis

Allergic contact dermatitis is an inflammatory reaction within skin initiated by antigenic stimulation of a delayed cellular immune hypersensitivity response. Following skin exposure, potentially allergenic chemicals may complex in various ways with tissue protein; such complexes are called a hapten–protein conjugates. If receptors on the surfaces of specialized epidermal cells (Langerhans cells) "recognize" and bind these complexes, the allergen (hapten–protein conjugate) is processed and carried to regional lymphatic glands, stimulating the release of sensitized lymphocytes (T-cells) into the circulation. These sensitized lymphocytes recognize the allergen wherever they encounter appropriately formed hapten–protein conjugates in skin and initiate a complex series of biochemical events. The resulting inflammatory reaction is often more intense than, but sometimes indistinguishable from, an inflammatory irritant reaction. Most potential allergens encountered in MWFs are relatively weak and, unlike the allergenic resin in poison ivy or oak, usually require months or even years of exposure before sensitization actually develops. Once sensitization has occurred, only minimal exposure may be necessary to provoke or sustain dermatitis. Biocides added to MWFs are the most common causes of allergic contact dermatitis, although sensitization has occasionally been reported from antioxidants, fragrances, rust inhibitors (chromate), or metallic salts (nickel, chromate, cobalt) leached from machined metals [4]. Some biocides containing potential sensitizers are listed in Table 1. The risk of sensitization increases when undiluted biocides are handled or added to MWFs at higher than recommended concentrations.

III. CLINICAL FINDINGS

The cutaneous changes of irritant contact dermatitis caused by MWFs are often mild, consisting only of dryness and chapping. With more severe irritation, cutaneous changes progress through various stages of clinical redness (erythema), scaling, and crusting; occasional small blisters (vesicles) may be present, particularly on the sides of the fingers. Clinical changes accompanying allergic contact dermatitis are often more severe and associated with more extensive erythema, swelling, and larger vesicles or blisters. Itching is usually the predominant symptom, although varying degrees of stinging or burning may occur as well. While allergic contact dermatitis is usually more severe than irritant contact dermatitis, it is impossible to differentiate the two on the basis of appearance alone. A unique clinical form of irritant contact dermatitis, contact folliculitis, is characterized by erythematous papules or pustules surrounding hair follicles. Negative bacterial cultures distinguish this from folliculitis caused by infectious organisms.

Both irritant and allergic contact dermatitis occur on skin surfaces where maximal exposure takes place and, in the case of MWFs, almost invariably involve the hands and/or forearms. Dermatitis may be accentuated in the web spaces between the fingers; involvement of the backs of the hands is common. When contact occurs almost exclusively from handling wetted parts, the palmar surfaces may be preferentially affected. Contact dermatitis from MWFs does not usually affect skin surfaces covered by clothing unless the clothing becomes noticeably wetted and saturated for extended periods of time.

IV. DIAGNOSTIC TESTING

A. Patch Tests

The patch test is the procedure most frequently utilized to establish a diagnosis of contact dermatitis. This test is appropriately performed only when allergic sensitization to a constituent of MWFs is suspected. Since the overwhelming majority of cases of contact dermatitis from MWFs are due to contact irritation rather than allergy, a significant potential for misuse and misinterpretation of a patch test exists. As it is likely that many workers with contact dermatitis from MWFs will

TABLE 1 Potentially Allergenic Biocides Used or Added to Metalworking Fluids and Their Patch Test Concentrations

Some common trade names[a]	Chemical names	Patch test concentrations[b]
Bioban CS-1135	4,4-dimethyloxazolidine/ 3,4,4-trimethyloxazoline	1%
Bioban CS-1246	1-aza-3,7-dioxa-5-ethylbicyclo(3,3,0) octane	1%
Bioban P-1487	4-(2-nitrobutyl) morpholine/ 4,4,-(2-ethyl-2-nitrotrimethylene) dimorpholine	1%
Busan 85	potassium dimethyldithiocarbamate	1%
Captax, Dermacid, Mertax, Thiotax	mercaptobenzothiazole	1%
Dowicide 1	o-phenylphenol	1%
Dowicide A	sodium o-phenylphenate	1%
Dowicil 75 or 200 Quaternium-15	1-(3-chloroallyl)-3,5,7,triaza-1-azoniaadamantane chloride	2%
Formalin	formaldehyde	1% aq
Bioban GK, Busan 1060, Grotan BK, Onyxide 200, Triadine 3	hexahydro-1,3,5-tris-(2-hydroxyethyl)-s-triazine	1%
Grotan HD	N-methylol chloroacetamide	0.1%
Hibitane	chloroacetamide	0.2%
Kathon 886 MW	5-chloro-2-methyl-4-isothiazolin-3-one/2-methyl-4 isothiazolin-3-one	0.01% aq
Omadine (zinc) Omadine (sodium)	salts of 2-pyridinethiol-1-oxide	1%
Onyxide 500, Bronopol Bioban BNPD	2-bromo-2-nitropropane-1,3-diol	0.25%
Ottafact	p-chloro-m-cresol	1%
Ottasept Extra	p-chloro-m-xylenol	1%
Proxel CRL	1,2 benzisothiazolin-3-one	0.1%
Tris Nitro	2-(hydroxymethyl)-2-nitro-1,3-propanediol	1%
Vancide TH	hexahydro-1,3,5-triethyl-s-triazine	1%
Vancide 51	sodium dimethyldithiocarbamate/ sodium 2-mercaptobenzothiazole	1%

[a]List compiled from a number of worldwide medical and trade journals. May also be available under other trade names.
[b]Allergens prepared in white petrolatum except where otherwise specified; aq = aqueous.

undergo patch testing, health and safety personnel in machine tool industries must understand the performance, interpretation, and limitations of the patch test.

The test procedure is basically simple but has been carefully standardized; deviation from recommended guidelines may lead to false positive or false negative test results [5]. Since skin reactivity varies on different body surfaces, the upper and mid-back regions are recommended as the standard test sites; the upper outer arms are also acceptable. Suspected allergens, prepared at standardized concentrations (usually in white petrolatum or water) specifically for testing on the recommended sites, are placed on small inert disks and taped to the skin securely with adhesive tape. The most widely used test device is the Finn chamber on Scanpor tape, which consists of a series of aluminum disks premounted on special porous, hypoallergenic nonirritating tape. The tape is removed two days (48 h) after initial application. If a tested individual is allergic, an inflammatory reaction appears under the disk containing the substance to which the individual is sensitized. The intensity of any observed reaction is generally graded as follows: ? = doubtful reaction; 1+ = weak, nonvesicular, slightly infiltrated reaction; 2+ = strong, edematous, or vesicular reaction; 3+ = extreme, blistered, or ulcerative reaction. Since 30% to 40% of true allergic reactions may not be clearly positive until 3 or 4 days (72 to 96 h) after the initial test application, a second delayed reading, in addition to the initial 48 h reading, is necessary. While some authors maintain that a single 72 h reading is sufficient, this author's experience finds weak reactions extremely difficult to interpret with only a single 72 h reading and prefers two readings at 48 and 96 h. There is general agreement that patch test reaction intensities of 2+ or 3+ at any reading probably indicate true allergic sensitization (assuming that proper test concentrations and procedures have been utilized); however, a 1+ reaction is a common intensity of false positive irritant reactions which may occasionally be observed and does not always indicate true allergic sensitization. Most false positive irritant 1+ reactions are apparent at 48 h but decrease in intensity to a doubtful or negative reaction by 96 h. It is therefore usually safe to regard a 1+ reaction as indicative of allergic sensitization if the 1+ intensity persists for at least 96 h or appears at any delayed reading following an initially negative or doubtful reaction at 48 h. Performing two separate readings at 48 and 96 h after patch test application greatly facilitates the interpretation of these 1+ reactions.

For obvious reasons, a conclusion as to whether allergic sensitization has occurred to any constituent in MWFs can be reached only for constituents which were actually patch tested; no conclusions can be reached about constituents which were not tested. Only 20 substances (Table 2) are currently recommended and widely available in the United States for routine patch testing. This screening series has been designed to detect common contact allergens within the general environment and is not specifically designed to detect allergies to common sensitizers in MWFs. However, positive reactions from the routine screening antigens may sometimes indicate the presence of a sensitizer in MWFs, either from occasional

TABLE 2 Routine Screening Patch Test Series[a]

Benzocaine 5%	Formaldehyde 1% (aq)[b]
Mercaptobenzothiazole 1%[b]	Ethylenediamine 1%[b]
Colophony (rosin) 20%[b]	Epoxy resin 1%
p-phenylenediamine 1%	Quaternium-15 1%[b]
Imidazolidinyl urea 2% (aq)	p-tert-butylphenol formaldehyde 1%
Cinnamic aldehyde 1%[b]	Mercapto mix 1%[b]
Lanolin alcohol 30%	Black rubber mix 0.6%
Carba mix 3%[b]	Potassium dichromate 0.25%[b]
Neomycin sulfate 20%	Balsam of Peru 25%[b]
Thiuram mix 1%	Nickel sulfate 2.5%[b]

[a]Hermal pharmaceuticals, Oak Hills, NY. All allergens prepared in white petrolatum except where otherwise specified; aq = aqueous.
[b]Positive reactions may indicate hypersensitivity to constituent of metalworking fluids. See Sec. IV, Diagnostic Testing.

use of these substances in MWFs or cross-reaction with chemically similar substances. Positive reactions to mercaptobenzothiazole and mercapto mix usually indicate hypersensitivity to rubber, since they are most frequently used as accelerators in rubber; but they may also indicate contact allergy to mercaptobenzothiazole used as a biocide and corrosion inhibitor in MWFs. Carba mix contains several related carbamates also used as antioxidants in rubber; positive reactions usually mean hypersensitivity to rubber, but may indicate contact allergy to related carbamates used as biocides in MWFs. Positive reactions to formaldehyde may indicate hypersensitivity to any number of formaldehyde-releasing biocides used in MWFs; however, the allergen may also be the entire molecular structure of formaldehyde-releasing biocide, and contact allergy will be missed in this latter case if the whole biocide is not tested. Quaternum-15, a formaldehyde-releasing preservative frequently used in cosmetics, is a case in point. Also called Dowicil 200 in industry, it is a frequent sensitizer in its own right and cross-reacts with formaldehyde in only 25% of cases. Ethylenediamine hydrochloride is a sensitizing preservative found in some prescription topical medications, but it is also present in at least one microbiocide (Proxel CRL). Chromate and nickel are common sensitizing metals in the general environment; positive reactions may indicate hypersensitivity to machined metal leached into recirculating MWFs, and chromate may be directly added to some MWFs as a corrosion inhibitor. Colophony and balsam of Peru are derived from pine tree resins; positive reactions usually indicate hypersensitivity to fragrance but could indicate contact allergy to substances added to MWFs as deodorizers (e.g., pine oil). Cinnamic aldehyde, another common sensitizing fragrance, could also indicate hypersensitivity to a fragrance added to MWFs.

Wherever possible, patch testing should be supplemented with common sensitizers often encountered in MWFs (e.g., Table 1), prepared for testing at recommended concentrations. Supplemental patch test allergens specifically prepared to detect contact allergies caused by MWFs may be purchased from companies manufacturing patch test antigens abroad, but they are not widely available in the United States due to restrictions on importation. Alternatively, raw materials (e.g., biocides) may be supplied or prepared for testing by industries which either manufacture MWFs or use them; in such cases, these raw materials must be carefully prepared and thoroughly mixed at recommended patch test concentrations before testing. Where these options are not available, the actual MWF may be tested, but there is no unanimous consensus on appropriate patch test concentrations. Full strength concentrates of MWFs frequently produce false positive irritant reactions; if diluted to working concentrations (e.g., 1:30 to 1:40), the concentration of the allergen may be reduced below the threshold concentration needed to induce a positive reaction on the back where the test is performed, thereby giving a false negative reaction. Under actual working conditions, however, working dilutions of MWFs may still elicit allergic contact dermatitis reactions due to factors which enhance percutaneous absorption at sites of exposure. This author recommends a patch test concentration of 10% (prepared from clean undiluted MWF), with the caveat that even this concentration may occasionally produce a marginal false positive reaction. "Used," contaminated MWFs should not be patch tested, as they are more likely to cause false positive irritant reactions and run the risk of secondary skin infection at the patch test site from microbial contamination.

B. Miscellaneous

When a true positive allergic patch test reaction is obtained and the allergen can be traced to a particular MWF, the intellectual craving for "proof" that dermatitis was caused by MWF is more easily satisfied. However, the patch test alone is not infallible proof. For example, a positive patch test to nickel may be obtained, and trace amounts of nickel may be found in MWFs (presumably from leaching into the fluid from machined parts). But what if the dermatitis had occurred only on the soles of the feet, not on the hands or arms? Would it automatically be logical to assume that contact allergy to nickel in MWFs must be causing the dermatitis, considering that contact allergy to nickel is a common occurrence in the general population from sensitization to jewelry or metallic clothing fasteners, regardless of employment? Conversely, completely negative patch test results may be obtained. Is it justifiable to conclude that hand dermatitis was not caused by MWFs, considering that most cases are due to contact irritation rather than allergy, and negative patch test results are to be expected? The conundrum can be resolved by defining the role of patch testing in proper perspective, as only one of several criteria which should be considered in establishing causation. The following seven criteria have been proposed [6]: (1) dermatitis should be clinically eczematous; (2) noticeable skin exposure to substances capable of causing contact dermatitis should have occurred (e.g., MWFs repeatedly contacting skin); (3) dermatitis should occur on skin surfaces where exposure is principally occurring (e.g., hands and arms in the case of MWFs); (4) temporal relationship between onset of exposure and onset of dermatitis should be consistent with contact dermatitis (i.e., usually within the first three to six months of initial exposure or changes in work duties which increase the amount of exposure); (5) dermatitis improves when exposure ceases (e.g., vacations) and worsens upon reexposure (e.g., return to work); (6) no other diagnoses or causes are likely; and (7) patch tests indicate a specific allergen encountered in the work environment when contact allergy suspected (e.g., specific biocide in MWF). All of the proposed criteria have exceptions and only reasonable medical probability, not 100% absolute certainty, can usually be ascertained. It has been suggested that at least four of the seven criteria be satisfied before concluding that a particular worker has contact dermatitis caused by MWFs.

V. PROGNOSIS

Several independent follow-up studies of patients affected by occupational contact dermatitis have consistently demonstrated a surprisingly poor prognosis: approximately 25% clear completely, another 50% improve but have periodic exacerbations, while 25% persist unchanged or worse, despite protective measures or job modifications [7]. Contact dermatitis from MWFs is no exception. In one large follow-up study of 121 machinists diagnosed with contact dermatitis from MWFs, 78% of those who continued to work still had persistent dermatitis after 2 years, while 70% of those who had stopped working (no further exposure to MWFs) also had persistent dermatitis [8]. The reasons for this relatively poor prognosis are not well understood. It is often presumed that continued exposure to MWFs, type of contact dermatitis (irritant vs. allergic), and underlying endogenous factors have some influence on prognosis. However, no significant differences in prognosis were observed among machinists with contact dermatitis who either continued or stopped exposure to MWFs after onset of dermatitis, even after controlling for type of contact dermatitis (irritant vs. allergic) and endogenous factors.

VI. TREATMENT

The primary objective of medical treatment of contact dermatitis is simple and obvious: restore the skin to its previously normal appearance and function. Over-the-counter moisturizers (e.g., hand lotions or creams) may suffice for mild dryness or chapping, but should not be used on visibly inflamed skin; such preparations may actually worsen contact dermatitis from MWFs [9]. Once inflammation has occurred, the mainstays of treatment are topical corticosteroid medications which are simply rubbed onto the affected skin at an average frequency of four times per day. An abundant and sometimes bewildering number of preparations are available by prescription and differ principally by their respective potencies. In general, the higher the potency, the greater is the cost. Hydrocortisone, a low-potency topical steroid, is now available over the counter in 1% cream or ointment formulations. It is occasionally effective when contact dermatitis is mild and limited; it may be used by occupational health personnel as initial treatment for affected workers, but is not likely to be effective if contact dermatitis is moderate, severe, or extensive. Triamcinolone acetonide 0.1% cream or ointment is an intermediate strength steroid preparation available by prescription in both generic form and bulk quantity, making it an extremely cost-effective treatment for moderate contact dermatitis when large surface areas (e.g., arms) are involved and frequent applications required. Severe contact dermatitis usually requires high-potency prescription topical steroids; a few generic preparations are available to reduce cost. If palmar surfaces are involved, treatment may occasionally require one of the newer ultrapotent (and ultraexpensive) topical steroids, but they are seldom necessary elsewhere; unfortunately, no generic equivalents are available. Treatment of extensive, severe, or disabling contact dermatitis sometimes requires internal corticosteroid administration by mouth or injection.

Secondary treatment objectives include keeping an affected worker on the job, maintaining safe working conditions, and preventing recurrences. In this author's experience, most cases of contact dermatitis from MWFs can be effectively treated without taking the worker off the job. This approach usually requires that the prescribed topical corticosteroid be applied to affected surfaces frequently during the workshift, sometimes as frequently as every one-half to one hour, since working conditions tend to remove applied medications from the skin. Dirty or contaminated skin should be gently cleaned and rinsed before applying the topical steroid. Small containers of water and dry rags may be kept at the workstation for this purpose. If palmar surfaces are affected by vesicular (blistering) dermatitis, this approach is less likely to be effective and the affected worker will need to be removed from duties which necessitate exposure to MWFs, at least temporarily until a response to treatment is obtained. If allergic sensitization has actually occurred, it is extremely unlikely that dermatitis will ever be satisfactorily controlled unless all exposure is eliminated; this usually means changing to a new MWFs without the offending allergen or removing the affected worker permanently from job duties where exposure will occur.

Treatment must also take into consideration the maintenance of safe working conditions. Although protective gloves are often used in other industries as a means of reducing or controlling skin exposure, safety considerations usually prohibit this approach in most machine tool industries, where a glove accidentally caught by a rapidly turning machine part may pull a finger or hand into the machine, causing severe crush injury or amputation. Ointment (petrolatum)-based topical steroids are generally too greasy for frequent application to the hands and may endanger the worker if a firm grip on a handled part cannot be maintained; cream-based topical preparations are preferable during the workshift.

Workers with contact dermatitis from MWFs who remain at their usual jobs are prone to recurrences when treatment is stopped, since exposure to MWFs continues. Barrier creams specifically marketed for "wet work" have not lived up to their expectations and may actually aggravate contact dermatitis if applied to affected skin before complete healing occurs [10]. White or yellow petrolatum (petroleum jelly, Vaseline) is water impermeable and is theoretically the most effective "barrier" which could be applied; unfortunately it is greasy and may present a safety concern when applied to the hands, as already mentioned. For these reasons, this author prefers using an equal part mixture of 1% hydrocortisone cream and 1% hydrocortisone ointment, applied to the hands and arms during the workshift like a "barrier cream." The consistency of this mixture, halfway between a cream (too water washable) and petrolatum (too greasy) is an excellent compromise; it provides a temporary barrier, does not have to be applied as frequently as straight creams, and contains hydrocortisone which may inhibit any tendency for contact dermatitis to recur.

VII. PREVENTION

Preventive efforts should be directed primarily at those workers exposed to MWFs who have never developed contact dermatitis. Workers with contact dermatitis are best managed by medical treatment as outlined above; in some cases, some of the preventive measures discussed below may actually aggravate preexisting contact dermatitis rather than improve it.

Strategies for preventing contact dermatitis from MWFs may be based upon an eight-step program [11]. Points of emphasis within any preventive program will vary, depending on specific circumstances and conditions within the workplace. In all cases, prevention begins with recognition and acknowledgment that frequent or prolonged exposure to some MWFs may cause irritant contact dermatitis irrespective of manufacturers' claims or reassurances. Some of these reasons have already been discussed. Safety personnel need to identify the ingredients used in their MWFs and recognize the additional hazard of allergenicity from some of them, particularly biocides and corrosion inhibitors (see Table 1). Allergenicity is a greater concern where concentrated biocides are added to MWFs in the workplace to prolong their working lives; such practices may cause exposure to high concentrations of potential allergens, either directly when the concentrated biocides are handled or indirectly when recommended concentrations are accidentally exceeded in the recirculated MWFs following measurement error.

After recognizing that some MWFs may cause contact dermatitis, additional preventive measures may be implemented. Engineering controls focused on process containment have been very successful, but may be limited by cost and feasibility. Splashguards were one of the first successful preventive measures and are still useful today. The introduction of robotic equipment, which coincidentally decreases worker exposure to MWFs while increasing the quality and output of manufactured products, is also likely to decrease the incidence of contact dermatitis. Careful monitoring of recirculating MWFs is another important aspect; this should include sump filtering to remove contaminating oils and debris, filter changes at appropriate intervals, maintenance of concentration and pH within manufacturers' guidelines, careful addition of biocides at appropriate intervals to prevent rancidity, regulation of effective biocide concentrations, and a prompt replacement with fresh MWFs when irreversible deterioration has occurred. Where feasibility permits, selection of MWFs containing biocides without demonstrated allergenicity (Table 1) will lessen the risk of allergic contact dermatitis.

Personal protective measures have only a limited role in prevention of contact dermatitis from MWFs. Gloves are usually impractical and may become a safety hazard if caught in the moving parts of machinery; when handling concentrated biocides, however, glove wearing should be mandatory. Protective aprons may be helpful around equipment where splashing of MWFs onto clothing is likely to occur. Barrier creams have no proven benefit in preventing contact dermatitis from MWFs; experimental data suggests that they may actually exacerbate the irritant effects of MWFs under some circumstances [6].

As some degree of skin contact with MWFs is inevitable in most machine tool operations, personal and environmental hygiene efforts have added importance. Mild soap and water is sufficient to remove water-soluble MWFs from skin and should be used several times during the workshift; most oil-soluble MWFs can usually be removed in similar fashion. Organic solvents should never be used for this purpose. Where oil-soluble MWFs (or associated grease stains) cannot be easily removed with mild soap, waterless hand cleaners (which contain an organic solvent dispersed in a cream formulation) are often satisfactory and are substantially less irritating than straight organic solvents. Tenacious stains on the palms may require abrasive soaps (e.g., pumice); these products work by stripping the outermost stained stratum corneum away but may contribute to irritation from MWFs if used on the thinner, more sensitive skin of the forearms and dorsal hands. Waterless cleaners and abrasive soaps should never be used to clean skin already affected by contact dermatitis, as they will likely aggravate it. Regardless of the method by which the skin is cleaned, the single most important preventive measure (in this author's experience) is application of a moisturizer immediately after washing. Petroleum jelly is very effective but greasy and may make it difficult to grip objects firmly if not wiped completely off the palms; some workers find both the greasy feeling and slipperiness of palms objectionable and will not use it. Moisturizing lotions are least objectionable but may need to be applied more frequently than is often feasible to keep the skin from drying out under the usual rigors of machine tool operations. For these reasons, this author prefers heavy moisturizing creams, which need to be applied less frequently but vanish quickly enough that they seldom interfere with grip. Hydrophilic ointment is a generic moisturizing cream, available through retail pharmacy outlets, which I have found quite satisfactory for this purpose. In addition to personal hygiene, environmental hygiene provides some additional benefit where significant skin exposure may occur from contaminated work surfaces. Wetted parts should be wiped clean whenever possible prior to handling. Other work areas should be kept clean so that hands and arms will not be resting on surfaces coated with MWF residues.

Education efforts, aimed at promoting safe work practices and awareness of MWFs as potential causes of contact dermatitis form another important element of a comprehensive prevention program. Such educational efforts need to be directed at management, safety personnel, and machine tool workers covering all relevant aspects of contact dermatitis prevention. Simple manuals or booklets (pocket size) should be developed for easy reference. From a practical point of view, most educational training received outside the employees' work areas (e.g., classroom instructional pamphlets) will not be successful unless reinforced by knowledgeable safety personnel who visit employees' workstations. This obviously requires a serious commitment on the part of management.

There are no current federal or state regulatory requirements governing skin exposure to MWFs. Any such regulations would have to be based on well-characterized dose–response relationships between exposure times and cutaneous irritation, of which there are none. Given the dynamic nature of the machine tool industry and the wide range of variables which may interact to produce contact dermatitis, it is unlikely that any generic guidelines or regulations will be forthcoming. Machine tool manufacturing plants and manufacturers of MWFs may voluntarily post signs in work areas where MWFs are used, indicating that prolonged or frequent skin exposure to MWFs may cause dermatitis; this warning is usually indicated on a material safety data sheet (MSDS). A reminder to follow the company's recommended dermatitis prevention program may accompany such warnings. Manufacturers of MWFs may consider voluntarily listing the biocide on the MSDS even if it is present at a level less than one percent (current mandated level for noncarcinogens), since biocides are the most important causes of allergic contact dermatitis from MWFs and may induce sensitization at concentrations less than one percent.

Motivation is an often neglected element of prevention programs. Ethical considerations aside, management may be motivated by the simple knowledge that happy and healthy workers are more productive workers, and in the long run this increases profits and reduces workers' compensation costs. Motivating workers to adopt safe work habits is a complex issue at best. Incentive programs which reward departments for no injuries or lost work time over a defined period have been criticized on the grounds that they may intimidate workers from reporting real injuries or illnesses for fear of repercussions. It is doubtful whether caps, T-shirts, or other paraphernalia, often used to promote overall company morale, can be a serious motivational tool for preventing occupational illness such as contact dermatitis. In this author's opinion, any successful motivational approach must be based on a demonstrated sincerity for the worker's well-being and personalized on an individual level. Safety personnel must know their workers as individuals and what motivates other aspects of their lives; employees must feel welcome to express their individual health concerns with safety personnel.

Finally, the role of preemployment screening to prevent contact dermatitis from MWFs must be considered. Legal requirements generally preclude discrimination in hiring on the basis of preexisting disease, unless it is likely that a given job will aggravate the preexisting disease. It is unclear at this time how the Americans with Disabilities Act will affect any preemployment screening efforts. Nonetheless, it is likely that any new worker with preexisting hand or arm dermatitis will have this condition aggravated by skin exposure to MWFs. Therefore, preemployment physical examinations for positions in machine tool manufacturing plants should at a minimum include questions about prior hand or arm dermatitis and a complete skin exam for evidence of active skin disease. In addition, individuals with personal or family histories of atopic allergies ("hay fever" and related seasonal allergies, asthma, childhood eczema) have increased predispositions to developing irritant contact dermatitis, particularly in wet-work occupations, compared to nonatopic individuals [12]. While the actual probabilities that contact dermatitis will occur are unknown, newly hired workers should be screened for these traits, carefully trained in dermatitis prevention techniques as outlined above, and supervised closely for the development of irritant contact dermatitis within the first several months of employment.

Prevention of contact dermatitis may seem like an idealistic goal in the machine tool industry, where skin exposure to MWFs, cleaners, and solvents are still inevitable with today's technology, but it is a goal worth pursuing. The elements of a successful program as outlined above will require a coordinated effort at all levels of management and employment.

REFERENCES

1. C. G. T. Mathias and J. H. Morrison, Occupational skin diseases, United States, *Arch. Dermatol. 124*: 1519 (1988).
2. *Occupational Skin Disease in California (With Special Reference to 1977)*, California Department of Industrial Relations, Division of Labor Statistics, San Francisco, 1982.
3. E. M. deBoer, W. G. Van Ketel, and D. P. Bruynzeel, Dermatoses in metal workers (I). Irritant contact dermatitis, *Contact Dermatitis 20*: 212 (1989).
4. A. A. Fisher, *Contact Dermatitis*, Lea and Febiger, Philadelphia, p. 531 (1990).
5. A. A. Fisher, *Contact Dermatitis*, Lea and Febiger, Philadelphia, p. 9 (1990).

6. C. G. T. Mathias, Contact dermatitis and workers' compensation: Criteria for establishing occupational causation and aggravation, dermatitis, *J. Am. Acad. Dermatol. 20*: 842 (1989).
7. J. R. Nethercott and C. Gallant, Disability due to occupational contact dermatitis, *Occup. Med. State Art Rev. 1*: 200 (1986).
8. D. W. Pryce, D. Irvine, J. S. C. English, and R. J. G. Rycroft, Soluble oil dermatitis: A follow-up study, *Contact Dermatitis 21*: 28 (1989).
9. C. L. Goh, Cutting oil dermatitis on guinea pig skin (II). Emollient creams and cutting oil dermatitis, *Contact Dermatitis 24*: 81 (1991).
10. C. L. Goh, Cutting oil dermatitis on guinea pig skin (I). Cutting oil dermatitis and barrier creams, *Contact Dermatitis 24*: 16 (1991).
11. C. G. T. Mathias, Prevention of occupational contact dermatitis, *J. Am. Acad. Dermatol. 23*: 742 (1990).
12. E. Shmunes, The role of atopy in occupational skin disease, *Occup. Med. State Art Rev. 1*: 219 (1986).

15
Health and Safety Aspects in the Use of Metalworking Fluids

P. J. BEATTIE AND B. H. STROHM
General Motors Corporation
Detroit, Michigan

I. INTRODUCTION

Metalworking fluids are widely used industrially. The National Institute for Occupational Safety and Health (NIOSH) has estimated that over 6 million workers are exposed to mineral oil and approximately 1.2 million are exposed in metalworking fluid applications [1]. Exposure to ingredients found in metalworking fluids can be high. In a 1987 toxicology and health data call-in on antimicrobial agents commonly used in these fluids, the Environmental Protection Agency (EPA) found metalworking fluid applications to be in the high exposure category. In a typical large automotive metal processing operation, it is estimated that approximately 360,000 gal of straight oil and 621,000 gal of water-miscible metalworking fluid concentrate are used per year. Because of their high volume of use in applications, which can give rise to significant occupational exposures, understanding of the potential health effects of these materials is critical.

As has been discussed in previous chapters, metalworking fluids are complex mixtures. Even those classified as straight oils typically contain additives such as sulfonated or chlorinated compounds. Water-based (soluble oils, semisynthetic, synthetic) fluids, although diluted in use possibly as high as 1:200 in water, may have a multitude of chemical constituents—amines, borates, nitrates, polyalkyl glycols, esters, dyes, biocides—all of which may contribute to the toxicological profile of the fluid.

Information on the potential health effects of chemicals is gleaned from two sources: toxicology or animal test data and epidemiology or studies on human populations. Toxicology studies are typically conducted using a single chemical with controlled exposure levels, test animals, and environment. Metalworking fluids have typically not been tested as a whole because of their chemical complexity, but toxicology data do exist on many of the individual components. Results of these studies are then extrapolated to effects that might be observed in man. Epidemiology studies eliminate the need for species-to-species extrapolation, but introduce many other variables. Unlike the controlled conditions employed in animal experimentation, real world exposures to chemicals are often unknown, even in industrial settings. Exposures can occur via multiple routes and be of inconsistent durations and frequencies that may fluctuate over time. A number of epidemiology studies have been conducted on populations working with metalworking fluids. These data together provide a comprehensive picture of the potential health hazards of exposure to metalworking fluids and allow for the control of these hazards.

II. CONCEPTS IN TOXICOLOGY AND TEST METHODS

Descriptive animal toxicity testing under well-defined conditions of exposure forms the foundation of chemical hazard evaluation. These studies are the starting point for evaluating the health and safety ramifications of chemical exposure. Most descriptive toxicity testing follow guidelines prescribed by regulation and are designed to predict possible effects in humans. In general, animal models have been found to be very good predictors of possible human responses.

The single most important factor determining the toxicity of a chemical is dose. As first expressed by Paracelsus in the 1500s, all chemical substances are potentially harmful to living organisms. The level of exposure or amount of a chemical to which an individual is exposed will decide the effect.

The total dose received by an individual is a function of the exposure concentration, duration, and for intermittent exposures, frequency. Animal toxicity testing is classified into acute, subchronic, and chronic based upon duration of exposure. Toxicity is generally a function of the total dose. At lower concentrations of exposure, toxicity is more likely to be manifested by exposures of longer duration or greater frequency.

Acute effects, produced by short-term exposure to high doses of a chemical, often produce different effects than long-term, low-level exposure to the same chemical. Ethanol is a familiar example of a chemical which produces primary acute effects upon the central nervous system and long-term effects in the liver. Toxicity, or the severity of response, tends to decrease as the dose administered is dispersed over time. The effect produced by consumption of six one-ounce servings of alcohol in six hours is less than that produced by six one-ounce servings in one hour. If the dose is spread over a great enough length of time, there may be no observable effects (i.e., one-tenth of an ounce of alcohol per day for 60 days).

This decrease in severity of effect as the dose is fractionated over time occurs in large part because of detoxification and excretion of the chemical between successive doses. Diminished effects may also be observed when the cumulative effect or injury produced by each administered dose is partially or fully reversed before the subsequent dose is administered.

One of the most frequently monitored toxicological endpoints is acute lethality. Lethal dose studies act as a starting point for the design of subsequent long-term exposure investigations. The acute or single dose of a substance producing death in 50% of the test animals is termed the lethal dose 50, denoted LD_{50}. More relevant acute toxicity endpoints for evaluating potential metalworking fluid hazards measure skin, eye and respiratory irritation, and sensitization effects. Results from these studies are expressed with numerical scores indicating the degree of severity of irritation, tissue damage, or response.

Subchronic experiments generally expose animals for 90 days and attempt to mimic potential human exposure. For example, if inhalation is expected to be the primary route of exposure to humans, then the animals will be exposed to the test compound for 6 to 8 h a day, five days a week over the 90-day period. Subchronic experiments can be quite informative without also incurring exorbitant costs in conducting the experiments. Included here would be specialty studies to determine target organ toxicity, such as bioassays for evaluating neurotoxicity, reproductive effects, or teratogenicity (effects on the fetus).

Chronic assays expose animals to the material over their lifetime (approximately 2 years for rats and mice). The endpoint of primary interest in these studies is carcinogenicity or the potential of exposure to a chemical to cause cancer or tumor formation. Chronic studies are very unusual for metalworking fluids as a whole and even fairly unusual for the components, except for some of the biocides or oils.

Lastly, as a substitute for chronic studies, short-term bioassays are used as indicators of mutagenicity and potential carcinogenicity. These assays use bacterial or mammalian cell culture systems and can be useful when conducted in a battery of tests providing complementary information. They can be run quickly with minimal cost.

III. HEALTH EFFECTS ASSOCIATED WITH EXPOSURE TO METALWORKING FLUIDS

A. Acute Effects

A number of components that are regularly used in metalworking fluids have been associated with various acute disorders. Skin contact is prevalent in metalworking fluid operations and often difficult to control. Primary irritant contact dermatitis and allergic contact dermatitis are more common today than oil-induced dermatitis, known as folliculitis. Causative factors are influenced by the ingredients and nature of the metalworking fluid, concentration and pH, duration of exposure, and other factors such as age, skin type, previous exposure, presence of other skin disease, and personal hygiene. Additives such as amines, petroleum sulfonate, and some of the biocides have been associated with contact dermatitis. Skin sensitization, which is an allergic response to a chemical or a component in the material, has also been reported. It is often believed that sensitization is due to the biocide. Isothiazalones, formaldehyde, and mercaptobenzothiazoles have been reported to have sensitization potential. Metal allergy dermatitis may also occur. This is believed to be due to the solubilization of metallic ions from the metals and alloys being worked in the system. Nickel, chromium, and cobalt are three of the most common metal skin sensitizers [2]. The previous chapter addresses in detail the effects of metalworking fluids on the skin.

Many of the components in metalworking fluids have been tested in animals for acute, oral, and dermal lethality (LD_{50}), as well as for skin and eye irritation. Mineral oils are classified as relatively nontoxic, the LD_{50} being greater than 10 g/kg on oral exposure in rats and greater than 3 g/kg following dermal application in rabbits. They are also classified as mild to moderate skin and eye irritants. Most other components in metalworking fluids are classified as moderately toxic to nontoxic when evaluated in LD_{50} studies.

Besides potential skin and eye contact, inhalation exposure is also a common route of occupational exposure. Workplace aerosols in an industrial metalworking environment are produced as machining fluids continuously flood the cutting and grinding tools and part being produced. The chemical nature and particle size of the aerosol generated vary with the type of fluid, chemical composition, and operation. In general, water-based fluids produce smaller particles than oil-containing fluids. Particles smaller than 10 μm are inhalable and therefore have the potential to cause respiratory effects. Chan et al. [3] studied the size characteristics of straight oils, soluble oils, and semisynthetic and synthetic fluids in various machining operations over a 16-month period. They found that the respirable portion of the total aerosol collected ranged from 19.2 to 55.3% and that the machining operation and fluid type were the most critical variables.

Acute respiratory effects from exposure to metalworking fluids have been manifested as irritation or alteration of pulmonary function. Schaper et al. [4]

evaluated the sensory and pulmonary irritation potential of ten aerosolized metalworking fluids in mice. The animals were exposed by inhalation for 3 h to 20 to 2000 mg/m^3 of fluid. Six of the fluids represented new and used pairs. All of the fluids at some dose were capable of producing sensory and pulmonary irritation with little or no change in pulmonary histopathology. The irritancy potential was as follows: synthetics > solubles > straight oils. Irritation potential did not increase with use. No conclusions were drawn with respect to which components in the fluids were the most likely causative agents. Costa and Amdur [5] studied the effects of oil mists on pulmonary function in guinea pigs. They found little effect following short-term exposures of up to 200 mg/m^3 measuring acute irritant response. A study of the nasal and pulmonary toxicity of 20 aliphatic amines, some of which may be metalworking fluid additives, was conducted in mice [6]. Most of these compounds were found to be nasal irritants, typically at levels greater than current occupational exposure limits. Diisopropylamine and di-n-butylamine, however, did produce lower respiratory (pulmonary) toxicity at relatively low levels.

The only occupational study evaluating the acute respiratory effects of exposure to metalworking fluids was reported by Kennedy et al. in 1989 [7]. Pulmonary function was measured Mondays and Fridays, pre- and postshift, in a group of employees working with metalworking fluids. A 5% cross-shift decrease in FEV_1, a pulmonary function measure indicative of air flow obstruction, was measured in association with exposure to inhalable (less than 9.8 μm) aerosols at concentrations of >0.2 mg/m^3 of straight oils, soluble, or synthetic fluids. No decrement was reported from Monday to Friday, suggesting reversibility of the observed effect with removal of exposure. The authors speculate that the decrease in FEV_1 reported may be due to multiple causes, including biological contamination, because of the diversity and chemical complexity of the fluids studied.

B. Chronic Effects

More extensive data exist on the subchronic and chronic effects of exposure to metalworking fluids and their components. These studies typically focus on respiratory effects or cancer potential.

1. Respiratory Effects

A number of studies evaluate the respiratory effects of long-term exposure to oil mists in laboratory animals. Lushbaugh et al. [8] exposed rats, mice, rabbits, and monkeys to high doses of either motor oil or diesel oil varying between 100 and 365 days. Only the monkeys exhibited adverse effects manifested by an increased incidence of infections, pneumonia, and gastritis. In another study, dogs, rats, rabbits, hamsters, and mice were exposed daily to 5 and 100 mg/m^3 of oil mist for 12 to 26 months. At the high dose, lipid-containing alveolar macrophages and lipid granulomas were reported in dogs and rats, as well as evidence of morphologic changes in respiratory tissue [9]. Stula and Kwon [10] reported similar findings to Wagner et al. [9] following exposure of dogs, rats, mice, and gerbils to 5 and 100 mg/m^3 for up to two years.

Several epidemiological studies have also focused on the potential respiratory effects following chronic exposure to oils and metalworking fluids. Jones [11] studied workers exposed to oil mists in a steel rolling mill. An increase in linear striations in the lungs were observed in x-rays of some patients; the significance of this was unknown. In a later study, Cullen et al. [12] reported normal x-rays from oil-exposed steel mill workers who had respiratory complaints, with one case of lipoid pneumonitis and restrictive pattern of pulmonary function. Three separate studies evaluated employees working with straight oils as well as water-based fluids for alterations in pulmonary function and respiratory symptoms. All reported an increased prevalence of symptoms, such as cough and phlegm, without alterations in pulmonary function [13–15]. Other studies have reported no alterations in respiratory morbidity, mortality, or symptomatology following occupational exposure to oil mists [16,17]. Robertson [18] reported several cases of occupational asthma in patients exposed to metalworking fluids. Water-based fluids were more often associated with respiratory effects than straight oils. The only specific agent identified occurred in a case where the patient reacted to the pine oil reodorant in the soluble fluid. Hendy et al. [19] reported occupational asthma in association with exposure to soluble oils containing pine oil reodorant and colophony, both extracted from pine trees. Occupational asthma has also been associated with exposure to an aliphatic amine, dimethyl ethanolamine, although not in a metalworking fluid application [20].

2. Carcinogenicity

Certainly the health effect that has caused the greatest concern in using metalworking fluids is the potential for cancer. One of the initial concerns in this area arose with the report [21] that nitrosamines, which have been shown to be liver carcinogens in laboratory animals, could form in metalworking fluids that contained both nitrites and amines in combination. Additionally, there has been concern over the potential carcinogenicity of petroleum-based mineral oils. The nitrosamine problem was addressed relatively easily in the industry by simply avoiding the combination of amines and nitrites. The oil concern is more complex and involves a critical component of all but synthetic metalworking fluids.

One of the major difficulties in an evaluation of the literature on mineral oils is the poor definition of the material under study. In the past, the term "mineral oil" has been used to describe oils derived from coal, shale, petroleum crude oil, and even animal and vegetable sources. There has been little recognition of the vast differences in the production, uses, chemical, and physical and toxicological characteristics of "mineral oils."

Health and Safety Aspects

This was the case when, in 1973, the International Agency for Research on Cancer (IARC) cited various reports in the literature and stated that there was sufficient evidence of carcinogenicity of *some* mineral oils in experimental animals and humans [22]. IARC did, however, acknowledge that mineral oils vary in their composition, which may also affect their carcinogenicity.

In order to clarify the mineral oil issue, IARC convened a group of scientists to evaluate the carcinogenicity of mineral oils derived from petroleum crude oils, which are then further refined and used as base oils in fuels and lubricants [23].

a. Refining History and Carcinogenic Potential of Mineral Oils. The important factors in the production of lubricating oil products are the petroleum crude oil type, the manufacturing or refining process, and the formulation of the final product. Petroleum crude oils are classed as paraffinic or naphthenic. Lubricant refining and, correspondingly, product formulations have changed considerably over the years. Until about 1940, processing consisted of acid refining with clay finishing and subsequent dewaxing by chilling. Solvent refining (and solvent dewaxing) was first introduced in the United States and Europe in the 1930s. This is an extraction process which, following solubilization of the polycyclic aromatic hydrocarbons (PAHs), selectively removes olefins, naphthenes, and then paraffins, depending on the severity of the process. Hydrotreating, a newer, more severe process than hydrofinishing, was introduced in the 1960s. Through a catalytic hydrogenation process, the lubricant base oil is made more paraffinic by the saturation of olefins. The severity of the hydrogenation dictates the degree of conversion of aromatics to naphthenes.

In general, the trend has been toward more highly refined oils with removal of unwanted impurities including PAHs, constituents believed to be major factors in imparting carcinogenic activity to these products. Consequently, animal studies have been conducted on refined mineral oils derived from these newer processing techniques in order to evaluate carcinogenic potential; these are primarily mouse skin-painting studies. These studies are fairly common in toxicology and considered to be relatively accurate in predicting skin carcinogenic potential in man. Data from these animal carcinogenicity studies are the primary basis for the following conclusions drawn by IARC:

There is sufficient evidence of carcinogenicity for:
    Untreated vacuum distillates
    Acid-treated oils (which includes caustic neutralization, dewaxing, or clay treating)
    Aromatic oils
    Mildly solvent-refined oils
    Mildly hydrotreated oils
There is no evidence of carcinogenicity for:
    Severely solvent-refined oils

White oils (when administered by routes other than intraperitoneal injection)

There is inadequate evidence (one study) to evaluate severely hydrotreated oils or oils that have been mildly solvent refined with subsequent mild hydrotreatment; however, the data to date have indicated these are not carcinogenic.

Therefore, if the oil in a metalworking fluid has been severely refined, carcinogenic risk from this component would be unlikely. From this information, a question which naturally arises is: How is "mild" or "severe" refining defined? Unfortunately, there is considerable controversy as well as data gaps in defining these terms, and the Chemical Abstract Services number (CAS number) indicates only the final refining process—not the severity or earlier processing. Because of this, the Occupational Safety and Health Administration (OSHA) published in the *Federal Register* on December 20, 1985, "Hazard Communication; Interpretation Regarding Lubricating Oils." One of the purposes of this document was to define "mild" hydrotreatment. The critical factors for defining this process are pressure and temperature. OSHA has decided that an oil has been mildly hydrotreated if it has been processed at a pressure of 800 psi or less at temperatures of up to 800°F. Unfortunately, OSHA did not define any of the other refining parameters.

b. The Modified Ames Assay: Predicting Mineral Oil Carcinogenicity. Another recently developed technique for evaluating potential carcinogenicity of oils is the modified Ames assay based on the standard Ames bacterial mutagenicity assay. The modified Ames assay is reported to have a correlation coefficient of 0.92 for oils with median boiling points between 500° and 1070°F when compared with the results of the long-term mouse skin-painting studies [24]. A significant correlation has also been observed between the 3 to 7 ring PAH compounds and both mutagenic and carcinogenic potency [25]. The two major advantages to this type of assay versus conducting mouse skin-painting tests are time and money. The modified Ames can be run and evaluated in a few days, whereas mouse skin painting takes approximately 2 years of testing and another year or so to analyze the results, obviously at a much greater cost. In addition, the modified Ames can be used to quickly screen oils of unknown refining history in order to predict potential carcinogenicity.

c. Human Studies on Metalworking Fluids. Numerous epidemiology studies have evaluated the carcinogenic potential of occupational exposure to oil mists and metalworking fluids. The majority of cancer types reported have been skin (scrotal), respiratory, or gastrointestinal. Cases of scrotal cancer have been extensively reported in the literature, primarily associated with use of poorly refined oil in conjunction with poor personal hygiene [27]. Possible causal relationships between exposure to oil mist and cancer of the respiratory or gastrointestinal tracts is

Health and Safety Aspects

difficult to establish as tumors in these organs are common and many other factors, such as smoking and diet, could be responsible. Additionally, few of these studies provide analytical data on either the bulk fluid, including the critical question of refining history of the oil, or air contaminant levels associated with the effects reported, making difficult extrapolation of the results to other settings.

Waldron [28] reported an excess of bronchus and larynx cancer in machine operators exposed to oil mists, with no increase in toolmakers similarly exposed. Ronneberg et al. [29] observed an increase in lung cancer in men exposed to naphthenic oils with limited refining in cable manufacturing. The authors state that smoking could have contributed to their findings. Jarvholm et al. [30], and in a follow-up study Jarvholm and Lavenius [31], studied cancer morbidity and mortality in workers exposed to oil mists and cutting fluids and found no increase in lung cancer or respiratory disease and no increase in any cancer type.

Several studies have reported increases in gastrointestinal cancers following long-term metalworking fluid exposures. DeCoufle [32] observed a twofold increase in cancer of the stomach and large intestine (combined). This effect was only found in employees who had worked with metalworking fluids for more than five years, prior to 1938, with more than a 20-year latency period. Vena et al. [33], in a proportionate mortality ratio study conducted using union records, reported an excess of digestive system cancer, particularly liver and pancreas, and respiratory system and urinary bladder cancers in a plant where various types of metalworking fluids had been used. An increase in stomach cancer was observed in association with precision grinding exposures using soluble oils [34]. In a larger study [35], similar findings were reported with a significant excess of stomach cancer in grinding using soluble oils and pancreatic cancer in grinding and machining operations (straight oils). The authors speculate that additives or abrasives in the water-based fluids may be related to the increased cancer rates.

d. Carcinogenicity of Metalworking Fluid Components. Limited data are available on the carcinogenicity of individual components used in metalworking fluids. Aliphatic amines are commonly present in water-based fluids. Some subchronic and chronic toxicology testing has been conducted on mono-, di-, and triethanolamines. Long-term ingestion or skin contact of the ethanolamines to rats, guinea pigs, or dogs produced primarily liver and kidney lesions [36]. To date, none of these chemicals are classified as carcinogens. The data do indicate that diethanolamine is the most toxic chronically of the three ethanolamines and is currently being tested for carcinogenicity by the National Toxicology Program because of effects observed in rats and mice following 13-week exposures [37]. As mentioned earlier, aliphatic amines can react with nitrites in metalworking fluids to form nitrosamines, which are liver carcinogens in laboratory animals; therefore, this combination should be avoided.

Another common family of compounds in metalworking fluids that have of late become a carcinogenic concern are the chlorinated paraffins. These are included in fluids as extreme-pressure additives and antiwear agents. The NTP recently completed carcinogenicity testing of two of the chlorinated paraffins, a C12, 58–60% chlorine (Cl) product and a C23/24, 40–43% chlorine material. Rats and mice were fed the chemical over their lifetimes. The C12, 58–60% Cl was positive for carcinogenicity in both rats and mice; however, equivocal results were obtained following exposure to the C23/24, 40–43% Cl compound [38,39]. Therefore, the C12, 60% chlorine, chlorinated paraffins are now classified as suspect carcinogens. These compounds are not, however, absorbed through the skin [40], eliminating this potential route of exposure.

e. Reclaimed, Rerefined, and Used Mineral Oils. Reclaimed, rerefined, and used oils represent a growing class of mineral oils that are increasingly used in manufacturing in response to pollution prevention and waste minimization initiatives. The potential hazards of these materials result from contaminants associated with their initial use. In large part, the toxicity of these fluids has not been studied. The IARC monograph does not address reclaimed oils and has only a brief section on "used oils," which is not defined. McKee et al. [26] recently assessed the dermal carcinogenic potential and PAH content of refined oils in new and used metalworking fluids. None of the solvent-extracted base oils induced skin tumors, and similarly, the cutting fluids prepared from these oils were not carcinogenic in mouse skin-painting bioassays. Additionally, there was no evidence that use increased PAH content or carcinogenic potential.

IV. CONCLUSIONS

The complexity of metalworking fluids results in a complex health and safety profile for these materials. Key considerations include the toxicity of the starting components and reaction products (such as nitrites and amines that form nitrosamines), as well as potential for exposure through inhalation of mists or dermal contact. Additional factors for exposure include particle size data as well as the ability of the material to be absorbed through the skin.

Epidemiology studies provide useful insight on the potential health effects of exposure to metalworking fluids as a whole, but are often designed to evaluate effects that may be due to exposures from fluids used 20 to 30 years earlier. This is because most cancers exhibit long latency periods, often as long as 20 years. Determination of specific causative agents in these studies is difficult because of a lack of chemical data both on the bulk fluids as well as on employee exposures that might have occurred historically. It is clear that the chemical nature of metalworking fluids has changed dramatically over time, especially with respect to oils often contained in these products, since changes have occurred in refining practices, aimed at removal of impurities such as PAHs. This then makes extrapo-

Health and Safety Aspects

lation of the results from many epidemiology studies to current metalworking fluid applications difficult.

Toxicology testing on metalworking fluid components provides both qualitative and quantitative data on the potential health effects from exposure and can assist in interpreting results of epidemiology studies.

Therefore, based on current toxicology and epidemiology data, if oil-containing metalworking fluids are used and employee exposure cannot be avoided, it is suggested that the following types of oils be used where possible:

Severely solvent refined,
Severely hydrotreated,
Mildly solvent refined with subsequent mild hydrotreatment, or
Oils that have been negative for carcinogenicity in toxicology studies (mouse skin painting, modified Ames bioassay).

In all applications where metalworking fluids are used, good industrial hygiene practice must be maintained. Impervious gloves and protective clothing, barrier creams, and ventilation, as well as other industrial hygiene control measures should be considered.

In many cases, by controlling for skin irritation and dermatitis, risk of additional chronic effects can be reduced. And, of course, employees must be informed of the potential hazards of these products and trained to minimize exposure as with all chemical materials.

REFERENCES

1. *National Occupational Hazard Survey*, Vol. III, National Institute of Occupational Safety and Health, publication no. 78–114, pp. 216–229, Cincinnati, (1977).
2. C. Zugerman, *Occup. Med.: State of the Art Rev. 1*: 245 (1986).
3. T. L. Chan, J. B. D'Arcy, and J. Siak, *Appl. Occup. Environ. Hyg. 5*(3): 162–170 (1990).
4. M. Schaper and K. Detwiler, *Fund. Appl. Toxicol. 16*: 309–319 (1991).
5. D. L. Costa and M. O. Amdur, *Am. Ind. Hyg. Assoc. J. 40*: 673–679 (1979).
6. F. Gagnaire, S. Azim, P. Bonnet, P. Simon, J. P. Guenier, and J. de Ceaurriz, *J. Appl. Toxicol. 9*(5): 301–304 (1989).
7. S. M. Kennedy, I. A. Greaves, D. Kriebel, E. A. Eison, T. J. Smith, and S. R. Woskie, *Am. J. Ind. Med. 15*: 627–641 (1989).
8. C. C. Lushbaugh, J. W. Green, and C. E. Redemann, *Ind. Hyg. Occup. Med. 1*: 237–247 (1950).
9. W. D. Wagner, P. G. Wright, and H. E. Stokinger, *Ind. Hyg. J. 25*: 158–168 (1964).
10. E. F. Stula and B. K. Kwon, *Am. Ind. Hyg. Assoc. J. 39*: 393–399 (1978).
11. J. G. Jones, *Ann. Occup. Hyg. 3*(4): 264–271 (1961).
12. M. R. Cullen, J. R. Balmes, J. M. Robins, and G. J. Walker Smith, *Am. J. Ind. Med. 2*: 51–58 (1981).
13. H. Oxhoj, H. Andreasen, and U. M. Henius, *Eur. J. Res. Dis.* Suppl. 18: 85–89 (1982).
14. B. Jarvholm, *Eur. J. Res. Dis.* Suppl. 18: 79–83 (1982).

15. B. Jarvholm, B. Bake, B. Lavenius, G. Thiringer, and R. Vokmann, *J. Occup. Med.* 24: 473 (1982).
16. D. H. Goldstein, J. N. Benoit, and H. A. Tyroler, *Arch. Environ. Health 21*: 600–603 (1970).
17. T. S. Ely, S. F. Pedley, F. T. Hearne, and W. T. Stille, *J. Occup. Med. 12*(7): 253–261 (1970).
18. A. S. Robertson, D. C. Weir, and P. S. Burge, *Thorax 43*: 200–205 (1988).
19. M. S. Hendy, B. E. Beattie, and P. S. Burge, *Br. J. Ind. Med. 42*: 51–54 (1985).
20. M. Vallieres, D. W. Cockcroft, D. M. Taylor, J. Dolovich, and F. E. Hargreave, *Am. Rev. Resp. Dis. 115*: 867–871 (1977).
21. DHEW (NIOSH), Current Intelligence Bulletin 15, No. 78-127: Nitrosamines in Cutting Fluids, October 6, 1976, Rockville, MD.
22. *IARC Monographs on the Evaluation of the Carcinogenic Risk of the Chemical to Man*, Vol. 3, International Agency for Research on Cancer, WHO, Lyon, France (1973).
23. *IARC Monographs on the Evaluation of Carcinogenic Risk of Chemicals to Humans*, Vol. 33, International Agency for Research on Cancer, WHO, Lyon, France, pp. 87–168 (1984).
24. G. R. Blackburn, R. A. Deitch, C. A. Schreiner, and C. R. Mackerer, *Cell Biol. Toxicol.* 2(1): 63–84 (1986).
25. T. A. Roy, S. W. Johnson, G. R. Blackburn, and C. R. Mackerer, *Fund. Appl. Toxicol. 10*: 466–476 (1988).
26. R. H. McKee, R. A. Scala, and C. Chauzy, *The Toxicologist 8*(1): 162 (1988).
27. E. Bingham, R. P. Trosset, and D. Warshawsky, *J. Environ. Pathol. Toxicol. 3*: 483–563 (1979).
28. H. A. Waldron, *Br. J. Cancer 32*: 256–257 (1975).
29. A. Ronneberg, A. Andersen, and K. Skyberg, *Br. J. Ind. Med. 45*: 595–601 (1988).
30. B. Jarvholm, L. Lillienberg, G. Sallsten, G. Thiringer, and O. Axelson, *J. Occup. Med. 23*(5): 333–337 (1981).
31. B. Jarvholm and B. Lavenius, *Arch. Environ. Health 42*(6): 361–366 (1987).
32. P. DeCoufle, *J. Nat. Cancer 61*(4): 1025–1030 (1978).
33. J. E. Vena, H. A. Sultz, R. C. Fiedler, and R. E. Barnes, *Br. J. Ind. Med. 42*: 85–93 (1985).
34. R. M. Park, D. H. Wegman, M. A. Silverstein, N. A. Maizlish, and F. E. Mirer, *Am. J. Ind. Med. 13*: 569–580 (1988).
35. M. Silverstein, R. Park, M. Marmor, N. Maizlish, and F. Mirer, *J. Occup. Med. 30*(9): 706–714 (1988).
36. Cosmetic Ingredient Review Panel, *J. Am. Col. Toxicol.* 2(7): 183–235 (1983).
37. R. L. Melnick, M. Hejtmancik, A. Peters, M. Ryan, A. Singer, L. Mezza, and R. Persing, *The Toxicologist 10*(1): 154 (1990).
38. *Toxicology and Carcinogenesis Studies of Chlorinated Paraffins (C12, 60% Chlorine) (CAS No. 63449-39-8) in F344/N Rats and B6C3F1 Mice*, Tech. Report NTP TR 308, National Toxicology Program, Research Triangle Park, NC (May 1986).
39. *Toxicology and Carcinogenesis Studies of Chlorinated Paraffins (C23, 43% Chlorine) (CAS No. 63449-39-8) in F344/N Rats and B6C3F1 Mice*, Tech. Report NTP TR 305, National Toxicology Program, Research Triangle Park, NC (May 1986).
40. J. J. Yang, T. A. Roy, W. Neil, A. J. Krueger, and C. R. Mackerer, *Toxical. Ind. Health* 3(3): 405–412 (1987).

16
Government Regulations Affecting Metalworking Fluids

WILLIAM E. LUCKE
Cincinnati Milacron
Cincinnati, Ohio

I. INTRODUCTION

At the end of World War II, American chemical and pharmaceutical industries enjoyed a well-deserved reputation for their contributions to the war effort, which included synthetic rubber, plastics, nylon, DDT, sulfa drugs, and penicillin. In the years after the war, new successes followed in synthetic fabrics, wonder drugs, and agricultural chemicals. Raw material feedstocks were cheap. Competition from the giant German cartels was eliminated as part of the peace treaty. Popular magazines carried "gee whiz" articles describing the everyday items that could be made from oil, water, and air. Synthetic materials began to replace natural materials in clothing, automobiles, and houses. The list of wonders promised for the future was almost endless.

In the early 1960s three unrelated events began a process of change in how the public perceived the chemical industry and their products. Rachel Carson published *Silent Spring*, which questioned the widespread use and safety of pesticides. Thalidomide, a sedative commonly used in Europe, was found to be a teratogen, preventing normal development of the limbs of a fetus exposed during a critical period of pregnancy. At the same time, new technologies in the analysis of trace levels of chemicals, especially gas liquid chromatography using electron capture detectors, made it possible to find chemicals in food and water at levels of parts per million and parts per billion. Questions were raised about the long-term effects of exposure to low levels of chemicals for which there were no satisfactory answers.

Public concern led to the development of a new science, toxicology, which studies the effects of exposure to chemicals on animals. Testing priorities focused on chemicals known or suspected to be toxic; approximately 50% of all chemicals tested for carcinogenesis gave positive results, especially when tested at very high-dose levels. By 1970, the perception of chemicals had changed from a source of unending miracles to a lurking menace from continuing exposure to low levels of toxins whose harmful effects would only show up years later as an incurable disease or in unborn children. The use of DDT and other pesticides was discontinued. Cyclamates were removed from soft drinks. The use of saccharin was allowed to continue only through special action by Congress. Nitrosamines were found in bacon, beer, baby nipples, and metalworking fluids. Companies dropped the word "chemical" from their corporate names. Dupont shortened their slogan "Better things for better living through chemistry" by two words.

The general alarm about the perceived chemical menace inevitably resulted in a rush of legislation in the 1970s and 1980s designed to protect human health and the environment. The Occupational Safety and Health Act mandated that chemical hazards in the workplace be identified and monitored. The Clean Water Act and the Clean Air Act imposed restrictions on industrial effluents. The Safe Drinking Water Act set standards for drinking water, including limits on metal and pesticide content. The Environmental Protection Agency (EPA) was created to administer these ambitious programs and soon found itself caught between conflicting demands and claims of industry and militant environmental lobby groups. Proposed rules were challenged in the courts and modified by decree, not always consistently.

Metalworking fluids have usually been at the edges of this activity, but they have never been far from the focus. The issue of nitrosamines in nitrite containing fluids directly involved the industry in health and safety questions. The ingredients in a fluid must be on the Toxic Substances Control Act (TSCA) inventory. Transport of the component chemicals or the fluids may be regulated by the Department of Transportation (DOT) under the Hazardous Materials Transportation Act (HMTA). Workplace hazards associated with a fluid must be disclosed under the Occupational Safety and Health Administration (OSHA) Hazard Communication Standard (HCS) on the product label and on a material safety data sheet (MSDS). Users must be informed of these hazards and how to work safely with them by their employers. All fluids are disposed of in some manner at the end of their useful life and will be subject to some degree of regulation at that time.

This chapter will summarize the regulations covering the formulation, transport, use, and disposal of metalworking fluids and will discuss ways to minimize product liability risks.

II. REVIEW OF THE REGULATORY PROCESS

Under the Constitution, Congress has the authority to draft and enact legislation which is administered and enforced by the executive branch. It is common for Congress to specify a particular agency to be charged with responsibility for a specific law. The direction given to the agency may be very specific in what is to be done and how it is to be done, or it may be a general authority to determine what should be done to achieve a goal and to develop the mechanisms and rules to allow the goal to be achieved.

The procedures followed by the agencies in developing the rules which implement the laws are established under the Administrative Procedures Act. In cases where the agency is not confident that they have all the facts on an issue, they will publish an Advanced Notice of Proposed Rulemaking (ANPR) in the *Federal Register*, presenting the issues as they see them and asking comments on specific questions and issues. A specific period for comments, usually 90 days, is established. All comments become part of a docket which is available for public inspection by interested parties. (When confidential business information is submitted as part of the comments, the public file contains a notation that information was submitted and possibly a sanitized summary of the information.) The agency is required to consider all comments made within the comment period. Comments received after the deadline may be considered at the agency's option, but they are not required to respond to or consider the comments.

After the comments have been reviewed, Proposed Rules are drafted by the agency and published in the *Federal Register*, along with a summary of the responses to the ANPR, if one had been published. A full rationale of the proposal is given, along with a number of analyses of the impact of the Rules required by law, a cost–benefit analysis, an economic impact analysis, small business impact, and compliance with the Paperwork Reduction Act. Again, a comment period, usually 90 days, is established for responses by interested parties.

Once the rule has been revised (if needed) in response to the comments, the agency will publish a Final Rule which is normally effective in 30 days, except as otherwise provided in the rule. Again, any comments received are presented along with agency responses.

This is the end of the formal process in creating regulations. It is not unusual, however, for those whose comments and suggestions were not favorably accepted by the agency to pursue the issue in the courts. At one time, industry groups and environmental groups would compete to be the first to file suit over a new rule to ensure that a "friendly" judge would hear the case. When OSHA published the rule covering workplace exposure to formaldehyde in 1987, the agency was sued by industry, labor, and the environmentalists. After over six years, the standard is only now going into effect.

The rules published in the *Federal Register* are actually changes in the various sections of the "Code of Federal Regulations." This is the official codification of all federal regulations. The Code is divided into 50 Titles, each dedicated to a specific agency or activity of the government. For the formulators of metalworking fluids, the relevant sections are 29 CFR Part 1910 (OSHA), 40 CFR (EPA), and 49 CFR (DOT). Title 40 is divided into sections according to enabling legislation; for example, Resource Conservation and Recovery Act (RCRA) is covered by Sections 190–299; Comprehensive Environmental Response Compensation and Liability Act (CERCLA), Superfund, and Superfund Amendments and Reauthorization Act (SARA) by Sections 300–399; water programs, effluent guidelines, and standards by Sections 100–699; and TSCA by Sections 700 to the end.

A new issue of the *Federal Register* is published every working day. A review such as this can be current only on the day it is written. Before accepting what is presented here as definitive, it would be wise to consult the appropriate sections of the current edition of the "Code of Federal Regulations" for any possible revisions, and the *Federal Register* for any pending or proposed changes. With that caveat, what follows is a summary of the major federal regulations that have impact on the metalworking fluid industry suppliers, formulators, and users.

III. TOXIC SUBSTANCES CONTROL ACT

The Toxic Substances Control Act (TSCA) was enacted by Congress in 1976, giving EPA the authority to oversee the introduction of new chemical products into the marketplace and to protect health and the environment from possible adverse effects of chemicals already in use. The provisions cover both pure chemicals and mixtures formulated from other chemical products. While the primary impact is on suppliers of chemicals to the metalworking fluid industry, fluid formulators and users are impacted both directly and indirectly. The direct impact is found in 40 CFR Part 747, Metalworking Fluids. This part of the regulations restricts the use of nitrites in connection with three specific chemicals which have a potential to form nitrosamines.

A. TSCA Inventory and Introduction of New Chemicals

Under the provisions of TSCA, any commercially available chemical must be included in the TSCA inventory. The initial inventory included all chemicals manufactured between July 1, 1974 and December 31, 1976. New chemicals introduced after that time have been or are subject to a premanufacturing notice (PMN) process before introduction to the market. The PMN process can be laborious, involving detailed process information about products and by-products, estimated production levels and sites, toxicity information, environmental impact estimates, and worker exposure during manufacture and use. The cost of filing a PMN has been estimated at $150,000.

Once the PMN has been filed, EPA has 90 days to request more data or to deny the addition of the chemical to the inventory. If EPA does not act in the 90 days, the manufacturer may begin production, but must notify EPA that production has begun.

EPA may extend the 90 day period if it is determined that the chemical may be subject to regulation under Section 5e or 5f of TSCA, more information is needed, or new information has been received during the review period. (Section 5 gives EPA powers to regulate chemicals which pose hazards to human health or the environment.)

For chemicals that are on the inventory, development of new uses are also regulated. Before such a chemical can be marketed for a new application, it will be subject to a Significant New Use Rule (SNUR). Information similar to that submitted for a PMN must be provided to EPA for possible rejection. For example, EPA determined in 1991 that nitrites had not been used in metalworking fluids since 1985 and that any future use would fall under the SNUR regulations.

B. Regulatory Authority Under TSCA

EPA does have the power to find a substance or class of substance to be hazardous. Hazardous materials may be regulated by one or more of the following requirements:

> Prohibition of manufacturing, processing or disposal, or the imposition of a maximum permissible amount to be processed.
> Limiting the concentration of the substance in manufacturing, use, or disposal.
> Requirement of warning and instructional labeling.
> Requirement of manufacturing records, testing, and monitoring to confirm compliance with other orders.
> Prohibition of use.
> Prohibition of disposal.
> Requirement of notification of users of the risk, including public notice of such risk and repurchase of hazardous material.
>
> EPA has the authority to require records and reports of
>
> The trade name and formula of each product.

Categories of use.

Total amounts of each product made and the amount going to each use category. (EPA is presently considering using this power to collect information on the environmental fate of chlorinated paraffins and α-olefins.)

A description of by-products.

All existing data concerning the environmental and health effects of each product.

The number of individuals exposed to the product and the duration of exposure.

Methods of disposal.

Obviously, some of these reporting requirements would affect users as well as formulators of fluids.

C. Reporting Under TSCA Section 8e

Under TSCA Section 8e, any person who manufactures, processes, or distributes in commerce a chemical substance, and who obtains information which reasonably supports the conclusion that such substance or a mixture presents a substantial risk of injury to health or the environment, shall inform EPA of such information within 15 working days of receiving the information. In cases of environmental releases involving substantial risks, reporting is required within 24 h. Failure to report can result in fines up to $25,000 and a sentence of up to six months for each violation. Each day of failure to report is treated as a separate violation.

The following discussion will focus on reporting of information under Section 8e, but the same guidelines will apply to environmental release reports.

A substantial risk to human health or the environment is defined as a risk of considerable concern because of the seriousness of possible effects and the fact or probability of its occurrence. These two criteria are weighted differently for different effects. For serious human health effects (such as cancer, birth defects, mutagenicity, death, incapacitation), little weight is to be given to exposure levels. The mere fact that a chemical is in commerce constitutes sufficient evidence of exposure. In cases of environmental effects (such as widespread and unsuspected distribution in environmental media or bioaccumulation) or emergency contamination incidents, the new information must involve or be accompanied by the potential for significant levels of exposure. Obviously, the use of words like "reasonably" and "significant" are ambiguous and open to interpretation. In view of the serious potential penalties for failure to report and the absence of penalties for filing an unnecessary report, prudence suggests reporting when in doubt.

As written, the regulations apply not only to companies, but to all employees capable of appreciating the significance of the substantial risk information. EPA recommends that employers set up a formal process for receiving information from the employee and assuming the liability. In general, a committee composed of technical, management, and legal personnel would formally receive the report and determine if the information is in fact reportable. The employee must be informed of this decision in writing. If a decision is made not to report, the employee may choose to report as an individual. In such a case, no retribution can be taken against the employee. However, the employer has the right to preview any submission to prevent disclosure of confidential business information.

If it is decided to inform EPA, a summary report and the data are submitted to EPA before the 15 day limit passes. Reporting cannot be delayed while confirmation is made. Further work can be done and reported subsequently. Whether or not a report is filed, all records must be maintained for 30 years in cases of health risk or 5 years for environmental risks.

The notice to EPA must include the following information:

- A statement that the information is being submitted in accordance with Section 8e of TSCA.
- The name and address of the company, organization, or individual filing the report.
- The name, job title, signature, and telephone number of the person filing the report.
- Identification of the chemical substance or mixture, including the CAS registry number, if known.
- A summary of the adverse effects being reported, describing the nature and extent of the risk involved.
- The specific source(s) of information together with a summary and the source(s) of any available supporting technical data.

If confidential business information is disclosed in the report, this information may be identified in the report, and a sanitized version may be included to be placed in the public files. It can be expected that EPA will require substantiation of such claims, but justification need not be included with the initial report. Subsequent reports, if any, should follow the same guidelines.

EPA will establish a file for the information, including separate files for the confidential and sanitized versions, if needed. The existence of the report will be made known to the public through several channels. Any other actions, including requests for more information, will be based on their judgment of the seriousness of the risk and the extent of any exposure.

D. Recordkeeping Under TSCA Section 8c

In cases where health or environmental effects are alleged but not supported by data, manufacturers or processors are required to collect and file supporting information under 40 CFR Part 717. There is no requirement to file a report unless supporting data are forthcoming. In cases involving known effects of a substance, there is no obligation to open a file. Again, files are to be kept for 30 years for allegations about health, 5 years for environmental allegations.

E. Reporting Under Section 4 of TSCA

In addition to these reporting requirements, EPA can request any data pertaining to specific chemicals or classes of products under 40 CFR Part 716. Each request for data is the subject of a separate *Federal Register* notice. In general, data on mixtures containing the chemicals are not subject to reporting, if antagonistic or synergistic effects are absent.

In summary, the provisions of TSCA have potentially the most significant impact on metalworking fluids as a class of any existing regulations, but are probably the least familiar to the industry.

IV. HAZARDOUS MATERIALS TRANSPORTATION ACT

While it is not generally realized that the transport of metalworking fluids can be regulated, it is not unusual to do so. Fluid concentrates are by nature alkaline, with pH values as high as 10–11. Such materials can attack aluminum metal, which is amphoteric (subject to attack by both acids and bases). Under the definitions established by the Department of Transportation, chemicals which corrode aluminum are classified as Corrosives and are subject to regulation. Since trailer bodies and airplanes are often made from aluminum, this is a reasonable restriction. In addition to the concentrates, many fluid additives are caustic and may be corrosive to skin.

The Hazardous Materials Transportation Act was enacted in 1974 and has been amended several times since. The actual regulations are contained in Title 49 of the "Code of Federal Regulations." In general, they include requirements for

> Provision for a proper shipping name for each hazardous material that conveys the hazard of the material being shipped. This name may or may not reflect the chemical composition of the material. Its purpose is to alert first responders to the nature of hazards that may be present at the scene of an emergency.
> Provision for packaging which meets minimum integrity standards.
> Provision for hazard communication, including specific labeling, marking, placarding of containers and vehicles, and shipping papers which carry emergency response information. Marking of the containers consists of placing the proper shipping name on the side and top of the container. Labeling is done by providing a sticker which identifies the hazard class of the material; e.g., corrosive.
> Restrictions on loading, unloading, handling, and transporting of hazards.
> Provision for a mechanism of incident reporting.

Hazardous materials are divided into nine classes:

Class 1, Explosives
Class 2, Gases
Class 3, Flammable liquids
Class 4, Flammable solids, spontaneously combustible materials, and materials that are dangerous when wet
Class 5, Oxidizers and organic peroxides
Class 6, Poisons and infectious materials
Class 7, Radioactive materials
Class 8, Corrosives
Class 9, Miscellaneous hazardous materials

For most regulated metalworking fluids, the hazard class will be Class 8 and the proper shipping name will be "Corrosive Liquid, n.o.s." (not otherwise specified). In 1991, DOT began an effort under Docket HM-181 to simplify the regulations and to align U.S. rules with international regulations.

Hazardous materials offered for transport must be properly marked and labeled. The proper shipping name must be on the container, along with an indication of the chemical(s) which lead to the hazardous nature of the material. In mixtures like metalworking fluids, this is not always obvious. A typical proper shipping name might be "Corrosive Liquid n.o.s., contains ethanolamines." In addition, a DOT label must be on the container. This will be a diamond-shaped label (Fig. 1), showing the hazard class in pictographic form.

Shipping papers must accompany any hazardous load and must identify each hazard specifically by the proper shipping name as determined by DOT. This is not the product name; a corrosive product must be shown as "Corrosive Liquid, n.o.s." along with the UN number and the identity of the chemical(s) which creates the hazard. Different products with the same proper shipping name may be listed on the same line of the shipping paper; shipping papers with the DOT names preprinted allow for easy preparation. Hazardous materials must be clearly indicated on the papers by special marking and color, or by grouping hazardous materials on separate sheets. The papers must be with the driver in the cab, either on the seat next to him or in a pouch on the driver's door. Note that placarding and shipping paper requirements apply if transport is by a company truck or by commercial carrier.

Any truck transporting more than 110 gal of hazardous materials must bear placards on all four sides identifying the nature of the hazardous material. Many trailers are fitted with flip-style placards and will not require a placard from the shipper. However, by law, the shipper is required to offer placards to the driver. It is recommended that a space be provided on the shipping paper for the driver to sign that placards were offered. This gives protection against a driver who might remove the placards once off your property to avoid areas prohibiting hazardous

Figure 1 State Emergency Response Commission (SERC) facility identification form.

Government Regulations

loads. If a hazardous material is already on the truck, placarding against the most hazardous material on the load would take precedence. Again, the driver's signature, showing that placards were offered, protects the shipper in case of an incident.

All employees involved in the transport of hazardous materials must be trained. This includes management, sales, dock personnel, and drivers. Drivers must be certified to haul hazardous cargo. Training courses are available on videotape, audio/slide presentations, or by off-site courses. The purpose of training is to ensure safe loading, handling, storage, and transportation of hazardous materials. Employees who load, unload, handle hazardous materials, prepare hazardous materials for transportation, are responsible for the safety of transporting hazardous materials, or operate a vehicle used to transport hazardous materials must receive training within 90 days after employment or a change in job assignment, and may not work unsupervised until training is completed.

A training program must cover four areas (the fourth applies only to highway transport).

- All employees must be able to recognize and identify hazardous materials in the workplace and understand the importance of the work they perform.
- Training must cover the necessary knowledge skills and abilities for the employee's job function. As an example, a shipping office employee would be trained focusing on 49 CFR Part 172 Subpart C.
- All employees who would handle or transport hazardous materials, or who might be exposed to hazardous materials in responding to an incident, must be provided with knowledge of hazards under normal conditions and under likely accident conditions.
- Drivers involved in highway transport of hazardous materials must receive training in accident avoidance and mitigation of the release of hazardous materials. Responsibility for determining the training needs has been left to employers.
- Training must be repeated at least every two years.
- Effectiveness of training must be documented by employee testing. Employers are required to develop test content, format, and methods that will comply with the regulatory goals.

V. HAZARD COMMUNICATION STANDARD

OSHA's Hazard Communication Standard (HCS), the workplace "right-to-know law," has been in effect since May 25, 1986. Prior to implementation of HCS, OSHA had identified certain chemicals and classes of chemicals as hazardous and had set limits on workplace exposure (29 CFR Part 1910.1000). For each chemical, a workplace exposure limit was established that is considered to be a level which will cause no harm, even given daily exposure for an entire working life. Limits are recommended by the American Conference of Governmental Hygienists (ACGIH) as threshold limit values (TLVs). Regulatory limits are adopted by OSHA as permissible exposure limits (PELs). Most limits are quoted as time-weighted averages for an eight-hour work day. Short-term exposure limits (STELs) may be set for 15 min weighted averages for more toxic materials. In some cases, a ceiling limit may be established which must not be exceeded at any time.

Common metalworking fluid components for which a PEL has been established are given in Table 1.

The Hazard Communication Standard itself (29 CFR Part 1900.1200) is intended to ensure that workplace chemical hazards are evaluated and communicated to employers and employees through labels, a material safety data sheet (MSDS), and training programs. Under the provisions of OSHA, the workplace provisions of any state or local right-to-know laws are preempted by the federal regulations.

A. Hazard Determination

Under the Standard, chemical producers or importers are required to evaluate their products to determine if they are hazardous. In making that determination, they must consider available scientific evidence. Any chemical for which a PEL or TLV has been established is considered hazardous by definition. There is no requirement that new tests be conducted, but a literature search must be conducted for each

TABLE 1 The List of Lists for Metalworking Fluid Components

Chemical	CAS no.	OSHA TLV	CERCLA RQ(lb)	SARA 313	RCRA waste code
Ethanolamine	141-43-5	3 ppm			
Barium compounds		0.5 mg/m^3	1000	X	D005
Diethanolamine	111-42-2	3 ppm[a]	1	X	
Hexylene glycol	107-41-5	25 ppm C[b]			
Morpholine	110-91-8	20 ppm			
Oil mist		5 mg/m^3			
Stoddard solvent	8052-41-3	500 ppm			
p-Chloro-m-cresol	59-50-7		5000	Proposed	U039
NDELA	1116-54-7		1		U173
2-Phenylphenol	90-43-7			X	
Copper compounds				X	
Glycol ethers			1	X	

[a]TLV was promulgated by OSHA, but overturned by the courts, and will likely be reinstated through normal rulemaking.
[b]Ceiling TLV.

chemical or component. A hazard is established by one valid scientific study which finds statistically significant evidence for any of the hazards listed in Table 2. Such studies must be cited in the MSDS, but any other studies which tend to refute that study may also be cited.

The rules covering hazard determination for mixtures are of obvious interest to producers and users of metalworking fluids. The hazards can be determined in one of two ways: If the mixture has been tested as a whole, the results of those tests must be used to determine whether the mixture is hazardous. If the mixture has not been tested as a whole, the mixture will be assumed to have the same hazards as any component present at 1% or greater. If a carcinogenic material is present at levels greater than 0.1%, the mixture must be treated as a carcinogen.

An employer has the option of accepting the information provided with the product or of doing an independent hazard determination. One problem with this approach is that complete formulation or toxicological information is needed for an adequate hazard determination, and the determination should be done by a toxicologist or industrial hygienist. Few metalworking fluid users have either. For that matter, few metalworking fluid suppliers have toxicologists or industrial hygienists on their staffs; they are dependent on the information given them by their suppliers for their own hazard determinations.

Producers or employers must describe in writing the procedures used to determine the hazards of chemicals. The procedures must be provided to any employee requesting them or to OSHA on demand.

TABLE 2 Chemical Hazards Regulated by OSHA

Physical hazards	Health hazards
Combustible	Carcinogens[a]
Compressed gases	Toxic chemicals[a]
Explosives	Highly toxic chemicals[a]
Flammables	Reproductive toxins
Organic peroxides	Irritants
Oxidizers	Corrosives[a]
Pyrophorics	Sensitizers
Unstable chemicals	Hepatotoxins
Water-reactive chemicals	Nephrotoxins
	Neurotoxins
	Hematopoietic toxins
	Chemicals that damage the lungs, skin, eyes, or mucous membranes

[a]Definitions for these hazards are given in the Regulations.

B. Written Hazard Communication Programs

Employers must develop, implement, and maintain a written hazard communication program for their workplaces which describes how the requirements for labels, MSDSs, and employee training will be met. In addition, it must include a list of all hazardous chemicals in the workplace using an identity referenced on the appropriate MSDS and the methods to be used in training employees in the hazards of nonroutine tasks and hazards associated with chemicals in unlabeled pipes in the workplace.

The hazard communication program must be made available to employees or to OSHA on request.

C. Warning Labels

Manufacturers must ensure that each container leaving their workplace is labeled, tagged, or marked with

The identity of the hazardous chemical,
The appropriate hazard warnings,
The name and address of the manufacturer.

These label requirements are independent of any labels required under the Hazardous Materials Transportation Act.

The final user may not remove or deface these warning labels unless the container is immediately marked with the required information.

Labels must be prominently displayed on the container and must be in English. Information may be presented in other languages as long as it is also present in English.

D. Material Safety Data Sheets

Since their introduction in 1985, the role of the MSDS has expanded greatly beyond the requirements of the Hazard Communication Standard. The simple two-page fill-in-the-blank form that OSHA suggested has been replaced by a comprehensive document addressing general health and safety, plus all other regulatory information which might affect the product. A modern MSDS may consist of more than 20 pages.

A recent survey by OSHA estimated that 16 years of education was needed to understand the average MSDS. While this may be acceptable in the chemical industry where employees have a formal technical education or job training, the average machine tool operator does not have these advantages and needs information in a form useful to someone with a high school education. At the same time, there will be readers who need information at a level appropriate for engineers or health professionals. Further, the MSDS must properly distinguish between the

Government Regulations

undiluted product as sold and the product as diluted for use. Clearly, the preparer of the MSDS for a metalworking fluid must tailor the information in each part of the document to meet the needs of a range of users. The Chemical Manufacturer's Association has submitted a proposed standard for MSDSs to the American National Standards Institute (ANSI), which addresses these issues.

In any event, under the Hazard Communication Standard, the MSDS must be in English and must include certain minimal information:

> The identity of the chemical. This must match the name on the container label and, in the case of metalworking fluids, will be the trade name.
> The names of all hazardous ingredients. CAS numbers and percentages of ingredients may also be included, at the option of the preparer. The identities of hazardous proprietary materials may be withheld, if it is noted that information has been withheld and a generic name for the substance is included. The rules covering trade secret information will be discussed in more detail below.
> Physical and chemical characteristics of the chemical.
> Physical hazards, including the potential for fire, explosion, and reactivity.
> Health hazards, including signs and symptoms of exposure and any medical conditions which are generally recognized as being aggravated by exposure to the product.
> The primary routes of exposure. For metalworking fluids, the primary routes of exposure would be through skin contact and inhalation of mists from the diluted product.
> Exposure limits (PEL, TLV, STEL, or other limits) that may apply to any component of the fluid. Table 1 shows PELs for those common fluid constituents for which limits have been established.
> Whether any fluid component has been listed in the National Toxicology Program *Annual Report on Carcinogens* or has been found to be a potential carcinogen in the latest editions of the International Agency for Research on Cancer *Monographs* or by OSHA.
> Any generally applicable precautions for safe handling or use, including hygiene practices and procedures for cleanup of spills and leaks.
> Control measures, such as engineering controls, work practices, or personal protective equipment.
> Emergency and first aid procedures.
> The date of preparation of the MSDS or the date of the last revision.
> The name, address, and phone number of the party preparing the sheet who can provide additional information in case of emergency.

Where no information can be found in any of these mandatory areas, this information must be stated; no blank entries are permitted.

The manufacturer must repeat the hazard determination at least every two years; any MSDS older than two years should be considered suspect. If new information about a significant hazard of a chemical is discovered, a new MSDS must be issued within three months.

An MSDS must accompany each initial shipment of a product to a new customer or distributor. In addition, an MSDS must be sent with the first shipment each calendar year for those products covered under the reporting provisions of SARA Section 313.

Employers must maintain copies of MSDSs for each hazardous chemical in their workplace and must ensure that they are available to employees in their work areas on all shifts. While not specifically incorporated into the regulations, OSHA's policy has been that sheets must be given to employees within 24 h of a request. MSDSs must also be available to OSHA on request.

While OSHA has defined and established the MSDS, the scope of the document has expanded greatly since 1985. Users now expect to find information on the regulatory status of a product beyond OSHA requirements. Inclusion of community right-to-know information was mandated by Congress under the terms of SARA. Information concerning RCRA, CERCLA, and DOT requirements are conveniently distributed by this medium, as well as information about state right-to-know disclosures.

E. Employee Information and Training

Employees must be provided with information and training about the hazardous chemicals in their work area at the time of their initial assignment and whenever a new hazard is introduced into their work area.

The information given to the employees must include:

The requirements of the Hazard Communication Standard
Any operations in their work area where hazardous chemicals are present
The location and availability of the employer's written hazard communication program, the required list of hazardous chemicals and the location of the MSDSs for the hazardous chemicals.

Minimal employee training must include:

Methods and observations that may be used to detect the presence or release of a hazardous chemical in the work area. This may include periodic monitoring, continuous monitoring devices, visual appearances, or odors.
The physical and health hazards of the chemicals in the work area.
Protective measures available to employees, including specific procedures, work practices, emergency procedures, and personal protective equipment to be used.
The details of the employer's hazard communication program, including an

explanation of the label system, the MSDSs, and how to obtain and use the hazard information.

F. Trade Secrets

Specific chemical identities of hazardous chemicals may be withheld from the MSDS, provided that:

The claim that the information is a trade secret can be supported.
Information concerning the properties and effects of the hazardous chemical is included.
The MSDS indicates that the information has been withheld.

The chemical identity must be made available under specified conditions. When a treating physician or nurse determines that a medical emergency exists and the identity of the chemical is needed for emergency or first aid treatment, the information must be provided immediately. The physician or nurse must sign a confidentiality agreement as soon as circumstances permit.

In nonemergency situations, the manufacturer must disclose the trade secret information to a health professional (physician, industrial hygienist, toxicologist, epidemiologist, or occupational health nurse), if

The request is in writing.
The request describes with "reasonable detail" one or more of the following occupational health needs for the information:
To assess the hazards of chemicals to which employees might be exposed.
To conduct or evaluate sampling to determine employee exposure levels.
To conduct preassignment or periodic medical exams of exposed employees.
To select or evaluate personal protective equipment for exposed employees.
To design engineering controls or other protective measures.
To conduct studies to determine the health effects of exposure.
The request must explain "in detail" why disclosure of the chemical identity is essential and why the need could not be met by disclosure of:
The properties and effects of the chemical.
Measures for controlling workers' exposure.
Methods of monitoring and analyzing worker exposure.
Methods of diagnosing and treating harmful exposures.
The request must include a description of the procedures to be used to maintain the confidentiality of the information.
The health professional must agree in a written confidentiality agreement not to use the trade secret information for any purpose other than health needs

or to release the information to any third party other than the OSHA, except as authorized by the supplier.

Provisions of the confidentiality agreement may include a restriction of the use of the information to the purposes indicated in the request, and may include provision for legal remedies in the event of a breach of the agreement, but may not include a requirement for posting of a penalty bond.

If the health professional determines that the information must be provided to OSHA, the owner of the information must be informed prior to or "at the same time as" the disclosure.

The manufacturer may deny the request for the information. The denial must be provided, in writing, to the requester within 30 days, must include evidence that the information is a valid trade secret, state the specific reasons why the request is being denied, and explain in detail how alternative information may satisfy the specific medical or occupational health need without revealing the chemical identity.

The requester, in turn, may refer the request and the denial to OSHA for consideration. OSHA will then consider if:

The trade secret claim has been supported.
The claim that the information is needed has been supported.
Adequate means to protect the confidentiality of the information has been demonstrated.

If OSHA determines that the information is not a bona fide trade secret or that the health professional has signed the confidentiality agreement and shown the ability to protect the information, the supplier will be subject to citation by OSHA. If OSHA determines that a confidentiality agreement would not provide adequate protection for the information, it may impose additional restrictions on disclosure to provide that protection.

VI. EMERGENCY PLANNING AND COMMUNITY RIGHT-TO-KNOW ACT

Just as the OSHA Hazard Communication Standard is based on the assumption that a worker has the right to know what chemical hazards are present in his workplace, the Emergency Planning and Community Right-To-Know Act, also known as Title III of the Superfund Amendment and Reauthorization Act (SARA), is based on the assumption that communities have a similar right to know what chemical hazards might be present in their neighborhoods. Beyond that premise, the Act requires communities and states to collect the information required from industry and to use it as a base for emergency response planning.

A. Chemical Hazards

Three classes of chemical hazards are defined under this law. *Extremely hazardous substances* are defined as any chemical identified by the EPA in 40 CFR Part 355 Appendix A.

Hazardous chemicals are defined as any chemical for which an MSDS is required under the Hazard Communication Standard by OSHA. These chemicals are subject to the requirements of Sections 311 and 312.

Toxic chemicals are those chemicals on the list published in 40 CFR Part 372.65. This list originally consisted of those chemicals which were considered hazardous by the states of New Jersey and Maryland. These chemicals are regulated under Section 313.

In general, extremely hazardous chemicals are not known to be used in blending metalworking fluids. However, since most fluids are required to have an MSDS, manufacturers, distributors, and users of metalworking fluids may be subject to the rules under Sections 311 and 312. In addition, some commonly used fluid components are listed under Section 313 (see Table 1); manufacturers and users may be subject to reporting requirements for those chemicals.

B. Emergency Response Groups

Under Section 301, each state has been required to establish a State Emergency Response Commission (SERC). This commission, and the others established under the Act, must include, at a minimum, representatives from each of the following groups or organizations: elected state and local officials; law enforcement, civil defense, firefighting, first aid, health, local environmental, hospital, and transportation personnel; broadcast and print media; community groups; and owners and operators of regulated facilities. The SERC was charged with the appointment of local emergency planning committees (LEPCs) and the supervision and coordination of the activities of those committees. Further, they were responsible for the designation of emergency planning districts.

Each LEPC was made responsible for the establishment of rules providing for public notification of committee activities, public meetings to discuss the emergency plan, public comments, response to such comments by the committee, and distribution of the emergency plan. Further, they were to establish procedures for receiving and responding to requests from the public for information which had been submitted to the committee.

Any facility which is required to respond to the reporting requirements of the Act are required to notify their SERC of this fact (Fig. 1). They are further required to designate a facility emergency response coordinator (FERC) and identify that individual to the SERC. The FERC is responsible for submitting all reports to EPA, the SERC, the LEPC, and the local fire department.

C. SARA Section 311, 312 Reports

Any facility which is required to prepare or have on hand an MSDS for a hazardous chemical is subject to the reporting requirements of 40 CFR Part 370.20.

Under 40 CFR Part 370.21, each facility is required to submit to the SERC, LEPC, and fire department Tier 1 data, a list of all hazardous chemicals on the premises, the chemical or common name for each chemical as given on the MSDS, and any hazardous component of each hazardous chemical identified on the MSDS. An exception to the last requirement is made under 40 CFR Part 370.28(a)(2), which provides the option of providing the information for the mixture itself, rather than the individual ingredients. This option is greatly to be preferred for metalworking fluids, given the complexity of the formulations and the proprietary issues.

The regulations do provide that an MSDS may be required for each hazardous chemical instead of the list of chemicals. This option seems to be preferred by most fire departments, even though this means that the same MSDS may be submitted by several facilities in the jurisdiction, resulting in duplication and increased needs for file space. When this option has been chosen, revised MSDSs must be submitted within 90 days of receipt from the supplier.

Each year, before March 1, any facility in Standard Industrial Classification Codes (SIC) 20–39 must submit a Tier 2 report (Fig. 1) covering all extremely hazardous or hazardous chemicals present at the facility during the preceding calendar year in amounts equal to or greater than 10,000 lb. The report must include the average and maximum daily inventories, location, and storage conditions for each chemical. Inventory levels are reported as codes, where a number representing an inventory range is entered on the form.

The fire department has the authority to inspect any facility submitting an inventory report and can obtain specific location information on hazardous chemicals at the facility.

These reports, and all other reports under SARA, are made available to the public. The LEPC is required to advertise annually that this information is available and to explain to the public how access may be made to the reports. Specifically, any person may obtain a copy of any MSDS from a facility or the inventory report data by submitting a written request to the LEPC. All information will be made available to the requester unless the facility operator has asked the LEPC to withhold from disclosure the specific location of any specific chemical. This is of minor importance to a metalworking fluid user, but has the effect of making a substantial part of the ingredient list for formulators part of the public domain.

D. SARA Section 313 Reports

40 CFR Part 372.65 identifies those chemicals and chemical categories which are subject to the provisions of SARA Section 313. Of immediate importance to metalworking fluid makers and formulators are two listed chemicals, diethanolam-

Government Regulations

Figure 2 SERC emergency and hazardous chemical inventory form.

ine and o-phenylphenol, and two chemical categories, barium compounds and glycol ethers. These substances are not more or less toxic than other unlisted chemicals, they are included because they were part of the initial lists specified by Congress in the original Act.

The glycol ethers are of importance because the EPA defines glycol ethers as any chemical with a molecular formula of the type

$$R\text{-}(OCH_2CH_2)_n\text{-}OH \quad \text{or} \quad Ar\text{-}(OCH_2CH_2)_n\text{-}OH$$

where R is any alkyl group, Ar is any aryl group, and $n \leq 3$ as a glycol ether. Under this definition, portions of many of the ethylene oxide adduct surfactants used to blend metalworking fluids are considered to be glycol ethers and are reportable.

o-Phenylphenol is a registered fungicide used to treat metalworking fluids. Its use now triggers a possible reporting situation and, although use levels are unlikely to exceed the threshold levels for reporting, the user does have the obligation to document uses and releases to defend a decision to report or not report.

E. Notification About Toxic Chemicals

Any person who manufactures, imports, or processes a toxic chemical and sells or otherwise distributes a mixture or trade name product containing the toxic chemical must notify each person to whom the mixture or product is sold in writing that

> The produce contains a toxic chemical and is subject to the reporting requirements of SARA Section 313 and 40 CFR Part 372.
> The name of each toxic chemical and its CAS number, if applicable.
> The percent by weight of each toxic chemical in the product.

This notification must be provided with at least the first shipment of the product to a customer and must be made annually with the first shipment made each year. If the product formula is revised to remove the toxic chemical or to change the percentage of the chemical, each recipient must be notified in writing with the first shipment received. When an MSDS is required for the product, the written notification must be attached to the MSDS or made a part of the MSDS. If notification is attached, the notice must include clear instructions that the notifications must not be removed and that the notification must be included in any copies of the MSDS.

If the distributor of the product considers the identity of the chemical to be a trade secret, the notice shall contain a generic chemical name that is descriptive of the toxic chemical. If the percentage of the chemical is a trade secret, an upper limit of content may be given. The upper bound must be no larger than necessary to protect the trade secret.

F. Reporting Requirements

Any facility that meets all the following requirements for a calendar year is a covered facility for that year and is subject to reporting. The requirements are that:

The facility has 10 or more employees.
The facility has a primary SIC code of 20–39.
The facility manufactured, processed, or otherwise used a toxic chemical in excess of the applicable threshold quantity.

The threshold quantities are 25,000 lb for manufacturers or processors and 10,000 lb for facilities where a toxic chemical is used. For the user of metalworking fluids, it is necessary to review the MSDSs for all fluids purchased during a calendar year, calculate the toxic chemical level for each fluid from the purchased amount and the weight percent on the MSDS, and total the amount for all fluids used. If the aggregate sum is greater than 10,000 lb, releases of that chemical must be reported.

For each reportable chemical or chemical category, the facility must fill out and submit Form R (Fig. 3). The basic information required is the amounts of each reportable chemical released to the environment for the year. Estimates are made for releases to waterways, the air, and to the land. For metalworking fluids, the most common form of release will be to waterways, either via a publicly owned treatment works (POTW) or an on-site treatment facility. Some releases to air or land disposal may occur under circumstances peculiar to a given site.

In the case of the barium compound category (or other reportable metal compound categories), the threshold calculation must be made using the total amount of all members of the category used, but the amount reported as released should reflect only the actual amount of barium (or other metal) released.

Reports must be filed each year prior to July 1 with the EPA, SERC, and LEPC. Again, the reports are available to the public. In some areas, there is competition between industry and environmental groups to be the first to be able to release the data to the press with the "correct" interpretation to guide the public reaction. The industry response emphasizes the reductions realized since the last report; the environmentalists focus on the gross number of pounds released and the trickery involved in calculating the reductions. The original intent of Congress was that EPA would use the data to calculate a mass balance for the toxic chemicals within five years (1992) which would be used to guide future regulations. This goal is far from realization.

G. Trade Secret Protection

Provision is made for withholding information as trade secrets by a complicated process involving review of the claim by the EPA, appeals by the applicant or the

	Form Approved OMB Number: 2070-0093	
(IMPORTANT: Type or print; read instructions before completing form)	Approval Expires: 11/92	Page 1 of 9

⊕EPA
United States
Environmental Protection
Agency

FORM R TOXIC CHEMICAL RELEASE
INVENTORY REPORTING FORM
Section 313 of the Emergency Planning and Community Right-to-Know Act of 1986,
also known as Title III of the Superfund Amendments and Reauthorization Act

TRI FACILITY ID NUMBER

Toxic Chemical, Category, or Generic Name

WHERE TO SEND COMPLETED FORMS:
1. EPCRA Reporting Center
P.O. Box 3348
Merrifield, VA 22116-3348
ATTN: TOXIC CHEMICAL RELEASE INVENTORY

2. APPROPRIATE STATE OFFICE
(See instructions in Appendix F)

Enter "X" here if this is a revision

IMPORTANT: See instructions to determine when "Not Applicable (NA)" boxes should be checked.

For EPA use only

PART I. FACILITY IDENTIFICATION INFORMATION

SECTION 1.	SECTION 2. TRADE SECRET INFORMATION
	Are you claiming the toxic chemical identified on page 3 trade secret?
REPORTING YEAR	2.1 ☐ Yes (Answer question 2.2; Attach substantiation forms) ☐ No (Do not answer 2.2; Go to Section 3)
19 ___	2.2 If yes in 2.1, is this copy: ☐ Sanitized ☐ Unsanitized

SECTION 3. CERTIFICATION (Important: Read and sign after completing all form sections.)

I hereby certify that I have reviewed the attached documents and that, to the best of my knowledge and belief, the submitted information is true and complete and that the amounts and values in this report are accurate based on reasonable estimates using data available to the preparers of this report.

Name and official title of owner/operator or senior management official

Signature | Date Signed

SECTION 4. FACILITY IDENTIFICATION

4.1

Facility or Establishment Name	TRI Facility ID Number
Street Address	
City	County
State	Zip Code
Mailing Address (if different from street address)	
City	
State	Zip Code	PUT LABEL HERE

EPA Form 9350-1 (Rev. 12/4/92) - Previous editions are obsolete.

Figure 3 U.S. EPA SARA report, Form R, p. 1.

Government Regulations

Page 2 of 9

EPA FORM R
PART I. FACILITY IDENTIFICATION INFORMATION (CONTINUED)

⊕EPA United States Environmental Protection Agency

TRI FACILITY ID NUMBER

Toxic Chemical, Category, or Generic Name

SECTION 4. FACILITY IDENTIFICATION (Continued)

4.2	This report contains information for: (Important: check only one)	a. ☐ An entire facility	b. ☐ Part of a facility

4.3	Technical Contact	Name	Telephone Number (include area code)
4.4	Public Contact	Name	Telephone Number (include area code)

4.5	SIC Code (4-digit)	a.	b.	c.	d.	e.	f.

4.6	Latitude and Longitude	Latitude			Longitude		
		Degrees	Minutes	Seconds	Degrees	Minutes	Seconds

4.7	Dun & Bradstreet Number(s) (9 digits)	a.
		b.

4.8	EPA Identification Number(s) (RCRA I.D. No.) (12 characters)	a.
		b.

4.9	Facility NPDES Permit Number(s) (9 characters)	a.
		b.

4.10	Underground Injection Well Code (UIC) I.D. Number(s) (12 digits)	a.
		b.

SECTION 5. PARENT COMPANY INFORMATION

5.1	Name of Parent Company ☐ NA
5.2	Parent Company's Dun & Bradstreet Number ☐ NA (9 digits)

EPA Form 9350-1 (Rev. 12/4/92) - Previous editions are obsolete.

♻EPA
United States
Environmental Protection
Agency

EPA FORM R
PART II. CHEMICAL-SPECIFIC INFORMATION

TRI FACILITY ID NUMBER

Toxic Chemical, Category, or Generic Name

SECTION 1. TOXIC CHEMICAL IDENTITY
(Important: DO NOT complete this section if you complete Section 2 below.)

1.1 CAS Number (Important: Enter only one number exactly as it appears on the Section 313 list. Enter category code if reporting a chemical category.)

1.2 Toxic Chemical or Chemical Category Name (Important: Enter only one name exactly as it appears on the Section 313 list.)

1.3 Generic Chemical Name (Important: Complete **only** if Part I, Section 2.1 is checked "yes." Generic Name must be structurally descriptive.)

SECTION 2. MIXTURE COMPONENT IDENTITY
(Important: DO NOT complete this section if you complete Section 1 above.)

2.1 Generic Chemical Name Provided by Supplier (Important: Maximum of 70 characters, including numbers, letters, spaces, and punctuation.)

SECTION 3. ACTIVITIES AND USES OF THE TOXIC CHEMICAL AT THE FACILITY
(Important: Check all that apply.)

3.1 Manufacture the toxic chemical:
a. ☐ Produce
b. ☐ Import

If produce or import:
c. ☐ For on-site use/processing
d. ☐ For sale/distribution
e. ☐ As a byproduct
f. ☐ As an impurity

3.2 Process the toxic chemical:
a. ☐ As a reactant
b. ☐ As a formulation component
c. ☐ As an article component
d. ☐ Repackaging

3.3 Otherwise use the toxic chemical:
a. ☐ As a chemical processing aid
b. ☐ As a manufacturing aid
c. ☐ Ancillary or other use

SECTION 4. MAXIMUM AMOUNT OF THE TOXIC CHEMICAL ON-SITE AT ANY TIME DURING THE CALENDAR YEAR

4.1 ☐ (Enter two-digit code from instruction package.)

EPA Form 9350-1(Rev. 12/4/92) - Previous editions are obsolete.

Figure 3 *(continued)*

EPA FORM R
PART II. CHEMICAL-SPECIFIC INFORMATION (CONTINUED)

Page 4 of 9

TRI FACILITY ID NUMBER

Toxic Chemical, Category, or Generic Name

United States Environmental Protection Agency

SECTION 5. RELEASES OF THE TOXIC CHEMICAL TO THE ENVIRONMENT ON-SITE

			A. Total Release (pounds/year) (enter range code from instructions or estimate)	B. Basis of Estimate (enter code)	C. % From Stormwater
5.1	Fugitive or non-point air emissions	☐ NA			
5.2	Stack or point air emissions	☐ NA			
5.3	Discharges to receiving streams or water bodies (enter one name per box)				
5.3.1	Stream or Water Body Name				
5.3.2	Stream or Water Body Name				
5.3.3	Stream or Water Body Name				
5.4	Underground injections on-site	☐ NA			
5.5	Releases to land on-site				
5.5.1	Landfill	☐ NA			
5.5.2	Land treatment/ application farming	☐ NA			
5.5.3	Surface impoundment	☐ NA			
5.5.4	Other disposal	☐ NA			

☐ Check here only if additional Section 5.3 information is provided on page 5 of this form.

EPA Form 9350-1 (Rev. 12/4/92) - Previous editions are obsolete.

Range Codes: A = 1 - 10 pounds; B = 11 C = 500 - 999 pounds.

&EPA
United States
Environmental Protection
Agency

EPA FORM R
PART II. CHEMICAL-SPECIFIC
INFORMATION (CONTINUED)

Page 5 of 9

TRI FACILITY ID NUMBER

Toxic Chemical, Category, or Generic Name

SECTION 5.3 ADDITIONAL INFORMATION ON RELEASES OF THE TOXIC CHEMICAL TO THE ENVIRONMENT ON-SITE

5.3	Discharges to receiving streams or water bodies (enter one name per box)	A. Total Release (pounds/ year) (enter range code from instructions or estimate)	B. Basis of Estimate (enter code)	C. % From Stormwater
5.3.__	Stream or Water Body Name			
5.3.__	Stream or Water Body Name			
5.3.__	Stream or Water Body Name			

SECTION 6. TRANSFERS OF THE TOXIC CHEMICAL IN WASTES TO OFF-SITE LOCATIONS

6.1 DISCHARGES TO PUBLICLY OWNED TREATMENT WORKS (POTW)

6.1.A Total Quantity Transferred to POTWs and Basis of Estimate

6.1.A.1 Total Transfers (pounds/year) (enter range code or estimate)	6.1.A.2 Basis of Estimate (enter code)

6.1.B POTW Name and Location Information

6.1.B.__	POTW Name	6.1.B.__	POTW Name
Street Address		Street Address	
City	County	City	County
State	Zip Code	State	Zip Code

If additional pages of Part II, Sections 5.3 and/or 6.1 are attached, indicate the total number of pages in this box [ ] and indicate which Part II, Sections 5.3/6.1 page this is, here. [ ]
(example: 1, 2, 3, etc.)

EPA Form 9350-1 (Rev. 12/4/92) - Previous editions are obsolete.   Range Codes: A = 1 - 10 pounds; B = 11
C = 500 - 999 pounds.

Figure 3 (continued)

EPA FORM R
PART II. CHEMICAL-SPECIFIC INFORMATION (CONTINUED)

TRI FACILITY ID NUMBER

Toxic Chemical, Category, or Generic Name

SECTION 6.2 TRANSFERS TO OTHER OFF-SITE LOCATIONS

6.2. ___ Off-site EPA Identification Number (RCRA ID No.)

Off-Site Location Name

Street Address

City | County

State | Zip Code | Is location under control of reporting facility or parent company? ☐ Yes ☐ No

A. Total Transfers (pounds/year) (enter range code or estimate)	B. Basis of Estimate (enter code)	C. Type of Waste Treatment/Disposal/ Recycling/Energy Recovery (enter code)
1.	1.	1. M
2.	2.	2. M
3.	3.	3. M
4.	4.	4. M

SECTION 6.2 TRANSFERS TO OTHER OFF-SITE LOCATIONS

6.2. ___ Off-site EPA Identification Number (RCRA ID No.)

Off-Site Location Name

Street Address

City | County

State | Zip Code | Is location under control of reporting facility or parent company? ☐ Yes ☐ No

A. Total Transfers (pounds/year) (enter range code or estimate)	B. Basis of Estimate (enter code)	C. Type of Waste Treatment/Disposal/ Recycling/Energy Recovery (enter code)
1.	1.	1. M
2.	2.	2. M
3.	3.	3. M
4.	4.	4. M

If additional pages of Part II, Section 6.2 are attached, indicate the total number of pages in this box ☐ and indicate which Part II, Section 6.2 page this is, here. ☐ (example: 1, 2, 3, etc.)

EPA Form 9350-1 (Rev. 12/4/92) - Previous editions are obsolete.

Range Codes: A = 1 - 10 pounds; B = 11
C = 500 - 999 pounds.

EPA FORM R
PART II. CHEMICAL-SPECIFIC INFORMATION (CONTINUED)

United States Environmental Protection Agency

Page 7 of 9

TRI FACILITY ID NUMBER

Toxic Chemical, Category, or Generic Name

SECTION 7A. ON-SITE WASTE TREATMENT METHODS AND EFFICIENCY

☐ Not Applicable (NA) - Check here if no on-site waste treatment is applied to any waste stream containing the toxic chemical or chemical category.

a. General Waste Stream (enter code)	b. Waste Treatment Method(s) Sequence [enter 3-character code(s)]			c. Range of Influent Concentration	d. Waste Treatment Efficiency Estimate	e. Based on Operating Data?
7A.1a	7A.1b 1 ☐ 2 ☐ 3 ☐ 4 ☐ 5 ☐ 6 ☐ 7 ☐ 8 ☐			7A.1c	7A.1d %	7A.1e Yes ☐ No ☐
7A.2a	7A.2b 1 ☐ 2 ☐ 3 ☐ 4 ☐ 5 ☐ 6 ☐ 7 ☐ 8 ☐			7A.2c	7A.2d %	7A.2e Yes ☐ No ☐
7A.3a	7A.3b 1 ☐ 2 ☐ 3 ☐ 4 ☐ 5 ☐ 6 ☐ 7 ☐ 8 ☐			7A.3c	7A.3d %	7A.3e Yes ☐ No ☐
7A.4a	7A.4b 1 ☐ 2 ☐ 3 ☐ 4 ☐ 5 ☐ 6 ☐ 7 ☐ 8 ☐			7A.4c	7A.4d %	7A.4e Yes ☐ No ☐
7A.5a	7A.5b 1 ☐ 2 ☐ 3 ☐ 4 ☐ 5 ☐ 6 ☐ 7 ☐ 8 ☐			7A.5c	7A.5d %	7A.5e Yes ☐ No ☐

If additional copies of page 7 are attached, indicate the total number of pages in this box ☐ and indicate which page 7 this is, here. ☐ (example: 1, 2, 3, etc.)

EPA Form 9350-1 (Rev. 12/4/92) - Previous editions are obsolete.

Figure 3 *(continued)*

Government Regulations

⇔EPA EPA FORM R
United States
Environmental Protection **PART II. CHEMICAL-SPECIFIC**
Agency **INFORMATION (CONTINUED)**

Page 8 of 9

TRI FACILITY ID NUMBER

Toxic Chemical, Category, or Generic Name

SECTION 7B. ON-SITE ENERGY RECOVERY PROCESSES

☐ Not Applicable (NA) - Check here if <u>no</u> on-site energy recovery is applied to any waste stream containing the toxic chemical or chemical category.

Energy Recovery Methods [enter 3-character code(s)]

1. _____   2. _____   3. _____   4. _____

SECTION 7C. ON-SITE RECYCLING PROCESSES

☐ Not Applicable (NA) - Check here if <u>no</u> on-site recycling is applied to any waste stream containing the toxic chemical or chemical category.

Recycling Methods [enter 3-character code(s)]

1. _____   2. _____   3. _____   4. _____   5. _____
6. _____   7. _____   8. _____   9. _____   10. _____

EPA Form 9350-1 (Rev. 12/4/92) - Previous editions are obsolete.

⊕EPA
United States
Environmental Protection
Agency

EPA FORM R
PART II. CHEMICAL-SPECIFIC INFORMATION (CONTINUED)

TRI FACILITY ID NUMBER

Chemical, Category, or Generic Name

SECTION 8. SOURCE REDUCTION AND RECYCLING ACTIVITIES

All quantity estimates can be reported using up to two significant figures.	Column A 1991 (pounds/year)	Column B 1992 (pounds/year)	Column C 1993 (pounds/year)	Column D 1994 (pounds/year)
8.1 Quantity released *				
8.2 Quantity used for energy recovery on-site				
8.3 Quantity used for energy recovery off-site				
8.4 Quantity recycled on-site				
8.5 Quantity recycled off-site				
8.6 Quantity treated on-site				
8.7 Quantity treated off-site				

8.8	Quantity released to the environment as a result of remedial actions, catastrophic events, or one-time events not associated with production processes (pounds/year)
8.9	Production ratio or activity index
8.10	Did your facility engage in any source reduction activities for this chemical during the reporting year? If not, enter "NA" in Section 8.10.1 and answer Section 8.11.

	Source Reduction Activities [enter code(s)]	Methods to Identify Activity (enter codes)		
8.10.1		a.	b.	c.
8.10.2		a.	b.	c.
8.10.3		a.	b.	c.
8.10.4		a.	b.	c.
8.11	Is additional optional information on source reduction, recycling, or pollution control activities included with this report? (Check one box)		YES ☐	NO ☐

* Report releases pursuant to EPCRA Section 329(8) including "any spilling, leaking, pumping, pouring, emitting, emptying, discharging, injecting, escaping, leaching, dumping, or disposing into the environment." Do not include any quantity treated on-site or off-site.

EPA Form 9350 - 1 (Rev. 12/4/92) - Previous editions are obsolete.

Figure 3 *(continued)*

public and ultimately, legal determinations by the courts. There is a provision in the law for fines of up to $25,000 for "frivolous" trade secret claims.

H. Citizen Suits

Under SARA Section 326, any person may commence a civil action on his own behalf against

> An owner or operator of any facility who fails to comply with Sections 304(c), 311, 312, or 313.
> EPA for failure to respond to a petition to add or delete a toxic chemical under Section 313(3)(1) within 180 days of receipt of the petition, or render a decision on a petition under Section 322(d) within 9 months after receipt of the petition.
> A state governor or SERC for failure to respond to a request for Tier II information under Section 312(e)(3) within 120 days after receipt of the petition.

Any state or local government may commence a civil action against a facility for failure to

> Provide notification to the SERC under Section 302(c).
> Submit an MSDS or list of chemicals under Section 311(a).
> Make available information requested under Section 311(c).
> Complete and submit an inventory form under Section 312(a) containing Tier 1 information.

Any SERC or LEPC may commence a civil action against a facility for failure to provide information under Section 303(d) or Section(e)(1). Any state may commence a civil action against EPA for failure to provide information to the state under Section 322(g). The courts may, if they choose, include the costs of litigation of the prevailing party in any settlement.

VII. RESOURCE CONSERVATION AND RECOVERY ACT

The Resource Conservation and Recovery Act (RCRA) is the legislation regulating the disposal of solid wastes. The initial reaction would seem to be that solid wastes and metalworking fluids would be mutually exclusive subjects, but the power to create regulatory definitions goes beyond the constraints of technical rigor. The regulations (40 CFR Part 261.2) first define the term "solid waste" as any solid liquid, or contained gas which is discarded, has served its intended purpose, or is a manufacturing by-product. Certain classes of materials are excluded from this definition (domestic sewage, Clean Water Act point source discharges, irrigation return flow, Atomic Energy Commission nuclear or by-product materials, and in situ mining wastes) because they are otherwise regulated; all other industrial wastes, along with garbage, refuse, and sludge are RCRA solid wastes. This definition would include metal chips and grinding swarf as well as spent fluids. (Under the exclusion for domestic sewage, fluids which are discharged for treatment by a publicly owned treatment works (POTW) are not RCRA wastes.)

RCRA wastes are hazardous if they are listed under 40 CFR Part 261 Subpart D, contain a listed waste or are present in a mixture that contains a listed waste, exhibit any characteristic of a hazardous waste, or are specific discarded commercial chemicals.

Listed wastes are divided into two groups: those from nonspecific sources and those from specific sources. The industries listed as specific sources are wood preservation, inorganic pigments, organic chemicals, pesticides, explosives, petroleum refining, leather tanning and finishing, iron and steel, primary copper, primary lead, primary zinc, and secondary lead. Any waste metalworking fluids generated by one of these industries has the potential to be considered as a hazardous waste.

Among the wastes from nonspecific sources are spent chlorinated solvents which had been used for degreasing operations. Thus, any metalworking fluid which had been contaminated by degreasing solvents would become a hazardous waste.

The characteristics of a hazardous waste are defined as ignitability (flash point < 140°F), corrosive (pH values less than 2.0 or greater than 12.5), reactive, or toxic. A reactive waste will readily undergo violent chemical change, react violently with water, generate toxic fumes when mixed with water or mild acids or bases, or is explosive.

A solid waste is toxic if, using the test methods described in Appendix II of 40 CFR Part 261, an extract (in the case of liquid wastes, the liquid itself is considered to be the extract) of the waste contains any of the substances listed under 40 CFR Part 261.24 (Table 1) at concentrations equal or greater than the value given in the table. A listed material that could be present as a component of a metalworking fluid is barium, with a regulatory level of 100 mg/l. In addition, listed materials that could contaminate a fluid in some applications are cadmium (1.0 mg/l), chromium (5.0 mg/l), lead (5.0 mg/l), tetrachloroethylene (0.7 mg/l), or trichloroethylene (0.5 mg/l). When these could be in the waste fluid, it should be tested before disposal.

A number of substances are further identified under 40 CFR Part 261.33(f) as toxic wastes. Two of these, N-nitrosodiethanolamine and p-chloro-m-cresol, are of interest with regard to metalworking fluids. The nitrosamine, of course, should not be present in fluids which do not contain nitrites. p-Chloro-m-cresol is used as a biocide in some metalworking fluid concentrates and could be present in wastes from those fluids.

This section of the regulations deals with discarded commercial chemicals and restricts their disposal by spraying on the land. It is intended to be applied to pure materials or mixtures in which the only active ingredient is the listed substance. There is no direct application to metalworking fluids, though this may be a point of confusion. Note that diethanolnitrosamine has never been available as a commercial chemical, either pure or in a mixture. Its appearance on this list would not seem to be relevant in this context and probably reflects more on congressional zeal than on congressional understanding of what was being regulated. Note also that in Table 1, p-chloro-m-cresol will be added to the list of toxic chemicals under SARA Section 313 and is still of regulatory concern to those using it.

A user of metalworking fluids should determine if waste fluids are hazardous so that proper disposal can be made. If the waste stream is excluded because it is discharged to a sanitary sewer, it does not fall under the jurisdiction of RCRA. Any regulatory issues will be raised by the POTW. If the user is in one of the specific source classifications, the waste stream is most likely hazardous. Otherwise, a determination can be made by either testing the wastes or from knowledge of the process. If the fluid concentrate contains a listed material, or if the process contaminates the fluid with a listed toxic material, the waste can be accepted as hazardous without testing. Similarly, if it can be shown that these conditions are not present, the waste can be determined to be nonhazardous without testing. (A treatment facility may still insist on testing.)

If the waste is in fact hazardous, an EPA identity number for the facility must be secured. This identifies the user as a hazardous waste generator. If other hazardous wastes are generated and a number has already been assigned, this requirement is fulfilled. If the user has more than one location generating the waste, each site must get an ID number.

As wastes are generated, they must be stored in designated areas in proper containers. Containers must be labeled to clearly indicate that the contents are hazardous wastes and to show the date that accumulation began. No wastes may be stored for more than 90 days unless the facility has also registered as a hazardous waste storage facility. Few users will be able to justify doing this.

Accumulated wastes must be removed by a certified hazardous waste hauler. Transfer of the wastes is documented by a hazardous waste manifest which shows

- The intended treatment/storage/disposal (T/S/D) facility to which the waste is being sent
- A manifest document number
- The DOT name and common name of the waste
- The total quantity being shipped, the number and type of containers
- Certification of the proper classification, description, packaging, marking, and labeling

File copies of the manifest are created for the generator, hauler, and recipient. A copy of the manifest is to be returned to the generator by the T/S/D which must include the date of disposal of the wastes and the method of disposition. Copies of the manifest must be kept on file by the generator for at least three years. The retention period may be extended in cases of unresolved enforcement actions.

All containers must be packaged, labeled, marked, and placarded in compliance with DOT regulations. Containers of over 110 gal must be marked with the generator's name, address, manifest document number, and the message "Hazardous Waste—Federal law prohibits improper disposal. If found, contact police, public safety authority or EPA."

Registered generators must file annual reports with EPA by March 1, summarizing all hazardous wastes shipped from the site for the preceding year. Copies of these reports must be kept on file for at least three years. This period can be extended in cases of unresolved enforcement actions.

A generator who has not received a copy of a hazardous waste manifest signed by the T/S/D within 35 days after shipping must contact the transporter or the T/S/D to determine what happened to the manifest and the wastes. If, after 45 days, the generator still has not received the manifest, it must submit an exception report to the EPA regional administrator. The report must include a copy of the manifest and a cover letter detailing the efforts made to locate the manifest and the wastes and the results of those efforts. Failure to comply can result in penalties, but the major threat is that should the wastes later be found to have not, in fact, been properly treated, ownership and liability still rests with the generator along with the responsibility for recovery, clean up, and final disposal. If other wastes are also present, the generator may be required to share in the costs of remediation for any other materials which might be involved.

A. Used Oil Disposal

Under the Hazardous and Solid Waste Amendments of 1984, EPA was directed to promulgate regulations that might be needed to protect human health and the environment from hazards associated with recycled oil. Over the next eight years, EPA made a number of attempts to address this directive, resulting in extensive controversy and litigation. One of the issues was the classification of "synthetic oils" and whether the proposed regulations should apply to synthetic metalworking fluids. Further complicating the question was the fact that a number of materials were included in the class of synthetic oils; these are nonpetroleum-based lubricants, water-soluble petroleum-based lubricants, and polymeric lubricants. The final determination was that oils become toxic or hazardous as a result of use and that similar uses make the used materials indistinguishable from used oils, *except in the case of segregated, water-based metalworking oils/fluids*. However, regulations covering used oils will apply to oily sludges resulting from the waste treatment of water-based fluids.

In general, the major restriction on oils from metalworking operations is a limit on chlorine content. EPA has made a rebuttable presumption that any waste oil with a chlorine content greater than 1000 ppm has been contaminated by chlorinated solvents and is a hazardous waste. Generators may rebut that presumption by documenting that the chlorine is actually part of the metalworking fluid formulation and does not result from mixture with degreasing solvents. Even when the rebuttal is successful, many T/S/Ds are reluctant to accept halogen-containing oily wastes.

VIII. BEYOND THE LAWS

Beyond matters of simple compliance with regulations is the area of product liability. One of the principles of civil law has always been that someone who suffers injury through the negligence of another is entitled to recover damages from that party. Recent expansions of this idea by the courts and state legislatures to include the concepts of strict liability and joint-and-several liability has resulted in situations where an injured party can collect damages from someone who was not the negligent party or who had only a minor part in the cause of the injury. The willingness of juries to award ever higher damages, especially punitive damages for alleged injuries, has made product safety an important part of product stewardship. One unsuccessful defense of a lawsuit may erase years of profits for a product or, for a small business, may mean the end of the business. The cost of the defense, whether successful or not, can be greater than the cost of the judgment or settlement.

Machinists are not immune to the general suspicion of chemicals and are likely to attribute an unusual disease or injury to their workplace exposure to the chemicals in metalworking fluids. Allegations against metalworking fluids range from allergic dermatitis to cancer. The causes of allergies can be determined by patch testing; causes of cancer or other diseases are not so easily determined. Successful defenses in such cases consist of presenting a technical argument to a jury without technical or legal training, often through the efforts of an attorney with no technical background, with interpretations of issues made by a judge, also without technical training.

Clearly, product safety should be given as much consideration during development as any other aspect of product performance. Potential hazards must be identified by testing of raw materials or by testing of the final product. Once a hazard is identified, it can be eliminated by reformulation. Where this is not possible, the label must clearly warn against the hazard, the conditions under which the hazard is present, and directions on how to avoid or minimize the hazard.

When reformulating to eliminate a hazard, it is not unusual to be confronted with a choice of conflicting priorities. When treating a rancid system, using a biocide to kill the microorganisms is obvious, but is one biocide safer or more dangerous than another? Is it better to use a biocide that acts through release of formaldehyde, a possible carcinogen, or to use a product that has no known long-term health effects but is corrosive to the skin and may injure those who must handle the biocide during addition? A rational system of making such choices is needed to guide the formulator and to give maximum protection to the final user. Once these choices are made, they must be communicated to the user in a way that convinces everyone that the choices were properly made and that the product is reasonably safe when used as intended.

It is important to document the process by which the hazards are identified and brought to the attention of the user; the details of the process are not as important as the consistency with which it is applied. The documentation of the hazard communication program required by OSHA will also serve to document the product safety review for product liability defense. The choice of warning phrases to be used on the MSDS should be made using a formal written procedure. The temptation to soften warning language to avoid a negative impact on product acceptance should be resisted. Maximizing sales of a product with potential for liability is always poor business practice.

IX. SUMMARY

As can be seen, using metalworking fluids in compliance with all laws is a complicated process for formulators and users. In fact, this discussion is greatly simplified in that it deals only with laws enacted at the federal level. When state, city, and county regulations are also considered, with conflicting requirements and inconsistent definitions of what hazardous materials are, it is almost impossible for

TABLE 3 Where to Call for More Information and Help

Indoor Air Quality Information Clearinghouse	800-438-4318
Storm Water NPDES Permitting Hotline	703-821-4823
RCRA (Hazardous Waste) Ombudsman	800-262-7937
Solid and Hazardous Waste (RCRA), Superfund (CERCLA), and Underground Storage Tanks	800-424-9346
Emergency Planning and Community Right-to-Know	800-424-9346 703-412-9810
Solid Waste Information Clearinghouse (SWICH)	800-677-9424
National Response Center (for reporting oil spills and hazardous substance releases)	800-424-8802 202-267-2675
TSCA Asbestos Information/Retrieval	202-554-1404
CHEMTREC (for assistance in transportation emergencies)	800-262-8200 202-887-1315

Government Regulations

anyone to be in compliance with all laws and regulations at any time. One can only attempt to make the best possible effort and remember that "Murphy" will always be with us.

In anticipation of a time when these general comments may not meet the reader's needs, Table 3 provides a list of federal hot lines where assistance can be obtained. A number of offices are included which deal with topics not covered here.

Glossary

Terms and definitions were contributed by the authors of the chapters to which they pertain.

Abrasion Removal of material by rubbing.
ACGIH American Conference of Governmental Industrial Hygienists.
Acid A substance that releases protons (H^+) in water, lowering the pH.
Acid number A numerical expression of acidity based on the milligrams of potassium hydroxide required to neutralize the free fatty acid in one gram of fat, wax, or resin.
Acute health effects Occur rapidly as a result of short-term, one-time exposures.
Adhesion Molecular attraction between two surfaces in contact.
Adsorption The uptake and physical bonding of the active substance contained in the lubricant on the surface of the solid metal, a mechanism for forming the intermediate layer to reduce metallic contact between two surfaces. This is different from a reaction mechanism in which the intermediate layer is formed through the transformation of the active substances into other chemical compounds (reaction products).
Agar (agar-agar) A polysaccharide extract from seaweed used as a base to prepare a gelatinous media for growth and enumeration of microorganisms.
Air entrainment Air that is held in suspension by a fluid and is slow to rise to the surface.

AISI   American Iron and Steel Institute.
Alkali   Hydroxides of ammonia, lithium, potassium, and sodium are called alkalies (*see* Base). Alkalies raise pH when added to water.
Alkalinity   The concentration of basic or alkaline components in a mixture, determined by titration with an acid.
Allotrophism   A property of metals whereby they exhibit multiple crystal structures at different temperatures. Under equilibrium heating and cooling conditions the change in crystal structure is completely reversible.
Alloy   A mixture of two or more elements, at least one of which is a metal, combined in a solid. The resulting alloy has chemical and mechanical properties that are different than those of the separate elements that make up the alloy.
Alum   The common name for aluminum sulfate or potassium aluminum sulfate, chemical substances frequently used in liquid waste disposal to separate out small particles of suspended matter.
Amide   Organic amides are formed by a condensation reaction of a fatty acid and an amine. Used as emulsifiers and corrosion inhibitors. The amine to fatty acid ratio may be expressed as 1:1 with little or no excess free amine, or 2:1 with some unreacted amine remaining.
Amine   A nitrogen containing organic compound, basic in nature, having the general formula NR_3 where R may be either a hydrocarbon or hydrogen.
Amphoteric   (1) Molecules containing both an acid group (COOH) and a basic group (NR_3). (2) Metals that are attacked by both acids and alkalies, e.g., aluminum and zinc.
Anionic surfactant   A surface active agent that when dissolved in water carries a slight negative charge. These tend to be sensitive to water hardness. Examples are soaps and sulfonates.
Annealing   A conditioning treatment for metal designed to produce a soft final structure, improve ductility, remove stress, and refine the microstructure. It usually involves austenitic transformation followed by slow controlled rate cooling inside a furnace.
Anode   The terminal of an electrolytic cell at which oxidation (loss of electrons) occurs.
Anodizing   An electrochemical treatment applied to aluminum to induce the formation of a hard, tenacious oxide layer. Such a layer imparts corrosion and wear resistance to the aluminum.
Aqueous   Referring to water; an aqueous solution of sugar would be sugar dissolved in water.
Arbor   Spindle on which a cutter is mounted.
ASTM   American Society for Testing and Materials.
Atopic   Inherited tendency to develop an allergy.
Austenite   The allotrophic form of iron that exists above 1670°F. Austenite can dissolve up to 2 wt % carbon and is the precursor structure for most heat treatments.

Bacteria Single-celled, microscopic organisms. They are widespread in our environment and are found in a variety of shapes (round, rod, or spiral). Most bacteria grow well on organic substances, although some survive wholly on inorganic material.

Bacteria, aerobic Bacteria requiring oxygen in order to grow and replicate.

Bacteria, facultative anaerobic Bacteria that can grow in either the presence or absence of oxygen.

Bacteria, obligate anaerobic Bacteria that must have an oxygen-free environment in order to survive.

Base A substance that reacts with protons (H^+) in water; raises the pH of aqueous solutions (*see* Alkali).

Biochemical oxygen demand (BOD) A measure of the amount of oxygen used by microorganisms during the process of decomposing waste. The BOD is usually measured in milligrams per liter or in parts per million of oxygen required.

Biocide Chemical added to a metalworking fluid formulation or mix to restrict the growth of microorganisms. This broad term includes bactericides and fungicides.

Biodegradable The capacity to be decomposed by microorganisms.

Blown oil A vegetable- or animal-derived oil chemically modified by oxidizing to increase viscosity for enhanced boundary lubrication.

Boring An operation used to enlarge a hole to an exact size using a single-point tool.

Boundary lubrication A thin layer of lubricant film which physically adheres to the surface by molecular attraction of the lubricant to the metal surface. Examples are fats, fatty acids, esters, and soaps.

Brinell An indentation-type hardness test which uses a 10 mm ball indentor under a load of either 500 or 3000 kg. The diameter of the impression made is measured and the hardness number calculated as a function of load divided by projected area.

Broach A cutting operation which combines both roughing and finishing in a single pass. The broaching tool consists of a straight shank with multiple cutting teeth, each taking the cut wider or deeper than its predecessor.

Built-up edge (BUE) A piece of work material adhering to a cutting edge of the tool.

Burn A change in the material caused by the heat of the metalworking operation; may be accompanied by discoloration.

Carbide tooling Consists of tungsten carbide particles (or combinations of other carbides) held together by cobalt or nickel binders.

Carbon equivalent The combined effects of the elements in cast iron that affect the formation of graphite. Usually considered to be the sum of carbon plus one-third silicon plus one-third phosphorous. Control of the carbon equivalent determines the solidification characteristics of cast iron.

Carbonitriding A heat treat process that enriches the surface of a low carbon steel with both carbon and nitrogen simultaneously. Upon quenching the case becomes hard while the low carbon core remains soft. A carbonitrided case is harder than a carburized case and is frequently applied to steels that have low hardenability and do not respond well to straight carburizing.

Carbon steel Steel containing less than 2% carbon and small (residual) quantities of other elements such as manganese, sulfur, and phosphorous.

Carburizing A heat treat process that enriches the carbon content at the surface of a low carbon steel. Upon quenching the high carbon surface becomes hard while the low carbon core remains soft.

Carcinogen Capable of causing cancer or malignant tumor formation.

CAS number Chemical Abstract Services Registry number. A number that uniquely identifies a chemical. Assigned by the Chemical Abstract Services.

Cathode The terminal of an electrolytic cell at which reduction (gain of electrons) occurs.

Cationic surfactant A surface active agent that when dissolved in water carries a slight positive charge. Examples are quaternary ammonium compounds.

Cemented carbide Pressed and sintered carbide particles in a binder, e.g., carbide tools.

Cementite The compound Fe_3C. In steel and iron this is the hard, wear-resistant microconstituent.

Centerless grinding A process for grinding cylindrical parts between two abrasive wheels, a grinding wheel that cuts the metal and a regulating wheel that controls the speed and horizontal movement of the part. The part is supported from underneath by a work rest blade.

Centertype grinding External grinding process where the cylindrical part is supported between two pointed metal shafts (centers) that are placed in depressions at each end of the part. The part rotates during the grinding process.

Centipoise Units of viscosity measurements (0.01 g/cm-s).

Centrifuge Used to separate oil and/or dirt from metalworking fluids. Incorporates a spinning bowl which receives the dirty fluid.

CERCLA Comprehensive Environmental Response Compensation and Liability Act, administered by the EPA (U.S.).

CFR Code of Federal Regulations, U.S.

Chelating agent A molecule capable of attaching itself to a metal ion by two or more linkages to the same molecule. A common example is ethylenediaminetetraacetic acid (EDTA).

Chemical oxygen demand (COD) The amount of oxygen required for total oxidation of matter in water.

Chloride A salt resulting from the neutralization of HCl, e.g., sodium chloride (NaCl).

Chronic health effects Adverse health effects resulting from long-term exposure, or persistent health effects.

Glossary

Climb milling   Cutting operation in which the rotation of the cutting tool is in the same direction as the workpiece movement. Consequently, the chip being formed starts out thick and becomes thinner as the tooth progresses through the workpiece.
CNOMO   Committée de Normalisation de la Machine Outiels. A consortium of French automobile manufacturers setting standards for the industry.
Coagulation   The clumping together of solids to make them settle faster. Coagulation is brought about with the use of certain chemicals such as polyelectrolytes.
Coalescence   The growing together or uniting process, such as when oil droplets agglomerate on a receptive medium.
COC   Cleveland open cup. Device used to measure flash and fire points.
Coining   The imprinting of a design or pattern onto a metal surface.
Composite material   Material that contains two or more different materials separated by a distinct interface.
Computer numerical control (CNC)   A programmable automation system that utilizes a computer to control a process with numbers, letters, and symbols.
Concentration   A measurement of the content of one or more materials in a metalworking fluid.
Conductivity   A measurement of the ability of a metalworking fluid mix to conduct an electric current which is dependent on the amount of dissolved ionic material.
Corrosion   Oxidation of ferrous or nonferrous metals; includes rust, staining, pitting, and etching. The process by which a refined metal returns to its natural state.
Coupling agent   A chemical additive that aids in emulsion formation by being mutually soluble in both phases.
Crater   A depression on the cutting face of the tool because of wear.
Cratering   A surface defect in painted parts characterized by the appearance of small, pin-hole craters in the paint film.
Crystal lattice   The spatial arrangement of atoms in a repetitive pattern.
Cutaneous   Pertaining to the skin.
Cutting fluid   A substance applied to a tool to promote more efficient machining.
Cutting speed   Tangential velocity on the surface of the tool or workpiece at the cutting interface.
Cyclone filter   A device that separates a mixture according to density through centrifugal force. Cyclones are used for cleaning metalworking fluids (separating dirt particles, chips, etc.) and have no internal moving parts.
Defoamer   A chemical additive that physically alters the surface tension of a fluid to reduce or eliminate foaming.
Deformation   The forming of ductile metal into a new shape.
Density   The mass of a material divided by its volume.
Dermatitis   Inflammation of the skin, evidenced by a rash, itching, blisters, or crustiness.

Dermis Layer of skin beneath the epidermis; the bulk of perceptible skin thickness.
Die A term generally applied to the female part of a tooling set used for metalforming.
Diffusion Movement of atoms or molecules within a material or across a mating surface.
DIN Deutsches Institut für Normung. German industrial standards organization.
Dirt Metal chips, fines, swarf, grinding grit, etc. found in a metalworking fluid mix.
Dirt volume The percentage of solids in the metalworking fluid mix that is separated from the mix by settling or centrifuging. High dirt volumes usually indicate either inadequate filtration or filter problems. A high dirt volume can affect the performance of the metalworking fluid and lead to such problems as residue, poor finish, poor tool life, and microbial growth.
Dislocation A defect in a crystal which interrupts the normal regular atomic array. The simplest form is an edge dislocation which is an extra plane of atoms. The motion of dislocations under stress accounts for the plasticity exhibited by metals.
Dissolved oxygen (DO) Elemental oxygen dissolved in water. Microbial activity will cause the DO to drop.
DOT Department of Transportation, U.S.
Down milling *See* Climb milling.
Draw bar An instrument used to apply a thin, even coating of fluid to a surface.
Draw bead Used in metal forming dies to control metal flow during stamping operations.
Drawing The reshaping of a sheet metal blank into an elongated round cup or square boxlike form.
Drilling A method for producing a hole using a cutting tool with flutes or grooves spiraling around the drill body which serve as a conduit for chip removal.
Ductile iron A form of cast iron which has been innoculated with certain elements that cause the free carbon to form as nodules or spheroids. Ductile iron has ductility approaching that of steel.
E-coat The application of paint through an electrodeposition process.
Eczematous An inflammatory condition of the skin exhibiting redness, blistering, scales, crusts, and/or scabbing.
Edema Swelling of body tissues due to the presence of abnormally large amounts of fluid.
EEC European Economic Community.
Effluent Waste water leaving a waste treatment process, relatively cleaner than before.
Electrodeposition The application of electric current through a paint or lubricant medium so that it is attracted to a metal surface with opposite charge.

Emulsifier Substances that prevent dispersed droplets coming together in emulsions by reducing interfacial tension. Emulsifiers typically have both oil-soluble and water-soluble portions to their structure.
Emulsion A disperse system consisting of several phases which arises through the mixing of two liquids which are not soluble in each other. One liquid forms the inner (or disperse) phase, distributed in droplet form in the carrier liquid (the outer or continuous phase). The emulsifiable metalworking fluids are frequently oil-in-water emulsions, i.e., oil forms the inner phase.
Endurance limit The level of stress at which a component will have infinite life under a cyclic or fatigue loading regimen.
EPA Environmental Protection Agency, U.S.
Epidemiology Science concerned with the study of the incidence and distribution of diseases in a population.
Epidermis Outer layer of skin.
Erythema Redness of the skin.
Ester A compound which may be formed by reaction of an organic acid with an alcohol. An effective boundary lubricant.
Extraneous oil Often called "tramp oil," is the oil or oil-like material in a fluid which is *not* from the metalworking fluid concentrate.
Extreme-pressure (EP) additives A compound (usually containing chlorine, sulfur, or phosphorus) which reacts with the surface of the metal or tool to form compounds (chlorides, sulfides, or phosphates) which have a low shear strength.
Facing Generating a flat surface perpendicular to the axis of rotation by machining.
Fat An animal- or vegetable-derived additive comprised of glycerol esters of fatty acids. Examples are lard oil, tallow, castor oil.
Fatigue A mode of metal failure which occurs under cyclic loading below a calculated yield strength. Components which fail by fatigue usually show a "beachmark" pattern on the fracture surface representing a progressive propagation of the crack under each stress cycle.
Fatty acids A family of organic acids, so-called because some of these chemicals occur in animal fats. These may either be saturated with only single bonds between the carbon atoms (stearic acid) or unsaturated with one or more double bonds between the carbon atoms (oleic acid).
Feed The rate at which the grinding wheel or cutting tool is moved along or into the workpiece.
Feed rate (drilling) The distance the tool moves per revolution.
Feed rate (milling) The maximum thickness of material removed per tooth.
Feed rate (turning) The distance the tool moves per revolution of the workpiece.
Ferrite The allotrophic form of iron that exists below 1670°F. Ferrite is essentially pure iron.

Ferrous Refers to iron. Steel, cast iron, and other iron-based alloys are called "ferrous metals."
FEV$_1$ A measure of pulmonary (lung) function defined as forced expiratory volume in 1 s.
FIFRA Federal Insecticide, Fungicide, and Rodenticide Act (U.S.).
Filter media Any porous material (usually paper, cloth, or screen) that traps solids when a cutting fluid passes through.
Finish Surface quality or appearance.
Finishing Final machining cuts taken to obtain the desired accuracy and finish.
Fire point The temperature at which a sustained flame will burn for at least 5 s, usually slightly higher than the flash point.
Flank Surface of a cutting tool adjacent to the workpiece surface.
Flash point The temperature at which a brief ignition of vapors is first detected.
Fluidized bed reactor Involves a process of passing a liquid upward through a bed of granular media at a velocity sufficient to expand the media and hydraulically suspend the individual media particles providing a large surface area.
Fly cutter A single-toothed milling cutter.
Foam Gas dispersed in a liquid causing an increase in the volume of the liquid. Usually seen as bubbles on the surface of the liquid which may break quickly or be quite stable.
FOG In waste treatment, the content of fat, oil, and grease.
Folliculitis Inflammation of the hair follicles. A particular type of skin irritation often caused by cutting oils ("oil boils"). Usually occurs on the forearms and thighs where oil-soaked clothing comes in contact with the skin. The cutting oil and grime from the work combine to plug the pores, resulting in inflammation, open sores, and often an acne condition.
Forging The shaping of metal through the use of impact strikes or pressure to plastically deform the material.
Fracture Failure, breakage, or fragmentation of a specimen or workpiece.
Free oil The amount of oil in a metalworking fluid mix that easily separates and floats on the top. Typically it is tramp oil.
Fungi Aerobic microorganisms consisting of either single-celled yeast or filamentous mold.
Galvanic corrosion A common type of corrosion process in which a potential difference through an electrolyte causes surface attack at the interface of two dissimilar metals.
Grains The individual crystallites which make up the structure of a polycrystalline body.
Gram (g) Metric unit of mass. There are 453.59 g in 1 lb (avoirdupois: U.S. or British); there are 1000 g in 1 kilogram (kg).
Gram stain Bacteria may be categorized by a staining technique developed by

Glossary

H. C. J. Gram in 1884. If their cell walls retain the dye, the organism is "gram positive"; if not, "gram negative."

G-ratio A measure of grinding performance defined as the volume of metal removed divided by the volume of grinding wheel worn away in the process.

Gray iron A form of cast iron where the free carbon is present in the form of graphite flakes. A freshly fractured surface has a dull gray appearance.

Gun drill A long drill with passages for coolant, used for deep holes.

Hardenability The relative ease that a martensitic transformation can be achieved in steel. One measure of hardenability is the cooling rate necessary to achieve martensitic transformation. High hardenability steels required water quenching.

Hardness (metals) The property of a metal to resist permanent indentation.

Hardness (water) The combined calcium and magnesium content of water. Usually expressed as parts per million (ppm) of calcium carbonate ($CaCO_3$).

Hematopoietic toxin A toxin that affects the production and development of blood cells, or attacks and destroys blood cells.

Hepatotoxin A toxin that attacks liver cells.

High-speed steel (HSS) Tool steels containing tungsten, molybdenum, vanadium, cobalt, and other elements; first used in machining at high speeds.

HLB Hydrophilic-lipophilic balance. A numbering system describing the water solubility/oil solubility of emulsion components. Low values (<10) indicate oil solubility while higher numbers (>10) indicate water solubility.

HMIS Hazardous Materials Indentification System (U.S.).

Honey oil A high-viscosity lubricant used for difficult, high-stress forming operations on heavy gauge metals.

Honing A fine finishing process using "stones" or sticks containing grit and an oscillating motion to create a crosshatch pattern.

Hydrocyclone *See* Cyclone filter.

Hydrodynamic lubrication Exists if the surfaces sliding over each other are separated by a coherent lubricating film of liquid.

IARC International Agency for Research on Cancer.

ID grinding An internal grinding operation performed to change the inner diameter (ID) of a cylindrical part.

Indexable inserts Cutting tools of various shapes and with multiple cutting edges, usually made of carbides. When one edge becomes dull, the insert may be rotated to expose another.

Influent Waste water entering a waste treatment process.

Interstitial A site within a crystal structure where a small void exists. Elements that have a small atomic diameter such as carbon and nitrogen can occupy these sites. These elements are called interstitial alloying elements.

In vitro Experiments with cells or tissues from living organisms conducted outside of the organisms, generally in a test tube or petri dish.

In vivo Experiments conducted with live animals.
Ion Molecular species with electron excesses (anions) or electron deficiencies (cations).
IP Institute of Petroleum, London.
Ironing A drawing process in which the thickness of the sidewall of the part is reduced.
ISO Abbreviation for International Standards Organization, located in Geneva, Switzerland.
Kinematic viscosity A function of both internal friction (viscosity) and density of the fluid.
Kurtosis The characteristic sharpness of the peaks and valleys of a surface finish. Sharp peaks and valleys have kurtosis greater than three while broad peaks and valleys have kurtosis less than three.
Land A straight section of a cutting tool behind the cutting edge.
Lapping A fine finishing operation where an abrasive slurry is introduced between the workpiece and a lapping tool made of a soft material such as cast iron.
Lay The predominant directional characteristic of the roughness of a surface.
LD$_{50}$ Acute or single dose of a substance producing death in 50% of the test animals within 14 days of exposure.
Machinability The capability of a material to be machined with relative ease with regard to tool life, surface finish, and power consumption.
Magnetic cutting fluid A magnetic field is used to cause these ferromagnetic fluids to penetrate the cut zone. Recommended for special environments such as high altitude, high vacuum, or low gravity.
Martensite A metastable body-centered tetragonal crystal structure existing in iron and steel and resulting from nonequilibrium transformation of austenite. It is the hard, strong, heat-treated form of steel. Similar transformations also occur in other, nonferrous alloy systems.
Metalforming An operation designed to alter the shape of metal without producing chips.
Metalworking A broad term used to refer to the shaping of metal by cutting, grinding, bending, stretching, or stamping.
Metalworking fluid A liquid used to cool and/or lubricate the process of shaping a piece of metal into a useful object. The term most often refers to a water-based fluid.
Micelle A cluster of surfactant molecules in solution held together by Van der Waals forces.
Microbicides Active substances used to protect water-mixed metalworking fluids against microbial attack. This broad term includes both bactericides and fungicides.
Microemulsion An emulsion of oil in water with emulsion particle size so small that the emulsion appears translucent to transparent.

Glossary

Micron A measure of length, one-millionth of a meter.
Microorganisms Minute living things, generally including bacteria, yeasts, molds, algae, protozoa, rickettsia, and viruses.
Milliliter (ml) Metric unit if liquid measure. There are 946.3 ml in 1 qt (U.S.); there are 1000 ml in 1 liter (l).
Milling Removing material with a rotary cutting tool.
Mist application Application in which the metalworking fluid is propelled in a stream of compressed air into the area where work is being performed.
Model A mathematical relationship which describes the behavior of a process.
Modulus of elasticity A measure of stiffness of a material which is expressed as the ratio stress divided by strain. A material with a high modulus will act more stiffly (or deflect less under the same load) than a material with a low modulus.
Mold Filamentous microorganisms composed of many cells; may grow in metalworking fluids, interfere with filtration, and clog pipelines. *See* Fungi.
MSDS Material safety data sheet. A form containing safety, regulatory, and physical information regarding any chemical.
Mutagen A substance capable of altering the genetic material in a living cell.
Naphthenic oil A petroleum oil with a significant content of hydrocarbons with saturated, ring-type structures. These oils are more easily emulsified but are more readily oxidized or degraded than paraffinic oils. Classification of an oil as either naphthenic or paraffinic is based upon a viscosity–gravity constant (VGC) determination.
Neat oil Hydrocarbon oil with or without additives, used undiluted.
Nephrotoxin A toxin that destroys renal (kidney) cells.
Neurotoxin A toxin that attacks nerve cells.
NFPA National Fire Protection Association (U.S.).
NIOSH National Institute of Occupational Safety and Health, U.S.
Nitriding A heat treat process that enriches the surface of a steel with nitrogen to form iron nitrides. The resulting case is hard as formed and does not require quenching.
NOEL No observed effect limit. The highest dose used in a toxicity test which produces no observed adverse effects.
Nonferrous Any metal that is not based on iron, such as copper and aluminum.
Nonionic surfactant A surface active agent that has no ionic character, carries no electrical charge in solution. In metalworking fluids they typically function as cleaners or lubricants.
Normalizing A conditioning heat treatment of iron and steel where the material is heated into the austenitic crystal structure and allowed to cool naturally in the open air.
NTP National Toxicology Program, U.S.
OD grinding An external grinding operation to reduce the outer diameter (OD) of a cylindrical part.

Oil emulsification The property of a metalworking fluid that determines its capacity for emulsifying (absorbing) or dispersing oil, typically tramp oil.
Oleophilic Is the "oil-loving" characteristic that polypropylene (and certain other materials) exhibit, making them useful for scavenging tramp oils from metalworking fluid mixes.
Organic compound Substances containing the element carbon; most contain hydrogen, and many contain oxygen, nitrogen, or other elements as well. Simple oxides of carbon are excluded.
Orthogonal Mutually perpendicular.
OSHA Occupational Safety and Health Administration, U.S.
Oxidation A reaction that removes electrons from the substance being oxidized.
Paraffinic oil A petroleum oil in which the majority of the molecules are saturated straight or branched chain hydrocarbons. These oils are more plentiful than naphthenic oils and are more stable, but they are also more difficult to emulsify. Classification of an oil as either naphthenic or paraffinic is based upon a viscosity–gravity constant (VGC) determination.
Passivation The formation of a tenacious, protective oxide film on the surface of a metal.
Pearlite A macroconstituent in iron and steel composed of alternate layers of ferrite and cementite in a lamellar structure.
PEL Permissible exposure limit (OSHA).
pH The negative log of the hydrogen ion concentration of an aqueous solution. Can be expressed on a scale of 1 (acid) to 14 (alkaline). Pure water has a pH of 7.
Phosphate A salt resulting from neutralization of phosphoric acid, e.g., sodium phosphate (Na_3PO_4).
Pigment A term used to denote the solid lubricants sometimes used in metalforming lubricants. Examples include calcium carbonate, talc, mica, graphite, and molybdenum disulfide.
PMN Premanufacture notification to EPA (U.S.).
Polar additive A molecule with positive and negative charges isolated to different portions of the molecule. Many emulsifiers and lubricants are polar in nature. Polarity aids in emulsion formation and causes attraction to the metal surface.
Polyelectrolyte An organic polymer containing multiple, electrically charged sites. Useful in waste treatment.
Polymer A high molecular weight chemical compound consisting essentially of repeating structural units.
Positive filter A type of filter using some type of filtering material (paper, cloth, wire screen, etc.) to remove particulate from a metalworking fluid.
POTW Publicly owned treatment works (sewage).
Precipitate A reaction causing a solution to form an insoluble compound that settles to the bottom of the liquid. Also, the product of this process.
Precipitation hardening Also called age hardening, involving the precipitation

of submicroscopic intermetallic compounds from a supersaturated solid solution to strengthen the alloy.
Punch A term generally used for the male part of a tooling set used in metalforming.
R_a The average surface roughness computed as the arithmetic mean of the absolute value of the distance between the baseline to the maximum peak or valley height.
Rake, negative The face of the tool along which the chip travels is inclined forward from the cutting edge forming a greater than 90° angle with the freshly cut surface. Such a rake is useful in machining high-strength materials in order to reduce chipping of the tool.
Rake, positive The face of the tool along which the chip travels is inclined backward from the cutting edge, forming a less than 90° angle with the freshly cut surface.
Rake angle The angle between the tool face and an imaginary line perpendicular to the freshly cut surface.
Rancidity The condition in which a metalworking fluid develops a foul odor. Typically caused by high levels of anaerobic bacteria.
RCRA Resource Conservation and Recovery Act, administered by the EPA (U.S.).
Ream A cutting operation used to enlarge and finish a previously formed hole. The reaming tool consists of one or more cutting elements arranged along the longitudinal axis.
Rebinder effect Modification of the mechanical properties at or near the surface of a solid, attributable to interaction with a surfactant.
Recycling The process used to clean and restore used metalworking fluid mixes for reuse.
Refractometer An optical instrument that measures the refractive index of a metalworking fluid. Used to determine concentration.
Residue That part of a metalworking fluid mix left after the evaporation of water.
Reverse osmosis (RO) A separation process similar to ultrafiltration, but using higher pressures and tighter semipermeable membranes. This reverses the natural process of osmosis causing water to flow from the more concentrated to the more dilute solution side of the membrane.
Rockwell An indentation-type hardness test which uses a variety of indentors and loads. The hardness number is based on the depth of penetration of the indentor.
Roughing Machining without consideration of surface finish.
Roughness Short-range deviations in the surface plane of a workpiece because of interactions between the cutting tool or forming process and the workpiece material.
RTECS Registry of Toxic Effects of Chemical Substances, published by NIOSH.

Rust Hydrated oxides of iron.
Saponification number A measure of the fat or ester content of a material. Expressed as milligrams of potassium hydroxide required to hydrolyze one gram of sample.
SARA Superfund Amendment and Reauthorization Act, U.S.
Semisynthetic fluid A metalworking fluid with moderate to low content of mineral oil, usually 5% to 30%. Generally contains a significant amount of water, 30% to 60%. These are sometimes called "preformed emulsions."
Sensitization In metallurgy, the precipitation of chromium carbides at the grain boundaries of stainless steel caused by heating the material in the temperature range of 1100 to 1500°F. The loss of chromium from the matrix of the steel causes a marked reduction of corrosion resistance.
Sensitizer In dermatology, a chemical which causes a delayed-type allergic reaction in people or animals after repeated exposure to the chemical.
Shank That portion of a tool by which it is held, such as the shank of a drill held in a chuck.
Shear plane A narrow zone along which shearing takes place in metal cutting.
Sintered carbide *See* Cemented carbide.
Skewness The relative comparison of peaks and valleys of a surface; a measure of the symmetry of the profile about the mean line. If the peaks are generally higher than the valleys are deep, the surface has positive skew. If the valleys are deeper than the peaks are high, the surface has negative skew.
SME Society of Manufacturing Engineers, Dearborn, Michigan.
Soap An emulsifier prepared by neutralizing a fatty acid with an alkaline material such as an amine or sodium hydroxide.
Solid-film lubricant Solid lubricants, such as graphite or molybdenum disulfide.
Soluble oil A metalworking fluid with high oil content (50% to 80%) and little or no water content. As sold it consists solely of oil, emulsifiers, oil-soluble lubricants, corrosion inhibitors, etc. When mixed with water it creates an emulsion that is milky in appearance.
Solution annealed A precursor heat treatment applied to precipitation-hardening metal alloys to form the supersaturated crystal structure from which the strengthening precipitates can form during aging.
Specific energy (U) In grinding, the power required to remove one unit volume of material per unit of time.
Specific gravity The ratio of the mass of any volume of material to the mass of an equal volume of some reference material, usually water, at a standard temperature. Also called the "relative density."
Specific metal removal rate (Q′) The volume of metal removed per unit of time per unit of effective grinding wheel width.
Sperm oil Ester-type boundary lubricant from the head cavity of the sperm

whale. Since 1971 this material has not been available in the United States because of the endangerment to the whale population.
Spindle Rotating shaft for tool holders.
Stainless steel An alloy of iron containing at least 11% chromium, and sometimes nickel, that resists almost all forms of rusting and corrosion.
Stamping A variety of operations in which a part is formed from a flat strip or sheet stock through the use of a forming die set.
Steel A broad term for a variety of iron-based alloys.
STEL Short term exposure limit.
STLE Society of Tribologists and Lubrication Engineers; formerly ASLE, American Society of Lubrication Engineers.
Strain The elongation or stretching a body exhibits in response to an application of pressure.
Stratum corneum The outermost, hardened layer of the epidermis; the skin's principal physical barrier against penetration by chemical substances and microorganisms.
Stress The load applied to a body projected over the area on which it acts. Stress has the dimensional units of pressure.
Subchronic health effect Resulting from repeated daily exposure of experimental animals to a chemical for part of their life span.
Sulfate A salt resulting from neutralization of sulfuric acid, e.g., sodium sulfate (Na_2SO_4).
Sulfonate Sodium petroleum sulfonate; an emulsifier and corrosion inhibitor. Originally a by-product of white oil production, sulfonates may also be prepared synthetically. Molecular weights from 380 to 540 are generally most useful in metalworking.
Surface finish (or roughness) Fine irregularities measured in terms of height and spacing.
Surfactant A compound that reduces the surface tension of water, or the interfacial tension between two liquids or between a liquid and a solid. A surface active agent.
SUS Units of viscosity (Saybolt universal seconds).
Swaging The squeezing or compressing of metal bar stock or tubing so as to create a taper or point.
Swarf Metal fines and grinding wheel particles generated during grinding.
Synthetic fluid A metalworking fluid that contains no mineral oil. Some synthetics are totally water soluble (chemical solutions) while others are emulsions of water-insoluble, synthetically derived lubricants (synthetic emulsions).
Tapping A method used to cut or form threads inside of a predrilled hole.
TDS Total dissolved solids.
Temper Condition of an alloy, such as annealed, cold worked, heat treated.
Tempering A postquenching heat treatment that occurs when the metal is

reheated to an elevated temperature but below the austenite transformation temperature. The mechanisms of tempering involves the rejection of carbon from the martensite lattice.

Titration   A procedure for determining volumetrically the concentration of a certain substance in a solution by adding a standard solution of known volume and strength until the reaction is complete. The endpoint is usually detected by a color change of an indicator solution or by an electrical measurement.

TLV   Threshold limit value. The limit of exposure to a material at or below which workers should experience no health problems (ACGIH).

TOC   Total organic carbon.

Tolerance   Permissible variation in the dimensions of a part.

Tool life   A measure of the length of time a tool will cut satisfactorily.

Torque   The tendency to produce rotation.

Total Oil   Is the percentage of all oil or oil-like material present in the metalworking fluid mix. This value includes both product oil and extraneous Oil. Usually determined by acidifying the mix and centrifuging.

Toxicology   The study of the health effects of chemicals.

Tramp oil   That oil which is present in a metalworking fluid mix and is *not* from the product concentrate. The usual sources are machine tool lubrication systems and leaks.

Trepanning   Uses a single-point tool to produce a hole by cutting a circumferential groove in the metal and leaving a solid core.

Tribology   The study of interacting surfaces in relative motion which encompass the aspects of friction, lubrication, and wear. Derived from the Greek *tribos*, for rubbing.

TSCA   Toxic Substances Control Act, U.S.

TSS   Total suspended solids.

TTO   Total toxic organics.

Turning   Machining on a lathe or turning center with single-point cutting tools.

TWA   Time-weighted average.

Ultrafiltration (UF)   A separation technique in which a liquid is applied to a semipermeable membrane under moderate pressure. The liquid passing through the membrane (the "permeate") is cleaned and the waste stream (or "effluent") becomes more concentrated with dissolved solids.

UN number   An identification number assigned by the United Nations to hazardous materials in transportation.

Up milling   Opposite of climb (down) milling. The cutting tool rotates against the direction in which the workpiece is fed; the chip starts out thin and gets thicker as the cut progresses.

Vanishing oil   An evaporative type of solvent-based lubricant used to stamp or draw parts that will not be washed.

Vesicles   Blisters.

Vickers An indentation-type hardness test which uses a pyramidal-shaped diamond indentor and a range of loads. The diagonals of the indentation are measured and the hardness number calculated as a function of load divided by projected area.
Viscosity The internal resistance to flow exhibited by a fluid.
Viscosity index A means of expressing the relationship between viscosity and temperature.
VOC Volatile organic chemicals.
Waste treatable Ability to chemically or physically remove an additive from the water prior to disposing of the effluent.
Water hardness The combined calcium and magnesium content of water. Usually expressed as parts per million (ppm) of calcium carbonate ($CaCO_3$).
Waviness A surface feature of a workpiece involving regular, long-range deviations in the plane of the surface. Waviness is usually due to tool chatter, spindle eccentricity, or other unintended motions of the cutting tool or machinery.
Wear Loss of material from a surface due to rubbing.
Wetting agent An additive which reduces surface and interfacial tension of a fluid and facilitates spreading of the fluid over a surface.
White iron A form of cast iron which has no free or uncombined carbon present. All of the carbon is present as carbides. A freshly fractured surface has a sparkly, white appearance.
WHMIS Workplace Hazardous Materials Information System (Canada).
Wire drawing Pulling a metal rod through one or a series of dies to cause elongation and a reduction in diameter.
Workpiece A piece of material to be machined.
Yeast Mostly single-celled fungi. They are larger than bacteria and are either round or oval in shape.
Yield strength The stress level required to cause a measurable permanent distortion in a material. This is usually the basis (with appropriate safety factors applied) of the maximum allowable design stress for functional components which suffer bending.

Index

ACGIH, 434
Acid-alum, 378
Acid number, 194
Acute toxicity, 413, 414–415
Adhesives, 216
Agar plates, 252
Air entrainment, 199
Airless spray, 142
Alkaline cleaner, 143
Alkalinity, 213, 315
Allergic reactions, 398–399, 402, 403
Aluminum, 225, 228, 236
Aluminum alloys, 47–50
American Revolution, 7
Ames test, 418
Amide, 171, 179, 180, 181, 182, 186
Amines, 173, 180, 182, 186, 259, 261, 270, 412, 414, 419
Amphoteric metals, 225, 228
Anionic titration, 213
Annealing, 47

Anodizing, 49
Antifoams, 179, 181, 182, 195
API separator, 373

Bacteria, 247–270, 322, 345, 356–357
  aerobic, 234
  anaerobic, 225, 234–235
  generation time, 249
  growth curve, 250
  sulfate-reducing, 234–235
Barium, 444, 456
Barrier creams, 406
Base number, 194
Bessemer, H., 9, 10, 37
Biocide, 179, 181, 182, 214, 234, 357, 255–263, 399, 400, 402, 406, 407, 414, 444, 456–457, 459
  mixtures, 260–262
  registration, 255
  selection, 263
Biodegradable, 186

Bioreactor, 382–383, 384, 385–389
Bioresistance testing, 266–268
Bioresistant, 181, 186–188
Blanking, 144–146
Block on ring test, 200
BOD, 382, 383, 385, 389
Bodymaker, 150
Borates, 169, 173, 181, 182, 184, 186, 412
Boring, 76
Boron, 214
Brinell hardness, 32
Broaching, 81–82, 95–96
Built-up edge (BUE), 66, 67, 80
Burn, 330

Cancer, 416–420, 459
Can drawing, 144–154
Cans:
    aluminum, 149–150
    steel, 153–154
Carbon equivalent, 44
Carburizing, 46
Cartridge filters, 290–292
Cast iron, 37, 43–44
Catalase, 253, 254
Cationic titration, 213
Central systems, 12, 273–302, 345–346
Centrifuge, 266, 287–289, 358–360
CFR, 426
Charpy impact test, 35
Chip formation, 61–62
Chip geometry, grinding, 125–129
Chip rust tests, 239
Chip settling, 218
Chip size and shape, 274–279
Chip transport systems, 279–283
Chlorides, 316
Chlorine, 141, 169, 174, 175, 176, 177, 179, 180, 182, 183, 184, 412, 420, 458

p-Chloro-m-cresol, 456–457
Chromate, 236
Chromium, 236
Chronic toxicity, 413, 415–420
Citizen suits, 455
Cleaning, 139, 142–143
Cleanliness, 217
Clean-out schedule, 350, 352, 359
Cloud point, 195
Coalescence, 358–359, 360, 362, 374, 384
COD, 382, 385, 389
Coining, 135, 159
Concentrate stability, 168
Concentration control, 211–215, 318–320, 324, 344–345
Conductivity, 323
Contaminant, removal, 345–348
Contaminants, 308, 344–345
Contract fluid management, 335
Cooling mechanism, 67–68
Copper, 225, 228
Copper alloys, 50–52
Corrosion, 223–246, 327
    bacteria-induced, 225, 233–235, 237, 250
    bimetallic, 230, 242
    galvanic, 230, 242
    inhibitors, 169, 173, 179, 182, 185–187, 236–238, 261–262
    intergranular, 232
    prevention, 244–246
    stress, 233, 241
Corrosion tests, 139, 238–244
    acid atmosphere, 241
    AC impedence, 243
    chip, 239
    electrochemical, 243
    Herbert, 239
    polarization, 243
    salt spray, 241
Cost, 309

Index

Crater wear, 64
Crystal structure, metals, 26–31
Cubic boron nitride (CBN), 89
Cupping, 150
Cutting oil, 13, 14, 70

Defoamer, 179, 181, 182, 195
Deionization, 316
Density, 192
Department of Transportation (DOT), 424, 430–433, 457, 458
Depletion, 250–252
Dermatitis, 13, 332, 395–409, 414
Dipslides, 252–254, 322–323
Dirt, 321
Dissolved oxygen, 254, 323
Down milling, 77
Draw bead simulator, 209
Drawing, 147
Draw and iron, 148
Draw ratio, 147
Draw–redraw, 154–155
Drilling, 70–74, 82–88, 206
Drills, 6, 71–73
Drum filter, 296–299

E-coat, 144, 216
Effluent limits, 370
Electrochemical series, 231
Electrocoat, 144, 216
Electrodeposition, 142
Electroflotation, 376
Elongation, 157
Employee training, 438
Emulsifiers, 15, 171–172, 178, 180, 185, 186, 187
Emulsion breaking, 376, 378–379
Emulsion particle size, 178, 180, 182
Emulsion stability, 168, 194
End milling, 77

Endotoxins, 269–270
Environmental impact, 367–371
EPA, 255, 369, 411, 424, 426–430, 441–459
Epidemiology, 412
Equalization, 375
Ernst, H., 16–17
Erosion, 231
Ester, 174, 175, 179, 181, 182
Evaporators, 376, 384
Exposure limits, 437

Face milling, 77
Facing, 76
Fatigue, 36
Fats, 137
Federal Register, 425–426
FIFRA, 255
Filter design, 125–129
Filter media, 290–299
Filtration, 265, 273–302, 308
Fire point, 193
Flank wear, 64
Flash point, 193
Flocculation, 379
Flotation, 376, 379–381
Fluid application, 68–69, 141–142
Fluidized bed reactor, 382–383, 385–389
Fluid management, 341–353
Fluid selection, 306–309, 343–344
Flume design, 281–283
Foam, 169, 182, 329
  separators, 287
  tests, 197–199
Folliculitis, 414
Formaldehyde, 402, 459
Formaldehyde condensate, 256–263
Four-ball test, 200, 209
Fretting, 232
Friction pendulum, 203
Fungus, 170, 182, 248, 252

Galling, 209
Gerdes, J., 5, 156
Gloves, 405, 407
Glycol ethers, 444
Gram negative/positive, 248
Gravity separation, 291–292, 373–375
Grease, 13
Grinding, 7, 8, 11, 99–134
  efficiency (E), 113–116
  ratio, 106, 204
Grinding wheels:
  CBN, 111, 113, 123
  diamond, 123–125
  resinoid bond, 132–134

Hand washing, 407
Hard-water stability, 166, 168, 185, 186, 187
Harris, B. R., 15
Hazard determination, 434–435
Hazardous waste, 455–458
Health effects, 414–420
Heat treatment, 44–47, 49–50
Herbert test, 239
High-speed steel, 40
History, 3–19
HLB, 15, 168
Honey oil, 138, 183
Hutton, H. W., 14
Hydrocyclone separators, 287–289
Hydrotreatment, 417

IARC, 417
Industrial Revolution, 4–9
Iron, 225–228, 236
Ironing, 150
Irritation of eye, nose, or throat, 333
Izod impact test, 36

Kurtosis, 56

Lathes, 6, 7, 11, 74–78, 203
LD_{50}, 414
Legionnaire's disease, 269
Leonardo da Vinci, 4, 7, 136
Liability, 459–460
Limiting draw ratio, 148, 207
Lubricants:
  boundary, 140–141, 152, 153, 169, 172, 180, 182, 183
  extreme-pressure, 17, 137, 138, 140, 144, 169, 175, 176, 177, 179, 180, 182
  hydrodynamic, 141, 170, 171
  solid, 137, 141, 177, 184
  thick film, 151
  thin film, 151
Lubricity tests, 199–210

Machine tools, numbers, 191, 339
Magnetic fluids, 86
Magnetic separators, 289–290
Make-up additions, 300–301
Material safety data sheets (MSDS), 408, 424, 435, 436–438, 442, 444, 445, 455, 460
Mechanical properties, metals, 32
Membrane biological reactor, 386–389
Membrane filtration, 274
Merchant, M. E., 16–17, 61, 67
Metal allergy, 399
Metal chips, 61–62, 125–129, 218, 274–279, 325, 456
Metalforming, 18
Metalforming tests, 207–210
Metalworking fluids:
  composition, 165–188, 411–412
  functions, 64–70
  usage (volumes), 2, 411
Methyl chloroisothiazolone, 256–258
Microbial control, 214
  growth, 170, 182, 186

Microwaves, 265
Milling, 8, 17, 76–78, 93–95
Mist, 199
Moisturizing creams, 404–408
Mold, 248, 322
Molybdate, 182, 236
Molybdenum disulfide, 175, 176, 184
Monday morning odor, 252
Mushet, R. F., 9, 10, 12

Neutralization number, 194
NIOSH, 411
Nitriding, 47
Nitrite, 236, 419
Nitrosoamines, 424, 456
Nodular iron, 44
Nonionic surfactant, 180
Normalizing heat treatment, 47
Notching, 146, 160
Nozzle design, 116–125

Odor problems, 325, 328, 355
Oil, 136, 371, 417–420
   naphthenic, 136, 170–171, 178
   paraffinic, 136, 170–171, 178
   rejection, 211
   removal, 300, 354–356
Oil and grease, 371–372, 383, 385
Oleic acid, 170
OSHA, 418, 424, 433–440, 460

Paint softening, 218
Parting, 146
Passivation, 236
Pasteurization, 263–265, 357, 360, 362
Patch test, 399–404
Pendulum, 203
Permanent media, 296–299
Petroleum, discovery, 9

pH, 179, 213, 259–260, 270, 315, 320–321, 345, 357
Phenols, 256, 257, 260, 396, 444, 456–457
o-Phenylphenol, 444
Phosphate, 216, 316
Phosphorus, 141
Phosphorus additives, 169, 175, 177, 179, 180, 182
Physical properties, metals, 32
Piercing, 146, 160
Pigmented paste, 169
Pigments, 137
Pin and V-block, 200
Pitting, 232
Plastic, 218
Plate counts, 252, 322
Polyelectrolytes, 378–379
Pontiac fever, 269
Precipitation hardening (aluminum), 50
Prelube, 143, 184
Premanufacture notification (PMN), 427
Premixed fluid, 301
Product evaluation, 311–312
Pulmonary irritation, 415
Punching, 146, 159, 160

Q', 104

R_a (surface roughness), 55, 107
Radiation, 265
Rake angle, 62
Rancidity, 13, 179, 181, 328
Rapeseed, 171, 179
RCRA, 455–458
Rebinder effect, 68
Recycling, 309, 339–364
Redraw, 154
Refractometer, 211, 319–320
Residue, 169, 218, 334

Respiratory effects, 414–416
Reverse osmosis, 274, 317, 381–382
"Right-to-Know" Act, 440–455
Rockwell hardness, 32
Rubber, 218
Rust, 225–228

Saponification number, 194
SARA, 426, 440–455
Selective hardening, 46
Semisynthetic fluids, 17, 137, 166, 180–182
   formulation, 181
Sensitization, 398–399, 414
Sensory irritation, 415
Separators, oil/water, 373–376
Settling tanks, 285
Shaving, 146
Shear angle, 61
Silent Spring, 423
Skew, 56
Skimmers, 384
Skin, structure and function, 396–397
Skin irritation, 13, 332, 395–409, 414
Skin treatment, 404–406
Soap, 136, 179, 182
Soluble oil, 14, 15, 137, 153, 165, 177–180, 368
   formulation, 180, 187
Sonneborn, H., 13
Specific energy (U), 107
Specific gravity, 192
Specific metal removal rate, 104
Staining (of metals), 224, 234
Stainless steel, 236
Stainless steel alloys, 41–43
Stamping and drawing, 177, 183
Steel, alloying elements, 37–39
Stokes Law, 355, 358, 362, 373

Straight oil, 170–177
   formulation, 177
Stress/strain curve, 34
Stripper fingers, 151
Subchronic toxicity, 413
Sulfates, 316
Sulfonates, 13, 179, 414
Sulfur, 141, 169, 174–177, 179, 180, 182, 235
Sulfuric acid, 378
Surface finish, 52–58, 107, 330
Surface tension, 182, 219
Swaging, 160
Synthetic fluids, 17, 137, 153, 165, 182–183, 368–369, 382
   formulation, 183

Tapping, 78–81, 95
Tapping torque, 207
Taps, 78
Taylor, F. W., 12, 64, 108, 110
Temperature control, 129–134, 301–302
Temperature effects, grinding, 132–134
Tensile curve, 34
Tensile test, 34
Threading, 78–81
Thurston, R. F., 11
Tin-plated steel cans, 153–154
Titration, 194, 213–214, 320
Tolyltriazole, 173, 182
Tool life, 16, 17, 63–64, 331
Tool steel, 2, 10, 39–41
Tool wear, 63–64
Toxicology, 412–421
Trade secrets, 439–440, 445, 455
Tramp oil, 181, 182, 321–322, 325–326, 359
Triazine, 255–261
Trimming, 146
Troubleshooting, 326–334

TSCA, 424, 426, 427–430
Turning, 74–76, 88–93

Ultrafiltration, 274, 359, 376–378, 384, 385–389
Up milling, 77
Usage (volumes), 2

Vacuum filtration, 292–295
Vanishing oil, 183
Vapor degreaser, 143
Vegetable oil, 170, 171, 172, 182, 183
Vickers hardness, 32
Viscosity, 170, 177, 179, 192
Viscosity index, 193

Warning labels, 436
Waste minimization, 184–187

Waste treatment, 179, 184, 187, 188, 217, 363–364, 367–391
Water hardness, 168, 196, 314–315, 342–343
Water quality, 244, 307–308, 312–317, 325, 342–343
Water softening, 316
Wedge wire, 284–285, 296–299
Whitney, Eli, 8
Wire drawing, 3, 5, 155–159, 184
  formulation, 184
World War I, 13, 14, 15
World War II, 15, 40, 423

Yeast, 248
Young's modulus, 35

Zinc, 225, 228